ANIMAL COGNITION AND BEHAVIOR

Edited by

Roger L. MELLGREN

Department of Psychology
University of Oklahoma at Norman
Norman, Oklahoma, U.S.A.

N·H
P↘C

1983

NORTH-HOLLAND PUBLISHING COMPANY
AMSTERDAM • NEW YORK • OXFORD

ISBN: 0 444 86627 2

Publishers:
NORTH-HOLLAND PUBLISHING COMPANY
AMSTERDAM • NEW YORK • OXFORD

Sole distributors for the U.S.A. and Canada:
ELSEVIER SCIENCE PUBLISHING COMPANY, INC.
52 VANDERBILT AVENUE
NEW YORK, N.Y. 10017

Library of Congress Cataloging in Publication Data
Main entry under title:

Animal cognition and behavior.

    (Advances in psychology ; 13)
    1. Cognition in animals--Addresses, essays,
lectures.   2. Animal behavior--Addresses, essays,
lectures.   3. Learning in animals--Addresses, essays,
lectures.   I. Mellgren, Roger L., 1944-   .
II. Series: Advances in psychology (Amsterdam, Netherlands).; 13. [DNLM: 1. Ethology--Congresses.
2. Cognition--Congresses. W1 AD798L v. 13 / QL 785
A5979 1982]
QL785.A72 1983   156'.3   83-3998
ISBN 0-444-86627-2 (U.S.)

PRINTED IN THE NETHERLANDS

XL 4 1008

BOOK LOAN

Please RETURN or RENEW it no later
than the last date shown below

ADVANCES
IN
PSYCHOLOGY
13

Editors

G. E. STELMACH

P. A. VROON

NORTH-HOLLAND PUBLISHING COMPANY
AMSTERDAM • NEW YORK • OXFORD

CONTENTS

LIST OF CONTRIBUTORS

Numbers in parentheses indicate the pages on which the authors' contributions begin.

James Allison, Department of Psychology, Indiana University, Bloomington, Indiana 47405 (1)

William H. Baum, Department of Psychology, University of New Hampshire, Durham, New Hampshire 03824 (253)

Robert C. Bolles, Department of Psychology, University of Washington, Seattle, Washington 98195 (65)

Lynn D. Devenport, Department of Psychology, University of Oklahoma, Norman, Oklahoma 73019 (83)

Michael Domjan, Department of Psychology, University of Texas, Austin, Texas 78712 (319)

Roger S. Fouts, Department of Psychology, Central Washington University, Ellensburg, Washington 78926 (445)

Gordon G. Gallup, Jr., Department of Psychology, State University of New York, Albany, New York 12222 (473)

William C. Gordon, Department of Psychology, University of New Mexico, Albuquerque, New Mexico 87131 (399)

N. Jack Kanak, Department of Psychology, University of Oklahoma, Norman, Oklahoma 73019 (445)

George H. Kimball, Department of Psychology, University of Oklahoma, Norman, Oklahoma 73019 (445)

S. E. G. Lea, Department of Psychology, University of Exeter, Exeter EX4 4QG England (31)

Roger L. Mellgren, Department of Psychology, University of Oklahoma, Norman, Oklahoma 73019 (223)

Mark W. Olson, Department of Psychology, Coe College, Cedar Rapids, Iowa 52405 (223)

John P. J. Pinel, Department of Psychology, University of British Columbia, Vancouver, B.C. V6T 1Y7 Canada (285)

H. Ronald Pulliam, Department of Biological Sciences, State
    University of New York, Albany, New York 12222   (427)

Seth Roberts, Department of Psychology, University of California,
    Berkeley, Berkeley, California 94720   (345)

T. J. Roper, Department of Biological Sciences, University of
    Sussex, Brighton, Sussex, England   (127)

William Timberlake, Department of Psychology, Indiana University,
    Bloomington, Indiana 47405   (165)

Donald M. Wilkie, Department of Psychology, University of
    British Columbia, Vancouver, B.C. V6T 1Y7 Canada (285)

PREFACE

This book is an outgrowth of the simultaneous interaction of
a number of factors. My own uneasiness with what I perceived
to be the sterility traditional animal psychology and my wish
to change it and make it better prompted me to undertake this
project. The following paragraphs were written for a general
audience in an attempt to obtain financial support for a
symposium to be held at the University of Oklahoma and serve
to detail the ideas and biases that motivated me to do this
book.

Scientific Psychology is generally agreed to have been founded
in 1879 to study consciousness. By the turn of the century
many of America's most prominent psychologists had begun to use
lower organisms in their research. As part of the movement toward the use of animals in research many of these psychologists
ignored, or even denied, the study of conscious thought processes as being a relevant goal for a scientific psychology.
Under the influence of psychologists such as Clark L. Hull,
B. F. Skinner, E. L. Thorndike and John B. Watson, explanations
of cognitive processes were declared to be derivable from
simplier or more primitive concepts than the cognition itself.
For example, Skinner's book entitled Verbal Behavior was an
explanation of language acquisition and use based on Skinner's
"operant conditioning" work with rats and pigeons as subjects.
The key concept was the notion of a learned stimulus-response-
outcome relationship, sometimes called S-R psychology, which
was seen as the basis for language, and by overt implication,
cognition in general. In this regard Skinner was following the
precedent set by the Russian Physiologist, Ivan Pavlov, who,
working with dogs, demonstrated how reflexes could be conditioned and then used such conditioned reflexes to explain
language (called the "second-signalling system") and other
cognitive processes. Thus the school of thought known as
Behaviorism convinced many psychologists that the study of
consciousness could be accomplished by studying simple processes such as represented by conditioned reflexes, or the S-R
association, and ultimately by building-up larger structures
composed of such processes, consciousness, cognition and
language could be understood.

Contemporary thought is in the process of rejecting this view
of cognitive functioning. The rejection of Behaviorism is
coming about for a number of reasons, both internal to that
school of thought and external to it. Internally, those who

have attempted to study the "building-blocks" of conditioned
reflexes and S-R associations have had more and more difficulty
staying within the boundary of such simple explanatory concepts,
even when the experimental situation is highly structured and
as simple as possible. I include myself in this category.
Through the last ten years of teaching I have found myself
saying with increasing frequency, "The greater the degree of
study of a particular behavioral phenomenon, the more complex
become the explanations of that phenomenon." This is not to
say that conditioned reflexes and S-R habits are irrelevant to
understanding animal (including human) behavior. In fact,
these mechanisms have been extremely useful in a variety of
ways, particularly the areas known as "Behavior Modification"
and "Behavioral Therapy" where such conditioning and S-R formu-
lations have been directly applied to treating various problem
behaviors. The successes of these applications are well
established.

From a historical perspective what has happened in psychology
is something which happens in many disciplines. Theoretical
perspective may act like a pendulum with the influence of
Skinner and others causing the pendulum to swing toward a
reductionistic-mechanistic, environmentally-determined view of
behavior. The swing of the pendulum has reached its apogee and
is now beginning a swing back toward considering higher mental
functioning as a phenomenon worthy of direct study. By heading
away from the behavioristic framework the uniqueness of different species is more fully appreciated; a point which ethologists often made to behavioristic psychologists. Traditionally
behaviorism assumed that the principles of conditioned reflexes,
because they were the fundamental building blocks of all behavior, could be studied in any species and then applied to
other species. Although it may be true that principles of
conditioning hold across species, the fact that such principles are incapable of explaining all behavior (and some would
argue that they are incapable of explaining any behavior)
means that we must find different principles to explain the
diversity of animal behavior so evident in the ethological
literature.

We arrive at the crux of this proposal. The study of animal
cognition is changing rapidly to produce a new, more integrated understanding of mental functioning. This integrated
understanding involves combining behavioristic psychology,
ethology, economics, sociobiology and cognitive psychology
(not necessarily in order of importance). It is my contention that the future direction of the swing of the pendulum
is toward the viewpoint of a person who has a background in
all of the above areas. I plan to teach a course under the

title "Seminar in Psychology" (Psy 4973) with the semester
title being "Animal Cognition". My plan is to provide junior-
senior and graduate level students with reading materials and
seminar experience in these areas as they are relevant to
animal cognition. Toward the end of the semester six distinguished scientists representing various views of animal
cognition will be brought to campus for a two day symposium.
Funding is requested to bring about the symposium.

On the basis of the above proposal a seminar was indeed offered
at the University of Oklahoma in the Spring of 1982 and I
was able to bring six persons to campus and combined with myself and Lynn Devenport we held a two-day symposium on
Animal Cognition. It seemed a natural extension to produce
a book from the meeting, but in order to provide more complete
coverage of the topic I invited several individuals to contribute chapters who were not participants in the symposium. I
wish the time and money had been available for all of the
contributors to this volume to have been on campus but this
was not possible.

The symposium and the book were made possible through the
efforts of President William S. Banowsky and contributors
to the University of Oklahoma Associates Fund and Dean Jerome
Weber who judged my plans worthy of support. Several students
provided help and ideas before, during and after the symposium. They include Steve Brown, Robert Hale, Barbara King,
Donna Martin, Linda Misasi, Rebecca Ross, Rick Stevens,
Randy Stratton and David Whiteside. The tolerance and support
of my wife, Karen in all phases of this project is deeply
appreciated. The dedicated typing of Glenda Smith made the
book possible, and her willingness to work on it made my job
much easier and more effective.

ANIMAL COGNITION AND BEHAVIOR
Roger L. Mellgren, editor
© North-Holland Publishing Company, 1983

BEHAVIORAL SUBSTITUTES AND COMPLEMENTS

James Allison
Indiana University

I do not know whether animals have cognition -- a confession
that may seem agnostic, atheistic, or simply ignorant, depending much on the reader's own convictions about animal cognition.
But I do have some confidence in the rationality of animals,
even those as primitive as the snail. Do not leap to the wrong
conclusion: By some definitions of rationality, animals can behave rationally without having minds, souls, a strong chess
game, or the ability to figure the expected value of a lottery
ticket.

Consumer Demand Functions

As a case in point, consider the economists. Economists talk
a lot about maximizing utility -- a supremely sensible way to
behave, but a supremely difficult way to behave, even for the
Wunderkind with a Ph.D. in game theory. But asked to state one
factual test of rationality, the average economist will settle
for a fairly simple one: Does the organism obey the demand
law? That is, does the organism buy less of a particular commodity as its price rises, all else being equal? As the price
of coffee rises do we buy less coffee, allocating some of our
coffee money toward the purchase of a substitute, such as tea?
Such behavior can be derived mathematically from the hypothesis
that we allocate our limited resources so as to maximize the
utility gained from our expenditures (Awh, 1976; McKenzie &
Tullock, 1981).

I must quickly interject two related disclaimers. First, we
can obey the demand law without quite maximizing utility. Second, we can derive the demand law from theory that makes no
reference at all to utility (e.g., Allison, Miller, & Wonzy,
1979; Allison, 1981b). Perhaps it is fortunate that we can,
because we have come to understand within the past few years
that the creatures that obey the demand law range from the

big-brained human to <u>Stentor</u> <u>coeruleus</u>, a minute, no-brained
protozoan (Rapport & Turner, 1977).

The laboratory rat lies somewhere between these two extremes of
neural and psychological capability, and generally shows close
obedience to the demand law. The rat data in Figure 1 come
from an experiment that obliged the rat to work for its food
under various fixed-ratio schedules, each of which required a
certain number of lever presses as a behavioral price for each
small food pellet (Collier, Hirsch, & Hamlin, 1972, Experiment
2). The price of the pellet varied from one schedule to the
next, and each test session lasted 24 hr. Figure 1 presents a
re-analysis of the data that defines price in terms of the
number of lever presses paid per gram of food; price appears
on the horizontal axis, total grams of food consumed on the
vertical axis. Food consumption generally fell as the behavioral price of food increased, in accordance with the demand
law.

As a bonus, Figure 1 also shows that goldfish obey the demand
law. The fish curve comes from a re-analysis of an experiment
by Rozin and Mayer (1964) in which the fish, tested under various fixed-ratio schedules, nosed an underwater target for
waterproof food pellets dispensed to the center of a ring
floating on the surface.

Many additional examples appear in reviews published by Lea
(1978) and Allison (1979b). The examples cover a wide variety
of animals (including humans) performing many different kinds
of responses as behavioral payment for a wide variety of commodities, such as food, water, sucrose, saccharin, heat,
cigarettes, alcohol, cocaine, morphine and pentobarbital.

Those two reviews appeared during a time of some confusion
about what to do with interval schedules in terms of the economics of consumer demand. Under a typical interval schedule,
a clock specifies a series of food setup times over the course
of the session. Only when the setup comes due will the next
lever press result in the delivery of the food. But what is
the price of the food? Does it have to do with the time the
animal must wait between setups? What about the lever press
the animal must pay for each delivery of food? What about the
ineffective lever presses the animal happens to perform during
the interval between setups?

My own solution happens to give satisfactory results. It ac-
cepts the fact that the interval schedule sets in advance
only a minimum price, one the animal can achieve only by never

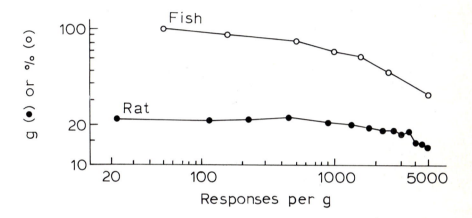

Figure 1. Mean food consumption by two rats tested in
24-hr sessions, and seven goldfish tested in 1-hr
sessions, as functions of the unit price of food
(adapted from Collier et al., 1972, Experiment 2,
and Rozin and Mayer, 1964, Experiment 3). Food
consumption is expressed in terms of grams for
the rats, and in terms of the percentage of the
amount consumed at the lowest price for the fish.

responding during the interval between setups. The minimum
price is the number of responses required per unit of the
reward, after the reward setup has come due. Each response
during the interval between setups raises the real price above
the procedural minimum. Because we typically have no way of
knowing how many responses the animal will pay out during the
intervals between setups, we can measure the real price after

the fact, but not before; the price actually paid is simply
the total number of lever presses emitted during the session,
divided by the total amount of the commodity consumed during
the session (Allison, Note 1; see also Lea & Tarpy, 1982).

Figure 2 illustrates this application to some of my own data
on interval schedules (Allison, 1980). Each rat got all of its
daily water during a daily 1-hr test session. The vertical
axis shows total licks at the water tube. The horizontal axis
shows the behavioral price, presses per lick, that emerged as
we tested the rats under a variety of variable-interval schedules. Each schedule required one lever press for access to
the water tube. Some schedules averaged 7 sec between setups,
some 14 sec; some schedules offered 25-lick setups, some offered 50-lick setups. The data points on the left-hand side of
the figure show the total number of licks that occurred under
a baseline condition that allowed free access to tube and
lever throughout the session. Like the group function displayed at the bottom of the figure, each of the six individual consumer demand functions conformed to the demand law.

Further analyses showed that as the interval between setups
increased from 7 to 14 sec. the rat raised the price of the
lick by performing more ineffective lever presses during the
time between setups. Price also increased as we reduced the
size of the setup from 50 licks to 25 licks, partly because
the rat performed more ineffective lever presses while waiting
for the smaller setup.

Some schedules invented by the fertile mind of George Collier
present similar problems, and yield to a similar solution.
The first published example used 24-hr test sessions. During
each session the schedule required x lever presses for access
to a food bin, like a conventional fixed-ratio schedule, but
allowed the rat to eat as much as it pleased; only if the rat
paused in its eating for several minutes did the bin close up,
whereupon the rat had to perform another x lever presses to open
the bin again (Collier et al., 1972, Experiment 1). Like the
interval schedule, the schedule that offers an optional magnitude of reward makes it impossible to specify the price of food
in advance. The price depends partly on x, the price of admission, which varies across sessions, and partly on how much
the rat chooses to eat after having paid the price of admission. The same uncertainty prevails in a restaurant that
charges a fixed amount of money for all the diner can eat.

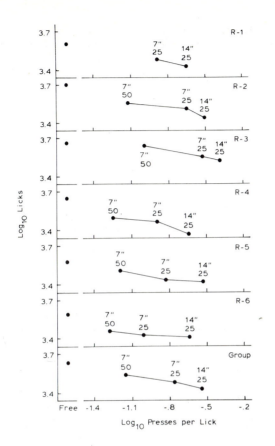

Figure 2. Total water licks as a function of unit
price, lever presses per lick, under variable-
interval schedules. Points on the left represent
the paired baseline condition with water and lever
freely available to the rat throughout the 1-hr
session. The schedules arranged 25- or 50-lick
setups at mean intervals of 7 or 14 sec (based on
data reported by Allison, 1980).

Just as we did with the interval schedule, we can measure price
after the fact by dividing the total number of lever presses
by the total grams of food consumed during the session.
Figure 3 reveals once again a satisfactory relation between
price and consumption. One curve shows that total food consumption fell as the price of food, presses per gram, rose.

The other curve shows that total meals, actually total bin admissions, fell as the fixed-ratio price of admission to the bin rose.

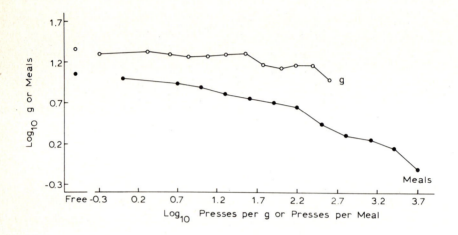

Figure 3. Mean food consumption by rats in 24-hr sessions (n = 3 except for the four highest prices, where n = 2). Open circles show the number of grams consumed as a function of the unit price of food, closed circles show the number of meals consumed as a function of the unit price of meals (adapted from Collier et al, 1972, Experiment 1).

In a later experiment, rats tested in 23-hr sessions performed x lever presses for each access to a water tube; the tube remained available until the rat stopped drinking for several minutes (Marwine & Collier, 1979, Experiment 1). A new analysis of the data appears in Figure 4, which refers to each access as a "bout". The figure displays on the horizontal axis both the a priori price of the bout, x, and the price of water, presses per ml, calculated after the fact. The group data, like the individual functions, conformed to the demand law: As their prices increased, total bouts and total water intake generally decreased.

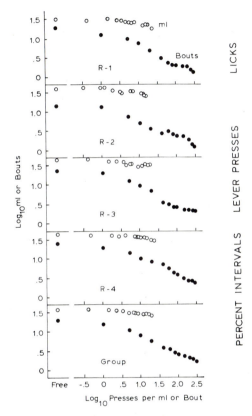

Figure 4. Total water intake and total drinking bouts by rats tested in 23-hr sessions as functions of the unit price of water and bouts (based on data reported by Marwine and Collier, 1979, Experiment 1).

In summary, the conventional fixed-ratio schedule, the variable-interval schedule, and the fixed-ratio schedule with an optional magnitude of reward all reveal the proper demand function when we define price in terms of the number of responses per unit of the commodity consumed: Total consumption falls as price rises. But do any of those functions really arise from the processes that are supposed to give rise to the

consumer demand functions we see in economics textbooks? When
the price of coffee rises humans consume less coffee, supposedly because tea or some similar commodity is readily available
as a substitute. But when the laboratory rat eats less in response to a rise in price, what does the rat substitute for
the missing food? Perhaps it substitutes its bodily stores of
food, the energy stored as body fat (Allison, 1981b). Perhaps
it reduces its daily activity, or its metabolic rate. Each of
these possibilities refers to an implicit substitution process, implicit but observable in principle. To see substitution more directly, we can turn to the literature on concurrent
ratio schedules.

Thanks to both the professional and the popular press, many of
us know that rats find root beer and Tom Collins mix mutually
substitutable (Alexander, 1980; Rachlin, Green, Kagel, &
Battalio, 1976). Imagine two levers arranged side by side, one
wired to a root beer dispenser, the other to a dispenser of
Tom Collins mix. The rat can drink either beverage at a particular behavioral price, presses per ml. As we raise the price
of either beverage while holding constant the price of the
other, the rat drinks less of the one, but more of the other
(Rachlin et al., 1976). The same authors later reported similar results with other beverages (Kagel, Battalio, Rachlin, &
Green, Note 2). That pattern of results would lead an economist to classify the two beverages as mutual substitutes,
like tea and coffee in the human economy. The pattern reflects
the kind of process that economists have in mind when they derive the demand law from the notion that the organism maximizes
utility. When the price of good A rises, we can maximize
utility by buying less of A and more of good B, a substitute
for A.

For a still clearer example imagine two levers arranged side by
side, one wired for cheap food pellets, the other wired for the
same food pellets at a higher behavioral price. The rational
shopper should buy all food at the cheaper "store", none at the
high-priced store. Rats generally fall short of this ideal,
but often come close; which of these two facts seems the more
remarkable depends on one's point of view.

Figure 5 shows an experimental example, a re-analysis of data
reported by Lea and Roper (1977). The figure plots food consumption as a function of price, presses per gram, for four
different pairs of prices. In the first pair on the left, one
lever charged 1 press per pellet, the other 8 presses per pellet. The remaining pairs charged 6 and 8, 8 and 11, and 8 and
16 presses per pellet. Note of each pair that the rats bought

some at the more expensive one. The upper curve shows total food consumption at the two stores combined and, like the pairwise curves, conforms to the demand law.

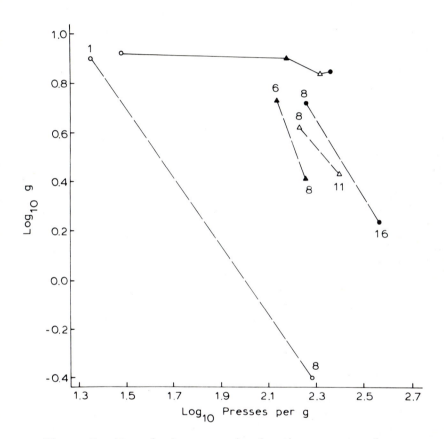

Figure 5. Mean food consumption by six rats tested in 1-hr sessions as a function of the unit price of food. Points joined by a broken line represent the amount consumed at one of the two stores. Points joined by a continuous line represent the amount consumed at the two stores combined. The four nominal price comparisons were 1 lever press per pellet against 8 presses per pellet, 6 against 8, 8 against 11, and 8 against 16. The unit prices plotted here are the prices actually paid, defined as total lever presses divided by total grams of food consumed in the session as a whole (adapted from Lea and Roper, 1977).

Why did the rats buy any food at the high-priced store? We
cannot explain their behavior as an artifactual consequence of
having to discover at the start of each session the cheaper of
the two stores, because the cheaper store had the same location
across sessions. Moreover, rats often sample the high-priced
store toward the end of the session (Shapiro & Allison, 1978).
Identical food pellets should be perfectly substitutable for
one another, in which case economic theory predicts a "corner
solution" (Awh, 1976), exclusive responding on the lever that
offers the cheaper pellets. The occasional sampling of the
other lever contradicts a number of theories from economics
and psychology (Allison, 1981a), and has not received the
attention it deserves. Perhaps the sampling behavior seen under concurrent ratio schedules has something to do with the
spontaneous alternation seen with the 8-arm radial maze (Olton,
1979).

Substitutes and Complements in Noncontingent Schedules

As mutual substitutes in the economic sense, one behavior rises
as the other falls. Next we examine some data on two phenomena,
schedule-induced polydipsia and autoshaped lever pressing, that
we might understand by viewing drinking and lever pressing as
substitutes for eating (Allison, Note 3).

Five rats got all of their daily food and water during a daily
2-hr session. Figure 6 plots total licks at the water tube
against total food intake. The open circle shows licks and
food intake under a paired baseline condition with food and
water freely available without limit throughout the 2-hr session.

The closed circles show what happened when we kept the rat from
eating its paired baseline amount. We managed that experimental suppression of eating by means of six fixed-time schedules
that delivered free food pellets, 45-mg each, at selected times
throughout the course of the session, with the water tube freely available. Three schedules delivered 2, 3, or 4 pellets
every 90 sec; three others delivered 1 pellet, 2, or 3 pellets
every 60 sec. Each schedule suppressed eating, in response to
which the rats became polydipsic, drinking substantially more
than they did under the paired baseline condition. Drinking
rose linearly as eating fell, as if the rats found drinking a
substitute for eating.

Each of the boxes used in that experiment had a retractable
lever between the water tube and the food receptacle; the lever
played no role there, as it stayed in its retracted position,
flush with the wall, throughout each session. Figure 7 shows

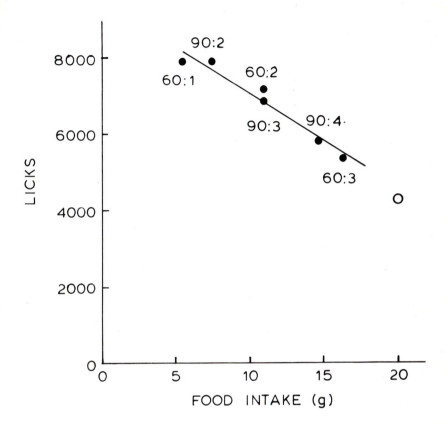

Figure 6. Total water licks as a function of total food
   intake under six fixed-time schedules (closed circles)
   and the paired baseline condition (open circles); group
   means for five rats tested in 2-hr sessions. The
   schedules delivered 1, 2, 3, or 4 45-mg pellets at
   intervals of 60 or 90 sec (Allison, Note 2).

what happened when we tested five new rats under an arrangement
that made use of the lever.

The 2-hr baseline sessions offered free access to food, water,
and lever. The open circle in the top panel shows baseline
licks and food intake; the open circle in the middle panel
shows baseline lever presses and food intake. After the baseline phase, we tested the rats with three different random-time
schedules of food delivery. Each schedule made 80 deliveries,
independently of the rat's behavior, at random intervals that

averaged 90 sec between deliveries; the three schedules dispensed 2, 3, or 4 45-mg pellets per delivery. Water was freely available throughout each session, just as it was in the first experiment. But our new procedure signalled each forthcoming food delivery by presenting the lever 10 sec beforehand, retracting the lever and delivering food simultaneously.

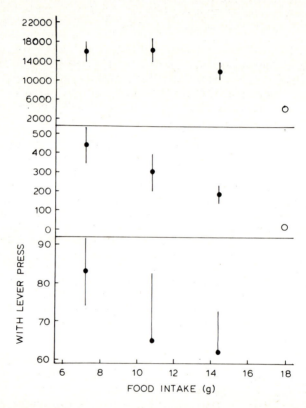

Figure 7. Total water licks (top panel), total lever presses (middle panel), and percentage of intervals with one or more lever presses (bottom panel) as functions of total food intake under three random-time schedules (closed circles) and the baseline condition (open circles); group means for five rats tested in 2-hr sessions. Each vertical line shows the standard error of the mean; standard errors of the basepoint means were too small to show graphically. The schedules delivered 2, 3, or 4 45-mg pellets at a mean interval of 90 sec (Allison, Note 2).

The closed circles in the top panel reveal a polydipsic response to our experimental suppression of eating; once again, drinking rose linearly as eating fell. The closed circles in the middle panel reveal the effect known as autoshaped lever pressing, substantially more responding than the baseline amount. Like drinking, lever pressing appeared to function as a substitute for eating, rising linearly as eating fell. The bottom panel shows another conventional measure of autoshaping, the percentage of intervals with one or more lever presses.

If drinking and pressing are substitutes for eating, they probably have different origins as substitutes. Drinking may, but lever pressing may not, have intrinsic value as a stand-in for eating. The substitute value of lever pressing appears to depend on the role of the lever as a predictor of food.

I say all of this because of a third experiment that presented lever and food independently of each other on a truly random basis, depriving the lever of its role as a predictor of food. In this third experiment our five new rats still became polydipsic, but showed no sign of autoshaped lever pressing. Thus, the substitute value of lever pressing appears to have an extrinsic character that may depend on associative learning. The substitutability of drinking probably does not depend on associative learning. Our experimental procedures deny the water tube any role as a predictor of food delivery, and a followup experiment has shown emphatically that the drinking response itself has little or no accidental contiguity with the next food delivery. Drinking typically rises to a peak soon after the delivery, and gradually decays to zero long before the next delivery.

Eating and drinking do not always function as substitutes; their function varies dramatically with the context in which we study them. I begin my illustration with a reminder: As mutual substitutes in the economic sense, one behavior rises as the other falls. But as mutual complements in the economic sense, the two behaviors rise or fall together. The classic economic examples of complementary goods are flour and shortening, key ingredients in the recipes of many bakery goods. If we cut our consumption of either in response to a rise in its price, we also cut our consumption of the one whose price stays constant, because we need less of it in our reduced production of bakery goods.

But the very same pair of goods can function as substitutes or complements, depending on our use of them. Used as footwear, left shoes and right shoes function as complements; for every left shoe that comes to hand, we need to find a matching right shoe to make one unit of footwear. Used as projectiles, left and right shoes function as substitutes; in flinging a slipper at the yowling cat on the fence we settle for the first slipper that comes to hand, left or right.

Figure 8 shows that eating and drinking may function as substitutes or complements. The circles at the top should look familiar; they come from our first experiment on polydipsia, and have already appeared in Figure 6. They show that if we suppress eating experimentally by dispensing food under various fixed-time schedules, always less than the rat eats normally, drinking substitutes for eating; as eating decreases, drinking increases. The squares at the bottom come from a similar experiment with five new rats that allowed free access to food, but limited access to a drinking tube on three different fixed-time schedules. Each schedule suppressed drinking experimentally by allowing less drinking than the paired baseline amount, the amount shown by the open square. Measured in a context that suppressed drinking, the two responses functioned as complements: As drinking decreased, so did eating.

It seems that if the rat has plenty of water but not enough food, eating and drinking may function as substitutes; if the rat has plenty of food but not enough water, they may function as complements. Perhaps the rat can reduce some aspect of hunger by drinking more water, but cannot eat dry food without incurring more thirst.

Labor Supply Functions

Labor supply theory deals with money and leisure as substitute goods. The left-hand panel in Figure 9 shows a standard economic analysis of how the worker's wage rate affects the amount of labor the worker is willing to supply (Awh, 1976; McKenzie & Tullock, 1981). The horizontal axis plots two variables, daily hours of work and daily hours of leisure. The work scale runs from left to right, the leisure scale from right to left. Because the analysis assumes that the person has a fixed number of hours available each day for work plus leisure, every hour of work deprives the person of one leisure hour.

The vertical axis shows daily dollars earned. The lines fanning out from the work origin on the left represent various wage rates, dollars per hour; the greater the slope of the wage rate line, the higher the wage rate. For any particular amount of labor, a high wage rate gains more daily dollars than a low wage rate; and the longer the person works at any particular wage rate, the greater the daily gain.

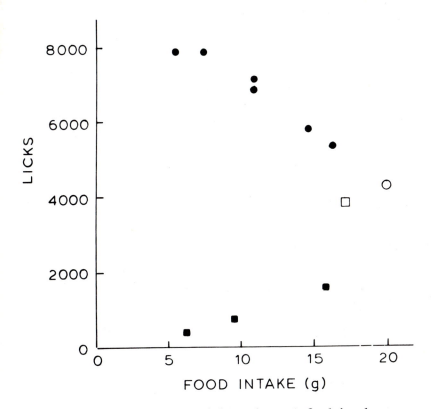

Figure 8.   Total water licks and total food intake
   under fixed-time schedules (closed symbols) and the
   baseline condition (open symbols); group means for
   rats tested in 2 hr sessions. Circles represent
   five rats whose schedules delivered 1, 2, 3, or
   4 45-mg pellets at intervals of 60 or 90 sec with
   water freely available (see Figure 6); squares
   represent five other rats whose schedules arranged
   5-, 10-, or 20-lick setups every 90 sec with 45-mg
   food pellets freely available (Allison, Note 2).

The curves tangent to the wage rate lines are indifference
curves that represent the substitutability of money and leisure. Every point on an indifference curve specifies a unique combination of two goods, daily dollars and daily leisure
hours, and all points on a particular curve hold equal attraction for the worker.

In relation to the leisure origin on the right-hand side, each indifference curve slopes downward; the downward slope means that the worker is willing to trade some daily dollars for more daily leisure, or leisure for dollars. That is the sense in which money and leisure are mutually substitutable.

Notice that any point on a higher indifference curve should hold more attraction than any point on a lower curve. For example, at five hours of leisure any higher curve offers more daily dollars than any lower curve. It follows that the worker should climb any particular wage rate line to the highest indifference curve attainable on that line. The worker should climb to a unique point of equilibrium, the point at which the wage rate line is tangent to an indifference curve. If the worker were to supply more labor or less than the amount marked by the equilibrium point, the worker would fall on some lower indifference curve, and would therefore fall short of the amount of utility allowed by the constraints of the prevailing wage rate.

By drawing a broken line through the equilibrium points in Figure 9, we draw the backward bending labor supply curve of economic theory. The theoretical curve says that at the highest wage rates, a drop in the wage rate should increase the supply of labor; at intermediate wage rates, a further drop in the wage rate should decrease the supply of labor. Throughout the entire range, as the wage rate drops we should also see a steady decline in daily dollars earned.

I should say a few more words about why the labor supply curve bends as it does, sloping downward amongst the higher wage rates, but upward amongst the lower ones. Let us begin with the highest wage rate shown in Figure 9. As we reduce the wage rate we reduce the income gained from an hour's work, but we also reduce the price of an hour's leisure in terms of lost work income. The price of an hour's leisure is the income lost from the hour not worked. The income effect would lead us to work longer so as to maintain our purchasing power. But the lower price of leisure would have the opposite effect; we should buy more of this cheaper leisure, and work less. If its income effect outweighs its price effect, the drop in the wage rate will lead us to work longer, as shown in the downward-sloping part of the labor supply curve. Leisure is now cheaper, but not all that cheap, because the wage rates are still pretty high. But if its price effect outweights its income effect, the drop in the wage rate will lead us to work less, as shown in the upward-sloping part of the curve. The logic seems plain enough: As the wage rate approaches zero the income gained from an hour's work approaches zero, and the

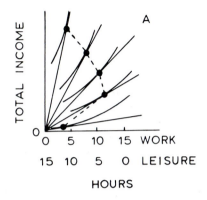
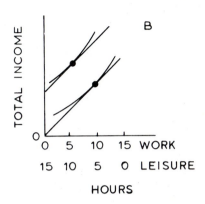

Figure 9. Daily income as a function of daily hours
of work or leisure, assuming 15 hrs. available each
day for work plus leisure. The slope of each line
is the wage rate, dollars per hr. Convex curves
are indifference curves that represent the substitutability of money and leisure. In Panel A
the broken line through the points of equilibrium
is the theoretical backward bending labor supply
curve. Panel B illustrates the theoretical effects
of nonlabor income; the intercept of the wage rate
line shows the amount of nonlabor income.

price of leisure therefore approaches zero. If we derive
practically no monetary benefit from work we have little reason to work, and every reason to use our time for leisure instead.

The right-hand panel in Figure 9 shows the predicted effects
of free income on labor supply. Both workers get the same
wage rate, but one starts the day with free income from some
source: interest on a bank account, dividends, food stamps,
welfare payments, social security payments, an inheritance, or
any other source besides the worker's current labor. The
analysis shown in Figure 9 predicts that free income will cut
the supply of labor, and raise total income.

Economic studies of the American labor supply market generally
reveal only the top part of the theoretical labor supply curve

in the left-hand panel, the part that slopes downward (Watts & Rees, 1977). Perhaps American wage rates are too high to show the part that slopes upward.

I recently completed a small-scale study of labor supply in humans, the labor supplied by 38 faculty members in a department of psychology, with results similar to those shown by large-scale studies of labor supply in America. A panel of four psychologists rated the performance of each colleague over a three-year period in terms of three different areas of productivity that supposedly determine such weighty matters as tenure, promotion, and salary; the three areas were teaching, service, and research. Each of the three rating scales ranged from 1 (non- or counterproductive) to 4 (star quality); a rating of 2.5 signified satisfactory performance. In their evaluations of teaching the raters considered such variables as the number of courses taught, class enrollments, service on thesis committees, textbooks published, evaluations by students, creation of new courses, academic coulseling, honorary awards for teaching, and financial awards for the development of innovative techniques. Their evaluations of service to department, university, profession, and community considered service on administrative, policy, and personnel search committees, editorships, service on editorial boards, reviewing of manuscripts and grant proposals, election to office in professional organizations, organization of professional conferences, and service rendered to community mental health organizations. In evaluating research the raters attended to the quantity and quality of publications, papers presented at conferences, invited addresses and invited papers, research grants, and honorary awards in recognition of distinguished research.

I averaged the three ratings to form a single composite measure of productivity and used that number, P, as the measure of the amount of labor supplied by the person. To arrive at the person's wage rate I calculated the person's average annual salary over the same three-year period, and called that number $. The ratio $/P is the person's wage rate, dollars per unit of labor, over the three-year period.

Let us examine some possible relations between the wage rate, $/P, and its two constituent parts: Salary ($), and productivity (P). Consider two psychologists with different wage rates. The one with the higher wage rate might have the higher salary, in conformity with the theoretical labor supply curve. That person might also show less productivity than the other,

more, or the same productivity as the other, depending on the
relative weight of income and price effects. The structure of
the ratio $/P allows each of those three empirical possibil-
ities.

As a fourth possibility, the one with the higher wage rate
might have the same salary as the other, but less productivity.
Fifth, the one with the higher wage rate might actually have
the lower salary, but much the lower productivity.

To examine the empirical relation between wage rate and its
two constituents I ordered the 38 persons in terms of wage
rate, and formed seven non-overlapping wage rate groups with
five to six persons in each group. Each of the seven points
in Figure 10 represents the median salary ($) and the mean
productivity (P) of one wage rate group. I chose median salary
because the 38 salaries showed the positive skewness typical
of income distributions. The functional relation in Figure 10
bears a close resemblance to the downward-sloping part of the
theoretical labor supply curve. As the wage rate dropped productivity rose significantly, $F(6, 31) = 4.65$, $p < .005$, and
salary fell significantly; by the Kruskal-Wallis test,
$X^2(6) = 25.15$, $p < .001$. Trend tests of the relation between
$ and P showed that salary rose linearly as productivity fell,
$F(1, 31) = 19.98$, $p < .001$. The data revealed no significant
curvilinearity in the relation between salary and productivity,
$F(5, 31) = 1.58$, $p > .10$.

Further analyses have shown that the picture changes very
little if we weigh research more heavily than teaching, and
teaching more than service. When we assign weights of 5, 3,
and 2 to research, teaching, and service, the highest wage
rate group shows the only major change, a leftward shift that
indicates a further drop in productivity.

In summary, the data in Figure 10 conform to the downward part
of the theoretical labor supply curve, where income effects
outweigh price effects. They show no appreciable sign of the
upward part where price effects outweigh income effects. A
participant in the symposium suggests, perhaps accurately, that
data would reveal the missing upward slope if we added another
point, a point for the graduate students in the department
(Bolles, Note 4).

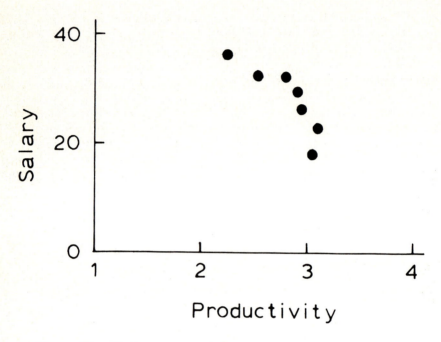

Figure 10. Median salary (thousands of dollars) and
    mean composite productivity in teaching, service,
    and research among seven groups of psychologists
    with different wage rates (n = 5 in each of the
    four highest groups, n = 6 in each of the three
    lowest groups).

I too have a hunch that a further drop in the wage rate would
finally reveal the full sweep of the theoretical curve.
Figure 11 shows why.

Figure 11 presents the results of an experiment that subjected
rats to an unconscionably wide range of wage rates (Kelsey &
Allison, 1976). Over a series of daily 1-hr sessions they
pressed a lever for access to sucrose solution under eight
different fixed-ratio schedules; each access allowed 10 licks
at the sucrose tube. Because the response requirement ranged

from 1 press of the lever to 128 presses the nominal wage rate, licks per press, ranged from 10 down to .08.

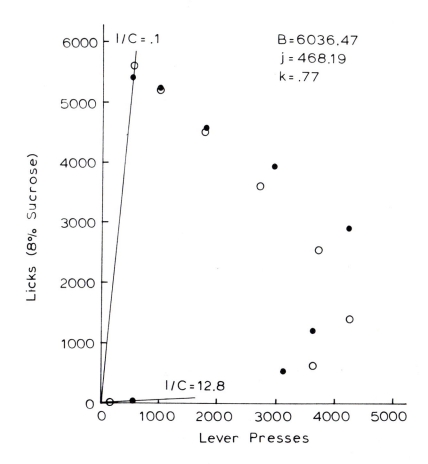

Figure 11.   Total licks (8% sucrose) as a function of total lever presses; group means for rats tested in 1-hr sessions. Each of the seven rats was tested with the six highest wage rates, five were tested at the next, and four at the lowest wage rate. Closed circles represent values observed, open circles represent values predicted by a conservation model (based on data reported by Kelsey and Allison & Boulter, Note 4).

Please ignore for the time being the eight open circles in Figure 11. Each of the eight closed circles represents total licks (our analog of income) and total lever presses (our

analog of labor) observed at one of the eight wage rates. The
data reveal a complete backward bending labor supply curve,
with total licks falling steadily as the wage rate fell.
Starting at the highest wage rate, total lever presses rose and
then fell as the wage rate fell.

My final figure shows another backward bending labor supply
curve from five rats that pressed a lever for all of their
daily water in 1-hr sessions under several fixed-ratio schedules; the results also confirm the theoretical effects of
free income (Allison & Boulter, Note 5). In Figure 12 the
three lines fanning out from the origin represent the wage
rates offered by three standard fixed-ratio schedules, ml per
press. The closed circle on each line shows the result observed for that schedule. A curve drawn through those three
data points would show a statistically significant bend, as
prescribed by economic theory; the greatest supply of labor
came at the intermediate wage rate.

The fourth line represents a free income payment. We started
that session by giving the rats about 4.5 ml of free water;
they had to earn all subsequent water by working at the
intermediate wage rate. The open circle on that fourth line
shows the results: Free water produced a significant decrease
in total lever presses, but a significant increase in total
water intake, in comparison with the same schedule that offered no free water. Thus, in agreement with economic theory,
the labor supply curve bent backward, and free income cut the
supply of labor and raised total income (see also Green &
Green, 1982).

The time has come to explain the 20 off-line points in Figure
12 and the 8 open circles in Figure 11. They represent predictions derived from a conservation model that has no
historical connection, and scant logical connection, with the
economic theory of labor supply shown in Figure 9.

The original model (Allison, 1976) assumed that the organism
conserves, or holds constant, the total amount of some dimension common to the two responses controlled by schedules of
the sort represented in Figures 11 and 12. Thus, each of the
two responses, pressing the lever and drinking, has some substitute value for the other; as the organism performs more of
either response, it will perform less of the other in attain-
ing the dimensional total specified by the model. Contrary
to economic theory, the organism does not want as much leisure and income as it can possibly get, but rather a fixed
amount of some dimension common to work and its proceeds.

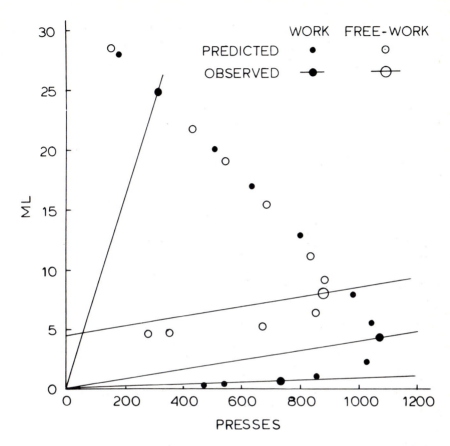

Figure 12. Total water intake as a function of total
  lever presses; group means for five rats tested in
  1-hr sessions, with free water (open circles) and
  without (closed circles). Large on-line circles
  represent values observed, small off-line circles
  represent values predicted by a conservation
  model (Allison & Boulter, Note 4).

To derive predictions from the model we need not specify the
dimension, and we need not assume that the same dimension
governs all cases. But energy serves as a convenient didactic
example: All responses, even cognitive ones, entail the ex-
penditure of energy. Maxwell's demon, standing at a porthole
between two gas-filled chambers of equal temperature, could
gate fast and slow molecules into opposite chambers, causing
one chamber to grow hotter, the other cooler, with no apparent
consumption of energy -- until we recognize that the demon's

cognitive function, in the act of measuring energy, itself
entails a consumption of energy (Monod, 1971). All responses
also take time, but we already know that organisms do not necessarily conserve the time allocated to the two responses
(Allison, 1976; Allison et al., 1979).

In the context of Figures 11 and 12, the original model predicts tht total drinking will decline linearly as total lever
pressing rises. It can deal with a bend in a labor supply
function (Allison, 1979a), but not very gracefully, and fails
badly in contending with the empirical function in Figure 11
(Allison, 1981b).

The newest conservation model (Allison, 1981b) does much better,
but at the cost of assuming an implicit substitution process.
This latest model assumes that when the behavioral price of
water grows very high, the rat substitutes its bodily stores of
water in place of the costly external water. The mathematical
version of the model generates the off-line predictions shown
in Figure 12, and the open-circle predictions in Figure 11.
Note that the theoretical predictions describe a backward bending labor supply curve that is practically linear at the highest wage rates -- a feature often seen in laboratory experiments that explore only the higher range of wage rates (Allison,
1981b).

We shall try to bring this implicit substitution process out
into the open. An experiment in progress provides three metal
tubes accessible through three holes in the chamber wall, a
water tube on the left, an empty tube in the middle, saccharin
on the right. The rat does all of its daily drinking in a
daily 2-hr session that starts with the presentation of two
tubes, empty and saccharin. Saccharin remains freely available throughout the session, a possible substitute for water.
By licking the empty tube instrumentally a certain number of
times, the rat earns 40 contingent licks at the water tube.
As we raise the instrumental requirement across sessions from
5 empty licks to 10, 20, 40, and 80, we decrease the wage rate
(water licks/empty licks), and raise the behavioral price of
water (empty licks/water licks).

As we decrease the wage rate, the rat generates an empirical
backward bending labor supply curve: The empty-lick total
rises and falls, and the water-lick total falls steadily.
Given that pattern of results, the new model leads us to expect a compensatory rise in total saccharin licks as the behavioral price of water rises. Preliminary results look promising, but it is too early to say how the model will fare

in handling the quantitative details of the data. Its mathematical form is

$$j\underline{S} + \underline{W} = \underline{B} - k\underline{E},$$

where $\underline{S}$, $\underline{W}$, and $\underline{E}$ signify total licks at the saccharin, water,
and empty tubes, $\underline{B}$ signifies the dimensional total, and $\underline{j}$ and
$\underline{k}$ signify the substitutability of a saccharin lick and an empty
lick for a lick at the water tube. The model says that the
weighted sum of total saccharin and water licks will fall
linearly as the empty-lick total rises. Its fate remains to
be seen.

Conclusion

If obedience to the demand law lays claim to rationality, many
would quickly accept the human's claim but reject, or accept
with great reluctance, an equally valid claim for rat or protozoan. Maybe the resistance comes from the conventional over-
tones of rationality in everyday usage of the term. Maybe we
should set it aside, along with the concept of response reinforcement (Allison, 1981b), in favor of scientific concepts
with broader applicability.

Biologists and economists alike have called attention to some
remarkable similarities between the animal foraging optimally
by the lights of ecological theory, and the human consuming
optimally by the lights of economic theory (McKenzie & Tullock,
1981; Rapport & Turner, 1977). Each of the two theories would
allow a prominent place for notions of substitutability and
complementarity. The optimal consumer chooses a market bundle
of two different goods, A and B, by buying a certain amount of
each. The consumer's specific selection represents an attempt
to maximize utility. The selection depends on the person's
budget, the relative price of A and B, and their mutual substitutability as represented by an indifference curve. An
optimally foraging predator chooses a combination of two different prey types, prey A and prey B. The specific selection
represents an attempt to maximize biological fitness. A predator too has budgetary constraints that depend on such variables as the time and energy available for foraging, and relative prey abundance. The selection depends partly on a fitness contour analogous to the indifference curve of economic
theory (Rapport & Turner, 1977, p. 368).

No one can yet discern the exact shape of the coming amal-
gamation, but several economists of an earlier era would have
taken some pleasure from its general form. They would surely
include Thorstein Veblen, who admired Darwin fervently and

longed for an approach to economics that would resemble
natural science. I might go so far as to credit Veblen (1898)
for an early statement of the conservation model:

> The modern scientist is unwilling to
> depart from the test of causal relation
> or quantitative sequence. When he asks
> the question, Why? he insists on an
> answer in terms of cause and effect.
> He wants to reduce his solution of all
> problems to terms of the conservation
> of energy or the persistence of quantity.
> This is his last recourse. (p. 377)

And why not?

Footnotes

I thank R. Aslin, A. Buchwald, D. Forrest, S. Hoffman, G. Lucas, R. Mack, K. Moore, I. Saltzman, and R. Shiffrin for various contributions to this paper. The work was supported in part by Grants MH31970 and MH34148 from the National Institute of Mental Health.

Reference Notes

1. Allison, J. Demand economics and conservation in ratio and interval schedules. Paper presented at the meeting of the Psychonomic Society, San Antonio, November 1978.

2. Kagel, J. H., Battalio, R. C., Rachlin, H., & Green, L. Demand curves for animal consumers. Unpublished manuscript, 1977. (Available from John H. Kagel, Department of Economics, Texas A. & M. University, College Station, Texas 77840).

3. Allison, J. Autoshaping and polydipsia: Lever pressing and drinking as substitutes for eating. Paper presented at the meeting of the Psychonomic Society, Philadelphia, November 1981.

4. Bolles, R. C. Personal communication, March 26, 1982.

5. Allison, J., & Boulter, P. Wage rate, nonlabor income, and labor supply in rats. Manuscript submitted for publication, 1982.

References

1 Alexander, T. Economics according to the rats. Fortune, December 1980, 127-132.

2 Allison, J. Contrast, induction, facilitation, suppression, and conservation. Journal of the Experimental Analysis of Behavior, 1976, 25, 185-198.

3 Allison, J. Remarks on Staddon's comment. Journal of Experimental Psychology: General, 1979, 108, 41-42. (a)

4 Allison, J. Demand economics and experimental psychology. Behavioral Science, 1979, 24, 403-415. (b)

5 Allison, J. Conservation, matching, and the variable-interval schedule. Animal Learning and Behavior, 1980, 8, 185-192.

6 Allison, J. Paired baseline performance as a behavioral ideal. Journal of the Experimental Analysis of Behavior, 1981, 35, 355-366. (a)

7 Allison, J. Economics and operant conditioning. In P. Harzem & M. D. Zeiler (Eds.), Advances in analysis of behaviour (Vol. 2): Predictability, correlation, and contiguity. Chichester, England: Wiley, 1981. (b)

8 Allison, J., Miller, M. & Wozny, M. Conservation in behavior. Journal of Experimental Psychology: General, 1979, 108, 4-34.

9 Awh, R. Y. Microeconomics: Theory and applications. New York: Wiley, 1976.

10 Collier, G. H., Hirsch, E. & Hamlin, P. H. The ecological determinants of reinforcement in the rat. Physiology and Behavior, 1972, 9, 705-716.

11 Green, J. K. & Green, L. Substitution of leisure for income in pigeon workers as a function of body weight. Behaviour Analysis Letters, 1982, 2, 103-112.

12 Kelsey, J. E. & Allison, J. Fixed-ratio lever pressing by VMH rats: Work vs. accessibility of sucrose reward. Physiology and Behavior, 1976, 17, 749-754.

13 Lea, S. E. G. The psychology and economics of demand. Psychological Bulletin, 1978, 85, 441-466.

14   Lea, S. E. G. & Roper, T. J.   Demand for food on fixed-
     ratio schedules as a function of the quality of concurrent-
     ly available reinforcement.   Journal of the Experimental
     Analysis of Behavior, 1977, 27, 371-380.

15   Lea, S. E. G & Tarpy, R. M.   Different demand curves from
     rats working under ratio and interval schedules.   Be-
     haviour Analysis Letters, 1982, 2, 113-121.

16   Marwine, A. & Collier, G.   The rat at the waterhole.
     Journal of Comparative and Physiological Psychology,
     1979, 93, 391-402.

17   McKenzie, R. B. & Tullock, G.   The new world of economics
     (3rd ed.).   Homewood, Illinois: Irwin, 1981.

18   Monod, J.   Chance and necessity.   New York: Knopf, 1971.

19   Olton, D. S.   Mazes, maps, and memory.   American
     Psychologist, 1979, 34, 583-596.

20   Rachlin, H., Green, L., Kagel, J. H. & Battalio, R. C.
     Economic demand theory and psychological studies of choice.
     In G. H. Bower (Ed.), The psychology of learning and
     motivation: Advances in research and theory (Vol. 10).
     New York: Academic Press, 1976.

21   Rapport, D. J. & Turner, J. E.   Economic models in
     ecology.   Science, 1977, 195, 367-373.

22   Rozin, P. & Mayer, J.   Some factors influencing short-term
     intake of the goldfish.   American Journal of Physiology,
     1964, 206, 1430-1436.

23   Shapiro, N. & Allison, J.   Conservation, choice, and the
     concurrent fixed-ratio schedule.   Journal of the Experi-
     mental Analysis of Behavior, 1978, 29, 211-223.

24   Veblen, T.   Why is economics not an evolutionary science?
     Quarterly Journal of Economics, 1898, 12, 373-397.

25   Watts, H. W. & Rees, A. (Eds.).   The New Jersey income
     maintenance experiments (Vol. 2):   Labor-supply responses.
     New York: Academic Press, 1977.

ANIMAL COGNITION AND BEHAVIOR
Roger L. Mellgren, editor
© *North-Holland Publishing Company, 1983*

THE ANALYSIS OF NEED

S.E.G. Lea

University of Exeter

"I need a microwave oven"

This chapter begins as a discussion of that simple statement. It must seem a bit of an oddity as a way of starting a chapter in a book on animal cognition and behavior. We do not suspect animals of needing microwave ovens, or even of feeling that they need them, still less of saying that they need them.

But my sample statement has a definite purpose. I have introduced it in an attempt to explain what a chapter on needs is doing in a book whose main focus is, after all, on cognition. And this is not the only chapter in that sort of area: those by Allison and Bolles also have a strong motivational component. Why are we breaking down the traditional partitions of psychology in this way?

Let us start by asking why a statement like "I need a microwave oven" should be interesting to a psychologist at all. I see at least three reasons.

1. It makes it clear that there is an irreducible cognitive aspect to motivational concepts like "need". You cannot "need" an artifact like a microwave oven, or a videorecorder, or a digital watch, unless you know both that such things exist, and what they are good for. Contrast this with the situation for a statement like

"I need something to eat"

"Children need a mother's care"

In such cases, we are readily prepared to suppose that such a need could have an instinctual or genetic basis, whatever we understand by that. The situation is of the sort that Pulliam

(chapter 13) considers as being dealt with by Lorenzian Fixed
Action Patterns. If, like God or Bolles (chapter 3), we are
in the business of designing animals, there is nothing to be
lost, and perhaps much to be gained, by making the individual's
experience of such needs independent of his or her individual
experience of the world. Of course such needs don't have to be
genetically determined, but they certainly could be.

2. A statement like "I need a microwave oven" strikes us as
actually or potentially phoney. It arouses our suspicions. We
are inclined to reply "Nobody actually NEEDS a microwave oven.
You simply WANT one." (I apologize to any reader who genuinely
and sincerely does need a microwave oven at this moment. If you
are in this sorry state of mind, please substitute your own
least favorite piece of consumer gadgetry.) In other words, to
think about a statement like "I need a microwave oven" is to
start thinking about what it actually means to NEED anything at
all, and that is a question that mainstream psychologists have
ignored for much too long.

3. The most important function of the statement "I need a
microwave oven", though, is that it starts us thinking about
the most important question in economic psychology. Before I
say what that question is, let me digress a little to say what
I mean by "economic psychology", for this too is a concept that
we shall need in the rest of this chapter.

"Economic psychology" is simply the attempt to explore the in-
terconnections between economics and psychology in a systematic
way, and come up with a coherent interdisciplinary account of
the phenomena to be observed at their interface. As an integrated field of study, it does not really exist in the English-
speaking world, but it does have quite a long, and continuous
history on the European continent. As long ago as the turn of the
century, the French social psychologist Gabriel Tarde published
a two-volume book entitled "La Psychologie Economique" (Tarde,
1902). I have recently translated a more recent French textbook into English (Reynaud, 1981), though what is probably the
best of the Continental texts is unfortunately in Swedish
(Warneryd, 1967). There are departments of economic psychology
in universities in France, the Netherlands, Sweden, and Germany,
and possibly elsewhere: the recent appearance of a Journal of
Economic Psychology in English suggests that at last the concept
is beginning to take roots in Britain and America.

Of course the component parts of economic psychology mostly do
exist in the English-speaking world. The study of human behavior in the workplace, of buying and selling, of spending and
saving, of gambling and investment--all these are common enough
in one form or another among American economists or psycholo-

gists, though too often they are left to a slightly despised
sub-species, the consumer or marketing scientists. What is
lacking is the attempt to bring then all together as a coherant
field of study which will involve concepts, methods and data
drawn from both economics and psychology, and from other disciplines as well.

Most Americans who have thought about the relation between
economics and psychology in recent years have been trying to do
one of two things. Either they have been trying to understand
economic behavior in terms of principles derived from the
psychological study of individuals, or else they have been trying to apply economic models to individual behavior. The first
is the approach of Allison (1981), Lea (1978), Scitovsky (1976),
and the one really well known American economic psychologist,
the late George Katona. The second is the approach of Rachlin
and Burkhard, 1978; Rachlin, 1981), Staddon (e.g., 1979), and of
the group of Chicago economists centering around G. S. Becker,
who have made something of a splash with an approach often referred to as "Time Allocation" or "The Economics of Time" (see
Becker, 1976). This contrast in the direction of applying
theories brings me back at last to my discussion of the statement "I need a microwave oven", and to that crucial question
for economic psychology, which I left the reader in suspense
about while I explained what economic psychology is.

In my opinion, the view of economic psychology as the simple
application either of economics to psychology, or of psychology
to economics, is much too simple (besides being likely to lead
to accusations of disciplinary imperialism from both sides). It
leaves unasked the crucial question, which can now be reviewed.

It is: What is the direction of causality in economic phenomena?

Does the behavior of the entire economy depend upon the preferences of the individual consumers within it, as the traditional economic doctrine of "consumer sovereignty" would have
us believe? Is the economy simply the aggregate of what we all
as individuals do? This is one version of the atomistic view of
society which has dominated Anglo-Saxon social thought right
back to Bentham, Mill and beyond.

Or does the kind of economy you live in, and your own position
within it, have a powerful, perhaps a dominant influence on the
behavior you show--on your personality--perhaps even on your
cognitions? Is "the economy" itself a vital actor in the economic drama? This is one version of the organic view of society
that has characterised Continental European thought right back
to Comte, Weber, and most of all, of course, Marx.

These two views give quite different accounts of a statement like

                    "I need a microwave oven".

Consider first the organic view, that individual psychology is a function of the economic situation. This position is some-times referred to as "economic determinism", though "vulgar Marxism" might be a better term. But despite its association with Marx, the version of this view that is easiest to use is associated with two American writers. Both of them have pro-duced ideas that lie somewhat outside the main stream of Anglo-American economics, and have been duly ostracised by their academic colleagues for doing so. Neither has been paid much attention by psychologists. I refer to Thorstein Veblen and John Kenneth Galbraith.

Veblen it was who gave us the phrase "conspicuous consumption", and who produced the first modern account of "status symbols". His best known book, "The Theory of the Leisure Class" (Veblen, 1899/1979), is a classic of sustained irony. It sets out in great detail Veblen's theory that many goods are consumed simply because they are expensive, their purchase therefore signals that the purchaser can afford them, and therefore has social status, position, or power. According to Veblen, there-fore, consumption of any commodity cannot be understood in terms of its physical characteristics. It is essential to take into account its social function--which is, of course, liable to change. In a classic but, it appears, too little read paper, Leibenstein (1950) showed how acceptance of this simple idea can have drastic effects on such basic applications of economic theory as the analysis of supply and demand. Not that we should really think of Veblen's ideas as all that simple: he works out their consequences in considerable detail and with considerable subtlety, not to say dry humour. But for our present purposes we need to take from Veblen one essential point. The amount of a commodity that people will buy, and the price they will be willing to pay, is not going to be a function of its physical characteristics and the unchanging human needs they meet. In-stead it will be a function of the entire current pattern of consumption in the society in question, and the social meanings attached to the different goods available within it (for more on this last point, see also Douglas and Isherwood, 1979).

Galbraith owes much to Veblen, a debt he acknowledges (e.g. Galbraith, 1975, p. 48). But with his usual gift for the tell-ing phrase that pinpoints a previously vague idea, he produces a clearer statement of the way particular social forces are held to change people's demand for goods, when he talks about "artificially created wants". Here the idea is that the need

for conspicuous consumption is only one way in which demand
for goods reflects something other than their inherent capacity
to satisfy human needs. Galbraith (1972) claims that there are
specific groups of individuals and societal forces working within society to modify demand in their own interest (and against
the interests of consumers). These interests include the ad-
vertising industry, of course, but also the entire "planning
sector" of the economy--the technologically sophisticated in-
dustrialized sector, financed by retained profits rather than
external share capital, where the investment needed to produce
a new commodity is so vast that the firm cannot risk consumers
being unwilling to buy it. They must therefore insure themselves by trying to find out consumers' wants and needs, and
either producing what consumers currently want, or, if that is
too inconvenient or expensive, quite as willingly and almost as
easily, changing what consumers want.

Both Veblen and Galbraith represent, as I have said, the organic
view of economic causation, in which economic behaviour is a
function of the society as a whole rather than of individual
psychology. They are true representatives of that view in that
both assert that the demand for goods is not a function of the
goods and of individual needs, or not a function of those variables alone. The alternative, atomistic view sees causation in
a different light: it sees the behaviour of the economy as being simply the result of aggregating a very large number of
individual behaviours, strictly caused by individuals' needs.
What can such an atomistic view say in response to the views we
have just discussed?

I make no secret of my sympathy for an organic approach to the
causation of economic behaviour in general, and for the views
of Veblen and Galbraith in particular. Nevertheless, the rest
of this chapter is devoted to trying to find out how good a job
we can do of finding an alternative account of statements like,

"I need a microwave oven".

an account that would lie within the mainstream English-speaking, atomistic tradition.

I have not yet characterized that alternative adequately. It
can be put shortly thus: People, and animals, have a relatively
small number of underlying needs. Commodities, the goods and
services that are traded in the market, or that (thinking of
the animal case) could be traded if a market was instituted,
are simply differing means of satisfying those underlying needs.
The use of these means depends on learning in two different
ways. As individuals, we each have to learn which goods satisfy
which needs--some of the ways in which this happends are

discussed by Bolles in this book. Secondly, the range of goods,
i.e. the range of ways of satisfying our needs, that is available at any particular time depends on a process of societal
learning, or rather invention--that is, it depends on the state
of our technology.

What follows from this approach is that the particular set of
goods for which an individual feels a need at any particular
time is subject to change. But it will not change in any
whimsical fashion, nor is it open to manipulation: it is a
function of individual knowledge and societal technological
advancement.

Let me summarize the questions which my simple statement

                    "I need a microwave oven"

has provoked. They are:

    1. How do we know that we need such artifacts?
    2. Can we distinguish "real needs" from "mere wants"?
    3. Can we relate our apparent needs for a great variety
        of commodities to a relatively small number of un-
        changing underlying needs?

Any one of these would be substance for a chapter. I propose
to concentrate on the third.

It has long been my view that if we can produce a successful
"underlying needs" model of economic behaviour, it will have
to be based fairly closely on the "new theory" of demand produced in 1966 by K. J. Lancaster of Columbia University.
Lancaster was attempting to deal with one of the most scandalous
omissions of the conventional theory of demand, its failure to
make any predictions concerning the substitutions that occur
between different goods when the price of one of them changes.
If (to adapt one of Lancaster's examples) the price of red
Cadillacs suddenly doubles, all other prices remaining constant,
it would not be unreasonable for there to be increases in the
demand for other goods. But we should expect these increases
to be felt more strongly in the demand for, say, black Cadillacs, and perhaps red Rolls-Royces, than in the demand for, say
Chevrolets or Volkswagens--let alone in the demand for china
tortoises, bow ties, butter or holidays in Bermuda. Demand for
all these other things might increase, but for purely personal
and idiosyncratic reasons: someone who has always wanted both a
red Cadillac and a holiday in Bermuda might decide to have the
latter rather than the former in a year in which red Cadillacs
have suddenly become relatively expensive. But the increase in
the demand for other colours of Cadillacs, or other ostentatious

red cars, would not be arbitrary in this way: it would be a reflection of the fact that, following a price hike for red Cadillacs, they represent a more efficient way of achieving the same ends.

Conventional economic theory can, to a limited extent, describe these different kinds of substitution: it can assess the pairs of commodities involved as showing different "elasticities of substitution: or "cross-price elasticities of demand". Lancaster's theory aims to describe them in a much more specific manner, and, in a certain sense, to predict them. Lancaster proposes that economic demand theory should be formulated, not in terms of demand for particular commodities, but in terms of demand for a relatively small number of underlying characteristics, which all commodities share to some extent. In fact any particular commodity can be represented as a vector of amounts of the characteristics. Lancaster insists that these characteristics should be seen as objective properties of goods, and not as anything psychological. The essential addition that an economic psychologist might want to make to his theory, therefore, is to suppose that each characteristic corresponds to the satisfaction of one of a limited number of needs.

The literature of economics (especially econometrics) and of economic psychology is beginning to contain a few empirical studies relating to Lancaster's theory. Lancaster himself includes a few early examples in the book in which he expounds the theory in greater detail (Lancaster, 1971). Most recent studies (e.g. Hensher's, 1981, investigation of Australian firms' requirements for sophisticated telecommunications equipment) rely on questionnaires or other verbal enquiry methods to discover what the underlying characteristics of the goods concerned might be, and typically they then investigate intended or predicted purchases rather than actual buying behaviour (e.g. Verhallen, 1982). While such studies are clearly a step in the right direction, they do not tell us whether Lancaster's theory can really be made operational at the level of overt individual behaviour. Nor are the characteristics that have been studied so far very convincing candidates for mapping into a set of fundamental underlying human needs: obviously it is not to be expected that they should be in the case of a study of industrial purchasing like Hensher's.

For some time now I have been trying to develop a variant of Lancaster's approach that would be usable in connection with animal behaviour. The analogy between performance of animals on certain schedules of reinforcement and the response of human consumers to variation in price and availability of goods is now well established (Allison, 1981; Hursh, 1980; Kagel, Battalio, Rachlin and Green, 19811 Lea, 1978). In particular, the

variation in quantities of reinforcers consumed as a function
of the fixed-ratio schedule on which they are made available
has been seen as analogous to the economist's demand curve,
which plots the variation in amounts of a commodity purchased
as a function of its price. The point of an animal analogue of
Lancaster's theory would be to predict the form the demand curve
should take, especially as a function of the availability or
otherwise of alternative reinforcers. We know (Lea and Roper,
1977; Rachlin, Green, Kagel and Battalio, 1976) that making substitutable alternatives available makes animals more sensitive
to increases in ratio schedules. The problem is to give an
account, preferably a quantitative account, of this intuitively
obvious phenomenon.

My proposed solution to this problem draws on ideas first proposed by Allison (1976) and Staddon (1979). It is important to
make clear, however, that I am using their theories in simplified forms. When, below, I refer to predictions from "Allison's
model" or Staddon's model", therefore, it should not be thought
that either Allison or Staddon can be held responsible for the
predictions I am making. I have isolated what seem to me to be
the key ideas from these two approaches, and pushed them further
(and perhaps in different directions) than their original
authors chose to. Not surprisingly, I can then demonstrate that
neither model gives a good account of some published data, or
of some new data that I shall present. That should not be interpreted to mean that the models' authors cannot cope with
those data, by adding extra detail to their models or restricting their field of application. In the case of Allison's model,
in fact, we see in chapter 1 of the present book how this is
done. My object is to show how, by combining the simplest versions of the two models we can arrive at a compound model which
both accounts for all the data at present to hand (and without
any free parameters, at that), and is related in an obvious way
to Lancaster's approach to demand theory. It thus has considerable integrative power, though I certainly would not claim at
this stage that it has better predictive power, or performance,
than the full-dress versions of the Allison or Staddon models
that their authors would no doubt wish to expound.

Of the two published models I wish to consider, Allison's is
perhaps the better known. It is also the simpler to explain.
In setting it out I shall modify the usual symbols, as some of
them are used with different meanings in some of the other
models I want to consider.

Allison's model is usually developed in terms of two responses,
thought of as consummatory and instrumental. I shall call
these responses C and I respectively; later, I shall also

want to consider a third response, which I shall call F, for
reasons that will become obvious. The rates of these three
responses I shall call c, i and f respectively. The essential
point of Allison's model is that the two responses C and I both
function as sources of some underlying quantity, which I shall
call X. We do not need to enquire at this stage what X might
be. It could be some desirable quantity, like energy input.
It could be some inherently limiting quantity, like time used
up. All that Allison specifies about it is that the rate at
which it is obtained (or performed, or generated, or used) is a
linear function of the rates of I and C. So if we let x be the
rate of X, we can write Allison's hypothesis as a very simple
equation:

$$x = ai + bc \qquad\qquad (1)$$

where a and b are unknown constants. As a matter of fact we
don't need both of them for Allison's model, only their ratio,
which Allison (1976) called k; but I am keeping them both in at
present, for compatibility with what comes later.

The first thing to note is that this model already takes us a
large step towards the goal of a psychological model that can
be related to Lancaster's demand theory. The quantity X has
just the properties of one of Lancaster's characteristics: it
is linearly related to actually available commodities (for
which, in the present context, read "responses"), and it is the
true determinant of behaviour when those responses are not--in
that it is the characteristic which is held to a constant rate,
not individual responses. It is this feature that give
Allison's model its usual name, the "conservation hypothesis".

Next we need to consider what predictions can be made from this
simple hypothesis. For all the models I shall consider here,
there are three areas where I shall look for predictions. They
are as follows:

1. The relation between different baseline measures of
    response rate;
2. The form of the demand curve, or of other curves
    derived from it;
3. The quantitative values of demand as a function of
    baseline response rates.

In the case of the conservation hypothesis, it happens that
Allison himself has concentrated on the second area of prediction. The other two are none the less areas where we can
apply the algebra of the model, but I should make it clear that
using it in this way would not necessarily meet with the appro-
val of its original author.

1.   Relation between baselines.

Where we have two responses under consideration, two different kinds of baseline performance can be observed. We can either make both responses available, and observe the rates of both, or we can make only one available and observe its rate in iso-lation. These are usually called "paired-baseline" and "single-baseline" conditions respectively. Throughout this chapter, I shall denote a response rate observed under paired baseline conditions by a subscript zero. Thus if we make both I and C available, the response rates we observe will be call i_0 and c_0 respectively. I shall denote single baseline rates by a subscript indicating the single response that was available, so if only the instrumental response I is available, its rate will be call i_i, and so forth. Note that these two conventions can be applied to characteristics as well as to directly observable responses, so the rate of the characteristic X under paired baseline conditions will be called x_0, its rate when only the instrumental response, I, is available will be called x_i, and its rate when only the consummatory response C is available will be called x_c.

The key to predicting the relation between baselines under the conservation hypothesis is to recognize that, because x is, on this hypothesis, held to the same value under all conditions.

$$x_0 = x_i = x_c \tag{2}$$

From these equations it is easy to work out relations between the paired and single baseline values of i and c, since straight-forward application of conservation shows that

$$ai_0 + bc_0 = x_0 \tag{3}$$

$$ai_i = x_i \tag{4}$$

$$bc_c = x_c \tag{5}$$

Eliminating x, a and b between Equations 3, 4 and 5 leads to the following easily tested relations:

$$i_0/i_i + c_0/c_c = 1 \tag{6}$$

$$(i_i - i_0)/i_0 = c_0/(c_c - c_0) \tag{7}$$

Equations 6 and 7 are in fact algebraically equivalent to each other and to the conservation hypothesis, so the truth of either one of them provides a necessary and sufficient condition for the truth of the conservation hypothesis, as applied to the

relation between baseline performance. Obviously, these relations make assertions about the way the two responses substitute for each other when one of them is made unavailable.
The greater the value of i_i-i_0 for example, the more response I
us used to replace response C when C is prevented. As a second
example of predictions from these equations, we can note that
they require both single baseline rates to be higher than the
corresponding paired baseline rates, since if this is not true
either i_0/i_i or c_0/c_c would be greater than one, so their sum
would be of necessity greater than one.

2.   The form of the demand curve.

What happens, according to the conservation hypothesis, when
we impose a fixed ratio schedule of reinforcement? Under these
conditions, the rate at which the consummatory response C can
be performed is constrained: it only becomes available when
some number (say r) of instrumental responses have been performed. For convenience let us suppose that the schedule is a
reciprocal one, so that the opportunity for the instrumental
response is also withdrawn whenever the consummatory response
is available (Allison, 1971). Then under the schedule it will
always be true that

$$i = rc \qquad (8)$$

If conservation holds, it will therefore be true that

$$arc + bc = x \qquad (9)$$

so that

$$c = x/(ar+b) \qquad (10)$$

Equation 10 gives the mathematical form of the demand curve
under conservation. A number of consequences, some of them
surprising, flow from this form. One that will concern us when
we come to consider some data is the following. The sum of the
rates of the instrumental and consummatory responses is given
by the expression:

$$i+c = (1+r)x/(ar+b) \qquad (11)$$

Differentiating this by r gives the following:

$$d(i+c)/dr = x(b-a)/(ar+b)^2 \qquad (12)$$

Since a and b are by definition constants, this quantity can
never be xero, unless it is zero at all values of r. It

follows that i+c is a monotonic function of r, regardless of the values of a, b and x.

3. Quantitative prediction of demand

Equation 10, the demand curve equation can be used to obtain quantitative predictions of the values of c (and hence of i, since i=rc) under any ratio. For this we need values for a, b and x, but by using Equations 3, 4 and 5, these can be obtained from the single and paired baseline values of i and c--provided, of course, that the predictions about the relations between those baselines, discussed above, are found to be accurate.

Now I want to turn to Staddon's model. From one point of view, this concentrates more at the level of overt behaviour: it asserts that what the organism is trying to achieve relates to the rates of observable responses, not those of unobservable characteristics. But it proposes its own underlying quantity, a variable which I shall call utility (to emphasize the link to conventional economic theory).

According to Staddon, the organism has certain ideal rates at which it would prefer to perform the instrumental and con-summatory responses I and C. Because these are ideal rates, we say that utility is maximized when they are achieved, and that any deviation from them will cause a loss of utility. Since they are ideal rates, presumably the animal will perform at these rates when both responses are freely available, that is, under paired baseline conditions; so we can represent the ideal rates by the symbols introduced above for paired baseline rates, i_0 and c_0. The only remaining question is how exactly utility will fall off if the animal must for some reason perform either I or C or both at something other than an ideal rate. Staddon adopts almost the simplest possible functional form for this, proposing the following equation, which we can call his utility function:

$$u = - k_1(i-i_0)^2 - k_2(c-c_0)^2 \qquad (13)$$

This function simply asserts that the further we go from ideal response rates, the less the animal will prefer the situation. Deviations of the two different response rates matter to dif-ferent extents, expressed by the constants k_1 and k_2. As in the development of Allison's model, we could if we liked drop one of these constants and use their ratio, but I am keeping them both in to help later developments.

Before we can consider the predictions of Staddon's model we need to know how the animal will respond when it cannot achieve ideal response rates. The answer is obvious: it will choose whatever available pattern of responding gives the greatest utility. This turns out to be all that we need to assume. From it we obtain the usual name for the model, the "minimum distance hypothesis". This name derives from the fact that maximizing u involves choosing a pair of response rates that are as near as possible to the ideal pair, in a space whose co-ordinates are the rates of the two responses rescaled by dividing i by k_1 and dividing c by k_2.

Let us therefore consider the predictions of the minimum distance hypothesis under the same three headings as we used for conservation.

1.   Relation between baseline rates.

We have already seen that under paired baseline conditions, the ideal rates of both I and C can be achieved. But what happens under single baseline conditions? Examination of Staddon's utility function makes it obvious that a very simple prediction can be made. Under a single baseline of instrumental responding, for example, the consummatory response C is unavailable and so its rate c is necessarily zero. But any change of i, the rate of the instrumental response, from its ideal (paired-baseline) value i_0 can only lead to a decrease in utility. Thus the minimum distance hypothesis asserts that single baseline response rates must be the same as paired baseline rates, a surprising and counter-intuitive prediction, and one that is inconsistent with data already available in the literature. It should be pointed out that the version of the model studied by Staddon (1979) does not make this prediction. We shall see later that the additional assumptions necessary to avoid it amounts to adopting a special case of the compound model to be advanced below.

2.   The form of the demand curve.

In order to discover what form the demand curve takes under the minimum distance hypothesis, we have to differentiate the utility function under the constraint, i=rc, imposed by a ratio schedule. This leads to the following expression:

$$c = (k_1 i_0 r + k_2 c_0)/(k_1 r^2 + k_2) \qquad (14)$$

This has many properties in common with the form we derived from the conservation hypothesis (Eq. 10), but it is not identical to it--there is no way of mapping the parameters of (10) into parameters of (14). And if we consider the sum of i and c, and differentiate as we did for the corresponding sum under the

conservation hypothesis, we come to quite a different conclusion: we find that this sum should have a maximum (or possibly a minimum) at some value of r, though the expression for
what that value of r will be is not simple or useful enough
for me to reproduce it here. Once again, however, the minimum
distance hypothesis makes a distinctively different prediction
from the conservation hypothesis, though in this case the weight
of published data lie on the side of minimum distance: in
Mazur's (1975) study of rats wheelrunning for water reinforcement, for example, the sum wheel+run shows a clear maximum.

3. Quantitative prediction of demand.

Unfortunately, it is impossible to make any prediction of the
actual level of consummatory and instrumental responding under
particular ratio schedules using the minimum distance hypothesis. Obviously, to make such a prediction we should need
to know the values of k_1 and k_2 (or to be precise, their ratio),
and because the relation between single and paired baseline
rates is so simple, it does not give any information about these
parameters. In this respect minimum distance is a markedly less
powerful hypothesis than conservation.

However, we have now seen two areas in which our two simple
models make different predictions. As regards the relation
between baselines, minimum distance predicts that single and
paired baselines will be the same, while conservation allows
them to be different (while making predictions about the
quantitative relations between the four response rates involved). As regards the form of the demand curve, or rather of
the relation between i+c and the ratio parameter r, conservation
predicts that this function will be monotonic, while minimum
distance requires it to have a maximum or minimum. In terms of
both intuition and extant data, conservation seems to have the
better of the contest in the first case, while minimum distance
does better in the second. This is enough to suggest that a
comprise between the two might be worth considering. I want
to propose the simplest possible compromise of this sort; I have
called it the "minimum characteristics distance hypothesis"
(Lea, 1981). For convenience we can shorten this to MCD.

The MCD model simply combines the features of the conservation
and minimum distance hypotheses. Like conservation, it assumes
that what matters to the animal is not response rates, but
underlying characteristics. However, it supposes that there
are two of these instead of just one, so we have the equation:

$$x_1 = a_1 i + b_1 c \qquad\qquad (15a)$$

$$x_2 = a_2 i + b_2 c \qquad (15b)$$

We now have four unknown constants (a_1, a_2, b_1, b_2) instead of two, though once again they are not all independent. But if the ratios a_1/b_1 and a_2/b_2 are different it will be impossible to conserve both x_1 and x_2 under the constraints imposed by (for example) a ratio schedule of reinforcement. Instead of supposing that characteristics are conserved, therefore, we suppose that they behave like the responses in the minimum distance hypothesis--that is, that the animal will get as close as it can to ideal rates of characteristics with two characteristics and two response rates, any desired characteristic rates can be achieved under paired baseline conditions, so it is reasonable to take paired baseline rates as ideal. This leads to the following utility function:

$$u = -k_1 (x_1 - x_{10})^2 - k_2 (x_2 - x_{20})^2 \qquad (16)$$

As in the minimum distance model, we then work out what the animal will do in any situation by finding what behaviour maximizes u.

We can consider the predictions of the MCD model under the same three headings as we used before.

1.   Relation between baseline response rates.

The MCD model places no restriction whatever on the relation between single and paired baseline response rates. Thus it is markedly less powerful than either conservation or minimum distance. However, it does turn out to make rather odd predictions for the case where either of the single baseline rates is less than the rate for the same response under paired-baseline conditions (recall that the conservation hypothesis sets an absolute ban on this happening).

2.   The form of the demand curve.

It is possible to work out what the rate of the consummatory response should be at any ratio schedule, in terms of the underlying parameters of the model, a_1, a_2, b_1, b_2, k_1 and k_2 (together with paired baseline response rates). The resulting equation does not look very tidy (Lea, 1981, Eq. 16), but it has an interesting form. Replacing various complex expressions in terms of the a's, b's and k's by constants A, B, C and D, we find that the demand curve can be expressed as:

$$c = (A + Br)/(1 + Cr + Dr^2). \qquad (17)$$

Now, if we simplify the demand curves under conservation and minimum distance (eqs. 10 and 16 respectively) in the same way, we find that (10) takes the form

$$c = A/(1 + Cr) \qquad (18)$$

while (16) takes the form

$$c = (A + Br)/(1 + Dr^2) \qquad (19)$$

The family relationship of the three models now becomes obvious: both minimum distance and conservation (eqs. 18 and 19) can be seen to be special cases of the MCD model (eq. 17). Conservation is obtained by setting B and D to zero, while minimum distance is obtained by setting C to zero.

3.   Quantitative prediction of demand.

It turns out that the untidy formula given by Lea (1981) for the demand curve under the MCD model can be improved considerably. As in the simpler models, response rates under baseline conditions are functions of the underlying parameters (the a's, b's and k's), and it is possible to use this fact to replace these parameters in the expression for the demand curve by baseline response rates. The resulting demand curve equation is still rather unpleasant looking:

$$c = \frac{c_c i_0 (i_i - i_0) + i_i c_0 (c_c - c_0)r}{i_0(i_i - i_0) + 2(i_i - i_0)(c_c - c_0)r + c_0(c_c - c_0)r^2} \qquad (20)$$

but it has one major advantage. It contains no free parameters. It can therefore be used directly, to obtain predictions of c (and hence of i) under any ratio schedule.

We can now see how to set about testing these various models. All we need to do, it appears, is to devise a situation where with two responses, we can observe paired baseline performance, single baselines of each response, and performance under a number of ratio schedules (preferably symmetrical ratios) linking the two. As I have already noted in passing, there are a number of published studies that meet this specification, notably those of Mazur (1975) on drinking and wheel-running and of Timberlake (1979) on drinking two different concentrations of saccharine. Much relevant data is also to be found in the earlier work by Allison and Timberlake that effectively started this whole field off. I have already referred to the typical data obtained in these experiments: the total of instrumental and consummatory response rates is a nonmonotonic function of the fixed ratio (or, at any rate, exceeds one or other of the single baseline rates), infirming the conservation hypothesis, but the single baseline rates exceed paired baseline rates, infirming the minimum distance hypothesis.

I want to present some data from a new kind of experiment, using a somewhat unconventional fixed-ratio schedule. As we have seen, what the schedule does is to force the two responses with which we are concerned to occur at proportional rates. There are a number of other situations that have the same general effect. Consider, for example, an animal faced with a diet that consists of a mixture of two foodstuffs. Our subject probably has a preferred ratio in which it would eat these two foods if they were freely available separately. Faced with them in a mixture of some other ratio, it must either eat too little of the constituent that is present in less than the preferred proportion, or too much of the under-represented one (or, of course, too little or too much of both). This is exactly like the situation of an animal under a fixed-ratio schedule, which must either leverpress more than it normally would, or else eat less than it normally would. In other words, a food mixture is effectively a fixed-ratio schedule. It may not be obvious which food corresponds to an instrumental response and which to a consummatory response, but that distinction is in any case completely unimportant the moment we start to think in terms of reciprocal schedules.

For my purpose, food mixtures have a considerable advantage over conventional schedules, in that it is possible to guess what the characteristics that might underlie overt behaviour would be. Such quantities as protein or energy intake suggest themselves immediately. Furthermore, the general approach of the minimum characteristics distance model has much in common with techniques, such as linear programming, that have been applied to diet problems in the literature of ecology and nutrition. And, finally, this is a chapter on needs: and if we cannot analyse needs in so simple an area as the makeup of an animal's diet, then I don't think we ever have much prospect of making an analytic approach to need work at all.

The experimental situation I have worked with is exceedingly simple. A number of rats (enough to use a fully balanced Latin square design to control for order effects between conditions) live in individual cages, with free access to standard laboratory diet and water. In addition, they have access on some basis to two other foodstuffs, both of which they will eat freely. These could be anything, but I have used sucrose (in the form of ordinary kitchen granulated sugar) and dried skimmed milk powder (again a kitchen grade, sold in Britain under the trade name "Marvel"). To begin with I used these unmodified,

but now I grind them up in a coffee mill to give equal particle
sizes. These substances are offered in cups firmly fixed to the
cage wall to prevent spillage. Paired-baseline conditions now
consist of giving two cups, one containing milk powder and the
other containing sugar. Single baselines consist of offering
just one substance, and fixed ratios consist of mixing the two
powders in some fixed proportion, and offering the mixture in
a single cup. A "session" consists of a 24-hour period (23.5
hours, to be precise, as servicing the experiment takes about
30 minutes each day) during which one of the possible conditions
is in force. At the end of this period, the amount of the substances that have been eaten is determined by weighing. The
rats, and their standard diet and water are weighed too. In
my first experiments, the rats had a new condition every day,
which made quite elaborate experiments very quick and easy to
run, but this obviously makes it possible that there will be
some carry-over effects, so in my most recent experiments I have
given the subjects a day or two off between conditions.

TABLE 1

Mean weights of sugar, milk powder and standard diet consumed by
12 rats in an experiment where standard diet and water were always available, while milk and sugar were made available for
23.5 hour sessions, in various combinations as follows: Paired
baseline: both sugar and milk, in separate containers. Single
baseline: either sugar or milk on its own. Mixture: milk
powder and sugar mixed together in the ratio 1:3.5 by weight,
delivered in a single container. All weights are in grams.

Consumption	Substances Offered			
	Sugar and Milk	Milk Alone	Mixture	Sugar Alone
Sugar	9.17	-	11.91	11.26
Milk Powder	5.29	6.57	3.40	-
Sugar + Milk	14.46	6.57	15.31	11.26
Standard Diet	5.58	13.83	6.17	9.92

Table 1 shows the result of one such experiment. In this case,
I used four conditions: single baselines of both milk powder
and sugar, paired baseline, and a mixture consisting of one
part milk powder by weight to 3.5 parts sugar (about equal parts
by volume). In this experiment, there were 12 rats, and they
had five "rest days" between each experimental day, so that on

any particular day only two of the rats were actually being tested--the rest just had standard diet and water available. Within a block of six days, all twelve rats were tested, three of them under each condition.

What happened is clear enough from the table. The rats ate more of both substances under single-baseline than paired-baseline conditions (this trend was actually significant only for sugar, though). And they ate more of the mixture than they did of either substance under single-baseline conditions (this was significant for both substances).

What are the implications of these results for the three models introduced earlier? We can consider them under the same three headings that were used to introduce the predictions made by those models.

1.   Relation between different baseline performance.

Recall that, according to conservation, the baseline response rates should be related by the following two equiations:

$$i_0/i_i + c_0/c_c = 1 \qquad\qquad (6)$$

$$(i_i - i_0)/i_0 = c_0/(c_c - c_0) \qquad\qquad (7)$$

Of the twelve rats in the experiment just described, it was the case for 11 that the sum $i_0/i_i + c_0/c_c$ exceeded 1.0 (the mean value of the sum was 1.72). Thus conservation failed to predict the relation between baselines satisfactorily.

The simple minimum distance model does no better, however. As we saw above, this requires single and paired baseline rates to be the same for both substances, and it is obvious from Table 1 that this did not happen.

The minimum characteristics distance model makes no prediction about the relation between baselines, so obviously these data can neither refute that model nor lend it any support.

2.   The form of the demand curve.

At first sight it is not obvious how the present data can be used to test predictions about the form of the demand curve, since demand data have to do with consumption at different "prices", i.e. at different fixed ratio schedules, and in this experiment only one such schedule was used, the 1:3.5 milk-powder:sugar mixture. However, the single baseline conditions provide implicit additional information about demand. Obviously, intake under single baseline sugar conditions, for example, must be very close to intake of a mixture containing a thousand parts of sugar to one of milk powder (or, come to that, one of anything else). Similarly, intake under single baseline milk conditions must be very close to intake of a 1000:1 milk-powder:sugar mixture. In other words, we can take the two single baseline performances as equivalent to performances under two additional fixed ratios, one very large and the other very small.

Given that, it is possible to use the present data to test the conservation model. It was shown above that, if conservation holds, the sum (i+c), that is, the total intake under mixture conditions, must be a monotonic function of the sugar:milk-powder ratio. It follows that, according to conservation, intake under the 1:3.5 mixture condition must lie between the intake levels under the two single baseline conditions (seen now as extreme ratio conditions). As has already been stated, this does not occur: mixture consumption was significantly higher than either single baseline. Thus the form of the total consumption curve, which is directly derived from the demand curve, is inconsistent with the conservation model.

The minimum distance model, on the other hand, predicts that total consumption will pass through a maximum (or possibly a minimum), and that is what happened in the present experiment. The minimum characteristics distance model allows for either a monotonic or a nonmontonic relation between (i+c) and r, and so once again it is impossible for it to be either refuted or supported by the data of this experiment.

As we saw earlier, all three models make more precise (and different) predictions about the algebraic form of the demand curve. With only one real set of mixture data, however, it really is impossible to test these predictions, except in a sense in which they reduce to the final separate type of prediction from the models, now to be considered.

3. The quantitative value of demand as a function of baseline response rates.

The conservation and minimum characteristics distance models (but not the simple minimum distance model) yield expressions for consumption at any sugar:milk-powder ratio as a function of single and paired baseline intake rates. However, because it has already been shown that the single and paired baseline rates are not in the relation to one another required by the conservation model, it is impossible to work out what the predictions

of that model are. This section can only concern itself, therefore, with the minimum characteristics distance model.

Equation 20 can be used for predicting consumption from the
baseline response rates, at any given fixed ratio. Using this
equation separately for each rat, the mean predicted intake of
the 1:3.5 mixture was 15.02 grams. The actual mean intake was
15.37 grams, and these two figures do not differ significantly,
indeed the agreement between them is impressive. However, there
is some reason to mistrust the mean predicted estimate: the
prediction equation involves the ratio of two observed quanti-
tatives, and means of such ratios tend to be biased. This can
be corrected by a technique known as "jackknifing" (Miller,
1974). The jackknifed mean predicted consumption was 13.71
grams. This also does not differ significantly from the observed mean, though it suggests a slightly less impressive fit
than the simple mean prediction.

From the one test that can give some information on it, therefore, the minimum characteristics distance model emerges un-
scathed, indeed with some credit. Nevertheless, it would be
better if it could be subjected to a more searching test. For
this purpose the results of a second experiment can be considered. The technique was very similar to that used in the
experiment just described, except that this one was actually
performed earlier, and there were a number of clumsinesses about
the way it was done. For instance, there were no "rest days"
between conditions, and the sugar and milk powder were not
ground up to constant particle size. Once again twelve rats
were used, but this time there were twelve different conditions,
single baselines of sugar and milk powder plus mixtures at ten
different ratios. The experiment was run as a Latin square
design spanning twelve days, so that on any one day each rat
was experiencing a different condition. In addition, the rats
experienced five days of paired baseline, one before the main
part of the experiment, and one after every three days of the
Latin square. The point of this design was to get, fairly
rapidly, a complete curve showing the relation of mixture consumption to mixture constitution.

Figure 1 shows what this curve looked like, and it also shows
the consumption of the two components of the mixture under each
condition and under paired baseline. Obviously, the results
were substantially similar to those of the previous experiment.
We can summarize them under the same three headings.

1.   Relation between baseline response rates.

Once again, the sum $i_i/i_0 + c_0/c_0$ was calculated for each rat.
Its mean value across rats was 1.40, which was significantly

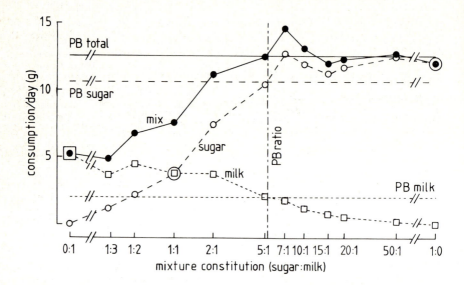

Figure 1. Mean daily consumptions in an experiment in which 12 rats each had access to one of 10 different mixtures of milk powder and sugar, or one of the pure substances, for a 23.5-hour session. The horizontal lines show mean consumption on intervening days when both substances were available (paired baseline condition). The vertical line shows the ratio of milk to sugar consumption under paired baseline conditions. Note that the scale on the horizontal axis is logarithmic.

greater than 1.0, contradicting the prediction from conservation. And once again, both single baseline intakes exceeded the corresponding paired baselines, though in this case the difference was significant for milk powder and not for sugar. This is sufficient to refute the simple minimum distance model.

2. The form of the demand curve.

The relation between mixture intake and mixture constitution appears from Figure 2 to be nonmonotonic, which is consistent with the minimum distance model (and the minimum characteristics distance model) but not with conservation. However, the difference between the peak intake (at a constitution of about 7 parts sugar to one of milk powder) and intake of more sugary mixtures falls short of significance, using Scheffé tests following analysis of variance of the Latin square data, so this aspect of the data cannot be used to reject either of the simple models.

With the large number of mixtures used in the present experiment, though, it is possible to use more powerful methods to investigate the form of the consumption curve in more detail. Recall that Equation 17, the demand curve under the minimum characteristics distance model has the general form:

$$c = (A + Br)/(1 + Cr + Dr^2) \qquad (17)$$

and that the conservation and minimum distance models are special cases in which, respectively, both B and D, and C only, are zero. It is not hard to test between these three models by using non-linear regression to fit this general form to consumption data, and looking at the values of A, B, C and D that emerge as giving the best fit. When this was done, separately for the data for each rat in the present experiment, it was found that for every one of the twelve rats, C was greater than zero, while for eleven of them both B and D were greater than zero. Thus both special cases can be rejected by signs tests: the form of the consumption data disagree significantly with both conservation and minimum distance models. Once again, though, these data do not give any evidence either way about the minimum characteristics distance model, since there is no more general model with which it can be compared.

3. Quantitative prediction of demand.

As was explained for the previous experiment, it is only for the minimum characteristics distance model that quantitative prediction of demand is practicable. Figure 2 shows the mean predicted mixture consumption (using the jackknife technique) plotted along with the observed data. I think it is fair to say that the fit is pretty good, although it is not obvious how we could test the significance of any deviations. It should be remembered that the prediction is made without the benefit of any free parameters.

The second line on Figure 2, and the second set of observed data, refer to the rats' consumption of standard diet, which in all these experiments is always available. This is an aspect of the procedure I have so far ignored, and it raises a whole series of further possibilities, and further possible

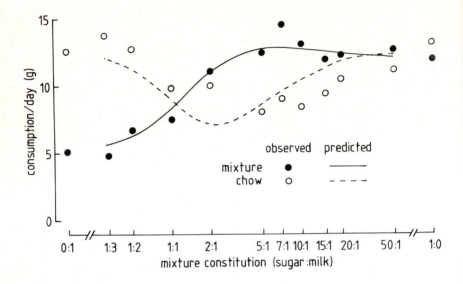

Figure 2. Solid line and filled circles show mean consumption of sugar-milk mixtures, as a function of
   mixture composition, compared with predictions from
   the Minimum Characteristics Distance (MCD) model.
   Dotted line and open circles show corresponding data
   for standard diet consumption. Each data point shows
   mean consumption for 12 rats who experienced the
   mixtures on different days.

models. In effect, the rats always had two responses available to them, eating the sugar/milk mixtures, or eating the
standard diet. If we are regarding eating mixtures as an
analogue of working under a fixed-ratio schedule of reinforcement, therefore, there is a sense in which these rats were
really working under two concurrent ratios. Furthermore, the
different conditions certainly generated different rates of
standard diet intake: Table 1 includes standard diet intake
levels under the different conditions of the first experiment
I described, and Figure 2 shows them at different mixture
constitutions for the second experiment. Obviously intake of
the "free" standard diet interacted with intake of the milk

and sugar in both these experiments. Might this not make a
major difference to the way in which the various models ought
to be tested?

What is involved here is taking into account the third response,
F, which was introduced at the beginning of this discussion of
models, but so far has not been used. In fact it turns out
that bringing F into consideration makes surprisingly little
difference to the predictions of the models so far discussed.
The minimum characteristics distance model was originally
developed to handle concurrent ratio schedules of different,
but partially substitutable reinforcers (Lea, 1981), so for
this model at least much of the algebra required is ready at
hand; and the other two models can be handled as special cases
of MCD, by choosing parameters appropriately. Here I shall try
to develop their implications in a more intuitive manner.

As regards the conservation model, it is true that the simple
predictions about relationships between baseline rates, and
prediction of demand data from baselines, both break down if
we suppose that the "free" response is relevant to the characteristic that is conserved. To find out what to put in their
place, we have to think in terms of the single characteristic,
x, being a linear function not only of the instrumental and
consummatory response rates, i and c, but also of the rate of
the "free" alternative response, which I shall call f:

$$x = ai + bc + b'f \qquad (21)$$

This condition has been stated by both Allison (this volume),
and by Staddon (1979), but both have used it in rather special
ways. Allison uses it only in the case where F is an unobserved
response, whose rate is to be inferred. Staddon similarly
treats F as an unobserved response, but for him the conserved
characteristic, X, represents the total time available, so that
F is really the category of "all other behaviour". In this
case the "conservation" conditions over the three responses I,
C and F is simply a way of stating the "budget constraint" that
all responses take time and time is limited. Staddon's original version of the minimum distance model in fact included this
budget constraint, and Staddon (1979) shows that, with the
budget constraint, minimum distance does not make the prediction, which I have shown to be false, that paired-baseline
and single-baseline response rates are equal. But there are
predictions of this extended conservation model which can be
used to test it in ways that neither Allison nor Staddon, so
far as I know, has explored. For the conservation condition,

$$x = ai + bc + b'f \qquad (22)$$

must be true under four special conditions. Under paired baseline, we must have

$$x = ai_0 + bc_0 + b'f_0 \tag{23}$$

Under the single baseline of the instrumental response, we must
have

$$x = ai_i + b'f_i \tag{24}$$

and under the single baseline of the consummatory response we
must have

$$x = bc_c + b'f_c \tag{25}$$

The fourth condition is the case where neither the instrumental
nor the consummatory response is available: In this case we
must have, simply

$$x = b'f_f \tag{26}$$

where f_f denotes "free" responding under this "non-availability"
condition. Furthermore, all these x's must be equal--that is
the meaning of conservation. Remembering that a can be set
arbitrarily to zero, we now find ourselves with four equations
(23-26), and only three unknowns (x, b and b'), since all the
other terms are observable. This is an equation system which
will have a solution only if certain relationships among its
coefficients are met. In the present case, it is the observ-
ables (i_0, c_0, etc) which are the coefficients, so we are able
to deduce a relationship among these terms which is predicted
by this extended form of conservation. This relationship
can be stated in a number of ways, none of them very intuitively
meaningful. Here is one:

$$c_c f_f (i_i - i_0) + i_i (f_0 c_c - f_f c_0) = 0 \tag{27}$$

For the first experiment I described, the value of f_f, the
standard diet consumption when neither milk nor sugar was
available, is readily obtained from the "rest days" between
conditions. Using data from the last rest day before each new
condition (so as to minimize any carry-over effects), I have
found a mean f_f of 23.0 grams, and have calculated the quantity
on the left-hand side of the equation just given (27). It is
less than zero for every one of the twelve rats. Thus even
this extended form of conservation is inconsistent with the present data.

There is an alternative way of extending the conservation model
to handle the case where two responses are available, and that

is to suppose that there are two characteristics, both contributed to by all three responses (instrumental, consummatory,
and free). But in this case it turns out that the prediction
about the general form of the demand curve is the same as for
simple two-response conservation, so this model can be dis-
counted on the basis of the non-linear regression results for
the second experiment I described.

It is even easier to see what happens under the minimum distance
hypothesis when we introduce free food. (I refer here to the
simple form of minimum distance discussed so far, without the
budget constraint.) According to this model, the three responses must all be assumed to be completely non-substitutable,
so the addition of the "free" response should make no difference whatsoever to performance of the other two responses,
under any condition. Thus all the predictions tested above, and
found to be false, remain predictions of the model.

For the minimum characteristics distance model, with three
responses available we have to consider three characteristics,
each influenced by the rates of all three responses, instead
of two. If this is done, it turns out that the addition of
the free response also makes no difference, in the sense that
there is no change in the equation (Eq. 20) predicting responding under ratio schedules (and hence, consumption under mixture
conditions) from baseline rates (Lea, 1981). It is to be
expected, of course, that addition of the free response will
affect both baseline and mixture consumption levels, but the
equation given above should still describe the relation between
them. So the good fit between observed and predicted mixture
consumption levels, found in both the experiments described
here, is still good evidence in support of the minimum characteristics distance model. Furthermore, it is possible to work
out an equation (which will not be given here because it is even
more untidy than the one for mixture consumption) for the rate
of the free response under different ratio schedules. Figure 2
includes both predicted and observed standard diet intakes and
intakes predicted using this equation, and the fit is again
reasonably good, though not as good as for the mixture intake.

So far, this has all been very abstract. We have considered
"characteristics" without asking what they might actually be.
Here I have to admit to the gravest weakness of the minimum
characteristics distance model. So far I have not been able
to find a way of working back from the response rates observed
to the underlying characteristics--in other words, to the
quantities a_i, b_k, b'_i, k_i etc. that appear in the underlying
equations of the model. So though the promising fit of this

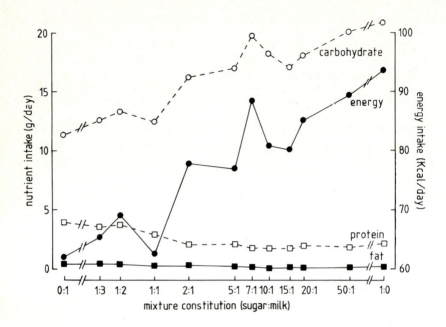

Figure 3. Mean intake of various dietary components, and
of energy, for 12 rats allowed access to milk/sugar
mixtures of various compositions.

type of model suggests that it is possible to analyse needs to
some extent. I cannot yet say what we are analyzing them into.

What we can do is look at some plausible candidates. One of the
advantages of working with food mixtures rather than conventional ratio schedules is that we can make some suggestions as to
what factors there might be which both (or all) responses concerned might be related to. Obvious possibilities are the conventional factors of nutrition, such as carbohydrate, protein,
and fat, or measures derived from them such as energy content.
Table 2 and Figure 3 show how these factors varied across conditions in the two experiments that have been discussed above.
These data emphasize the message that no obvious quantity can
be said to have been conserved in these experiments, but of
course we already knew that: the data presented above showed
that there could not be anything of this sort that would be
constant across conditions. The figures do add the information
that this is no mere mathematical quibble. All the obvious
candidates as characteristics vary substantially and systematically across conditions.

TABLE 2

Mean weights of carbohydrate, protein and fat consumed, and
mean total energy intake, for 12 rats in an experiment where
standard diet and water were always available, while milk and
sugar were made available for 23.5-hour sessions, in various
combinations as follows: Paired baseline: both sugar and
milk, in separate containers. Single baselines: either sugar
or milk on its own. Mixture: milk powder and sugar mixed
together in the ratio 1:3.5 by weight, delivered in a single
container. All weights are in grams; energy is in Kilocalories,
as calculated from the declared compositions of the various
foodstuffs.

Consumption	Substances Offered			
	Sugar and Milk	Milk Alone	Mixture	Sugar Alone
Carbohydrate	15.83	12.96	17.94	18.01
Protein	2.85	4.65	2.25	1.59
Fat	0.23	0.49	0.22	0.33
Energy (Kcals)	74.9	71.5	81.2	79.8

One of the messages of the success of the minimum characteristics distance model, however, is that we should not draw from
that fact the conclusion that these quantities are not those
determining behaviour. It is perfectly possible that the
characteristics are (say) protein, carbohydrate and fat. The
point is that, given that there are fewer independent responses
available than there are characteristics, none of these can be
held strictly constant except at the expense of the others.
The MCD model says that this does not occur: there is a compromise of some sort. That, perhaps, is the most general substantive conclusion one can draw from the present data.

But we still have to draw out their more general, methodological
consequences. I introduced these experiments as one attempt at
an analytic approach to needs. In my view, the experiments I
have described show that this approach is at least moderately
successful. At this kind of level, it seems that we can hope
to work up from very simple, perhaps physiological, basic ideas
about motivation to the level of real instrumental behaviors,
perhaps even economically important ones.

This conclusion obviously contrasts with the general bias which
I declared at the beginning of this chapter--my general feeling
that the more organic view of the economy has received less

attention than it deserves, and will probably prove to play at
least some substantial part in the explanation of individual
economic behaviour. Nonetheless I should like to lay claim to
at least some degree of consistency. It is just because I
suspect that an analytic approach to economic motivation will
not necessarily work, that I want to investigate it in detail.
It is something that should be tested, not assumed. Furthermore, we need to know what it looks like in an area where it
does work, in order to recognize when it is failing. Obviously,
there are other ways of analyzing need than the minimum characteristics distance model. But if we find in a substantial
number of cases that the MCD model gives a useful account of the
data, then cases where it fails will become prima facie cases
for the operation of some more global, organic factor in economic behaviour. Personally, I am convinced that serious
investigation of the individual's behaviour in the economy
would very soon produce many such cases.

References

1   Allison, J. Microbehavioral features of nutritive and
    nonnutritive drinking in rats. Journal of Comparative
    and Physiological Psychology, 1971, 76, 408-417.

2   Allison, J. Contrast, induction, facilitation, suppression,
    and conservation. Journal of the Experimental Analysis of
    Behavior, 1976, 25, 185-198.

3   Allison, J. Economics and operant conditioning. In
    P. Harzem and M. D. Zeiler (Eds.), Advances in Analysis of
    Behavior (Vol. 2): Predictability, correlation and
    contiguity. Chichester: Wiley, 1981, Pp. 321-352.

4   Becker, G. S. The economic approach to human behavior.
    Chicago: Chicago University Press, 1974.

5   Douglas, M. & Isherwood, B. The world of goods. London:
    Allen Lane, 1979.

6   Galbraith, J. K. Economics and the public purpose.
    Harmondsworth, Middlesex: Penguin, 1975.

7   Galbraith, J. K. The new industrial state, 2nd Ed.
    Harmondsworth, Middlesex: Penguin, 1972.

8   Hensher, D. A. Towards a design of consumer durable.
    Journal of Economic Psychology, 1981, 1, 135-164.

9   Hursh, S. R. Economic concepts for the analysis of be-
    havior, 1980, 34, 219-238.

10 Lancaster, K. J. A new approach to consumer theory.
    Journal of Political Economy, 1966, 74, 132-257.

11 Lancaster, K. J. Consumer demand: a new approach. New
    York: Columbia University Press, 1971.

12 Lea, S. E. G. The psychology and economics of demand.
    Psychological Bulletin, 1978, 85, 441-466.

13 Lea, S. E. G. Concurrent fixed-ratio schedules for dif-
    ferent reinforcers: a general theory. In C. M. Bradshaw,
    E. Szabadi & C. F. Lowe (Eds.), Quantification of steady-
    state operant behaviour. Amsterdam: Elsevier, 1981.
    Pp. 101-112.

14 Lea, S. E. G. & Roper, T. J.   Demand for food on fixed-
    ratio schedules as a function of concurrently available
    reinforcement.   Journal of the Experimental Analysis of
    Behavior, 1977, 27, 371-380.

15 Leibenstein, H.   Bandwaggon, snob and Veblen effects in the
    theory of consumers' demand.   Quarterly Journal of Economics,
    1950, 64, 183-207.

16 Mazur, J. E.   The matching law and quantifications related
    to Premack's principle.   Journal of Experimental Psychology:
    Animal Behavior Processes, 1975, 1, 374-386.

17 Rachlin, H.   Economics and behavioral psychology.   In J. E.
    R. Staddon (Ed.), Limits to Action.   New York: Academic
    Press, 1980, Pp. 205-236.

18 Rachlin, H & Burkhard, B.   The temporal triangle.   Psycho-
    logical Review, 1978, 85, 22-47.

19 Rachlin, H., Green, L., Kagel, J. H. & Battalio, R. C.
    Economic demand theory and psychological studies of choice.
    In G. H. Bower (Ed.), The psychology of learning and
    motivation (Vol. 10), New York: Academic Press, 1976,
    Pp. 129-154.

20 Reynaud, P. L.   Economic Psychology.   New York: Praeger,
    1981.

21 Scitovsky, T. I.   The joyless economy.   New York: Oxford
    University Press, 1976.

22 Staddon, J. E. R.   Operant behavior as adaptation to con-
    straint.   Journal of Experimental Psychology: General,
    1979, 108, 48-67.

23 Tarde, G.   La Psychologie economique (2 vols.).   Paris:
    Alcan, 1902.

24 Timberlake, W.   Licking one saccharin solution for access to
    another in rats: Contingent and noncontingent effects in
    instrumental performance.   Animal Learning and Behavior,
    1979, 7, 177-288.

25 Veblen, T.   The theory of the leisure class.   Harmondsworth,
    Middlesex: Penguin, 1979.   (Originally published, 1899).

26 Verhallen, T. M. M.   Scarcity and consumer choice behaviour.
    Journal of Economic Psychology, 1982, 2, in press.

27 Warneryd, K. E. Ekonomisk psykologi, 2nd Ed. Stockholm: Natur och Kultur, 1967.

ANIMAL COGNITION AND BEHAVIOR
Roger L. Mellgren, editor
© *North-Holland Publishing Company, 1983*

A 'MIXED' MODEL OF TASTE PREFERENCE

Robert C. Bolles

University of Washington

Nutritional Labels

If we look at the different nutritional needs of an organism, it is clear that one of the most important is the need for calories. And the priority of the caloric need is all the more marked in a small animal, such as the rat. It would be useful, therefore, to have innately recognized labels for high calorie foods. And we have something like that with the sweet taste, but that is surely not the whole story. On the one hand, for most animals in most habitats, sweets provide a relatively small part of the total caloric intake. Foraging in a grocery store may be different, but foraging in nature usually turns up a relatively small amount of sweet substances. An opportunistic ominivore, such as the rat and presumably the human, is well advised to recognize fatty substances; they are the richest caloric foods. Again, they might not be very wide spread in nature, but it is advisable to recognize fatty substances if one is an opportunist. We find that the rat does, indeed, not only key on sweets and fats, but has a decided preference for such substances.[1]

An animal that has to balance its diet would seem to need a label for protein. It ought to be able either to innately recognize or to learn to recognize protein foods, particularly those that have a high protein content. But there is a paradox. If we offer an animal an artificial food such as casein or powdered egg white, which are extremely high in protein content, the animal does not seem to want to consume these excellent foods. Nobody likes 100% protein; that is the paradox. It is easy enough to say that animals do not like pure protein and that there is no label for recognizing it because pure protein does not occur naturally, but the question then arises, where does protein occur naturally? I suggest that it occurs in different ways and is recognized in different ways by different species of animals, depending upon their eating style. The whole problem of food recognition is, of course, quite different

if an animal is a specialist. A pure canivore, for example,
may find all its food neatly packaged in feathers or in fur,
and hence it need not be too concerned about how to select protein. All it has to do is unwrap its food package and consume
the contents. The protein problem is solved--along with the
calorie problem and all the other nutritional problems. But
the situation is entirely different for an omnivore that eats
a variety of foods. Here we would expect there to be a label
for high protein foods. Perhaps there is. High protein sources
tend to be oily or fatty substances such as seeds, nuts, meat
and fish. All of these high protein foods have charactieristic
fatty properties, and perhaps then for the omnivore fat is the
label for protein.

We should also expect there to be labels for minerals and vitamins and other micro-components of the diet. But such labels
have been peculiarly difficult to isolate and identify. The
one mineral for which there is a well-known label is sodium,
which is known to us by the salty taste. We and a lot of other
animals have a specialized salt detector, which is useful to
pick up salt when it is available in the environment. Salt
deficiencies do occur, so we have a salt detector, and we have,
moreover, an appetite for salt which ensures that we can better
cope with the problem (Denton, 1973). The seriousness of the
problem is indicated by animals that eat greenery, particularly
mountain greenery. For example, mountain goats show a ferocious
appetite for mineral deposits and will fight to defend them
(Brandborg, 1955). Interestingly enough, one might anticipate
that a carnivore, such as a cat, would show little or no salt
appetite and be unresponsive to an induced salt need, because
of the fact that as long as it eats meat it is never likely to
encounter a salt deficiency. Like most of the other minerals,
salt regulation is only a problem on the deficiency side. If
there is a surplus of sodium ion in the body, it is simply
washed away, as long as drinking water is available. So we
should expect to find little of psychological interest in the
state of salt surplus. Accordingly, animals can be biased to
like salt, and to ingest it freely when it is available. This
bias helps protect animals from the real danger, which is salt
deficiency.

In contrast with the unambiguous salty label for sodium ion,
most minerals appear to lack appropriate taste labels. And
occasionally where there is a real need and even a specific
hunger, the specific hunger is inappropriate. A case in point
is the love of ice cubes shown by iron-deficient animals and
people (Woods & Weisinger, 1970). The craving for ice cubes is
diagnostic of an iron deficiency, both for pediatricians and
veterinarians, but the consumption of ice cubes does not solve

the problem. There are scattered reports here and there in the
literature of other specific hungers for minerals. For example,
Green (1925) reported that phospherous deficient cattle would
dig up and eat bones, a very uncattlish kind of behavior. But
by and large, the "wisdom of the body" philosophy that prevailed
in the 1920's, and which looked so attractive for so long, has
not been supported by recent evidence. We need many minerals
and vitamins in order to complete our nutritional requirements,
but we do not have specific hungers for most of them, and what
specific hungers we have are as often as not inappropriate
(e.g., the ice cube phenomenon). For many years it was thought
that there was a specific hunger for thiamin, but that possibility has now been dismissed (Rozin & Kalat, 1971). So taste
labels for specific microcomponents of the diet are peculiarly
inappropriate and/or nonexistent. All that emerges clearly
is the salt appetite. But perhaps the salt appetite leads
the animal not only to improving its chances of satisfying its
sodium requirement but its calcium, iron, and other mineral
needs as well. Perhaps we may assume that minerals tend to
occur together, so that the salt taste is a label for minerals
in general.

Innately Recognized Labels

Some labels are innately recognized. Sweetness is a good
example. There has been considerable discussion over the years
of whether the sweet taste was innately preferred by many mammals or whether it becomes preferred because milk is sweet.
That question would appear to have been answered now, at least
for the human organism. In a study by Steiner (1977), newborn
infants were tested prior to the time they had nursed, before
they had eaten anything. It was a simple experiment. The
experimenter moistened a finger in water and inserted it in the
infant's mouth, and observed no particular facial reaction.
When the same thing was done after the finger had been put in
quinine solution, clear facial rejection expressions were noted,
and when the finger had been dipped in a sugar solution, the
neonate smiled.

If we think of the sweet taste as a label for calories, then the
desirability of sweetness, its hedonic value, should increase
with an increase in hunger. And that phenomenon has been known
for some time (e.g., Young & Asdurian, 1957). The salty taste
is also evidently an innate label for the sodium ion (and perhaps also for minerals in general, as we have argued above).
And the preference for salt should again be expected to increase
as the need for the sodium ion increases. Indeed, that is how
one proves the innateness of the label. Krieckhaus and Wolf
(1968) had the most elegant demonstration of the phenomenon;

they first trained rats to press a bar to receive a salty
solution. Later, after the animals had been given formalin
injections to create a need for salt, they measured the extinction of the bar press response and found much more responding than in control animals. Krieckhaus and Wolf described it
as a "latent learning" experiment, because at the time the
animals were learning about bar pressing and the salt consequence they had no need for salt. But it is better thought
of as an incentive-shift study: when the salt need was introduced, responding for it increased dramatically. It would
appear then that sweetness and saltiness are innate labels for
calories on the one hand and sodium ion on the other. But not
all taste labels are innate. And the other side of the coin is
that some flavors may be highly preferred even though they are
not labels for any particular nutritional consequence. It is
this latter phenomenon we must now consider.

Preferred Tastes of Uncertain Significance

If we look at the kinds of substances people really like, we
find a strange bag. For example, there is coffee. Coffee is
one of the great, marvelous flavors. Who could deny that?
Well, actually, anyone drinking coffee for the first time would
deny it. Coffee is one of those things that Rozin (1976) has
called innately aversive. It is bitter and characterless; it
simply tastes bad the first time you encounter it. But by the
time you have drunk a few thousand cups of it, you cannot live
without it. Children do not like it, uninitiated adults do not
like it, rats do not like it: nobody likes coffee except those
that have drunk a fair amount of it, and they all love it. And
they will tell you it tastes good. They like a mediocre cup of
coffee, they relish a good cup of coffee, and they go into
ecstasies over a superb cup of coffee.

Alcoholic beverages affect us the same way. I dare say, anyone
who tastes beer for the first time is turned off, wine is
terrible, liquor is undrinkable. But we all learn to overcome
these initial adversities, and many of us come to love alcohol
after we have drunk a substantial amount of it. Nothing is
better than beer at the end of the day. Nothing tastes better
than wine at dinner. Some of us drink vodka martinis without
vermouth before dinner; it is pure alcohol, totally undrinkable,
and yet it is hard to get through the evening without it. Now,
it is easy enough to observe that the caffein beverages and the
alcoholic beverages all have pharmacological effects. One might
suppose that we learn to endure the bad taste because of the
pharmalogical effect; the pick up or the high is worth the bad
taste. But the argument is not convincing. It fails to per-
suade us because there are so many alcohol connoisseurs and so

many caffein connoisseurs, people who really delight in the
taste of the beverage. There comes about a total switch in the
hedonic value of the taste. It is not that we are willing to
endure it, rather, it is that we come to adore it. If you are
not personally well disposed toward cold vodka before dinner,
just try it for a couple of weeks; you will like it.

There are other substances that have less pharmacological effect, but which are still enormously enjoyed by those who are
familiar with them. Chocolate--is there anything that tastes
better, anything richer or more desirable than chocolate? Yet
when you think about chocolate, it is so bitter, it is usually
laced with sugar. But chocolate became popular, immensely
popular, throughout the western world several centuries ago at
a time before sugar was widely available. What I am saying is
that we do not necessarily like chocolate just because it has
been associated with sugar, we can also like chocolate because
it tastes so good. The Spaniards had a monopoly on cocoa, which
they prepared by grinding up roasted beans, and putting hot
water in it. They did that for a whole century before sugar was
known (Tannahill, 1973). Wars have been fought over cocoa. It
is simply a superb flavor, even though initially it is aversive
unless it is sweetened. Vanilla is another magnificent flavor.
Rats do not care for it very much, children rather like it, although most of them can take it or leave it. But after you
have had a lot of vanilla in your ice cream and in your cookies
(and again one must wonder whether the desirability of vanilla
is based upon its association with sweetness), vanilla becomes
one of the great tastes that nature provides. Does it have a
pharmacological effect? Is it a label for something? Does the
vanilla bean or the cocoa bean provide some as-yet-undiscovered
nutrient that we all tend to need more of? That would begin to
make sense of the thing.

Then there is pepper, beautiful pepper. If you start putting
pepper on your eggs, soon you cannot eat eggs without pepper.
It is addicting. The different peppers of the world provide
an unbelievable rainbow of delicious flavors. In Japan there
is sansho, hardly known outside of Japan, but just exquisite.
Why does it taste so good? And why is it that rats will not
eat pepper in their food? Why is it that children do not want
pepper on their food? Is pepper some kind of learned label?
What is it a label for? One does not consume enough pepper,
even in highly spiced food, to provide an appreciable amount of
vitamin A or anything else, some mysterious unknown nutrient,
to have a nutritional effect. We like pepper because it tastes
good, at least it does after we have become accustomed to it.
The pepper problem has been addressed in considerable detail
by Rozin and Schiller (1980) who focuses mainly on chili

pepper. Chili pepper really is innately aversive. It is horrible stuff, much too hot. It makes you cry. You want to spit it out, and drink something. It is terrible stuff. It is like alcohol and coffee, only much more painful. Yet after you have eaten chili pepper a few times you do not want to eat without it. Rozin documents the fact that in cultures where chili pepper is the accepted seasoning, e.g., in Mexico, one has the feeling of not having eaten unless one has tasted the horrible stuff. Many kinds of chilis grow in hot climates, and one rapidly becomes a connoisseur, so that a new pepper from a new chili becomes a great experience. Once when I was spending a few days in Los Angeles, I happened to drive down to San Diego and eat in a Mexican restaurant. I found their hot sauce so exquisite that I had to drive all the way down there again the next day. The question is how tastes that are innately so aversive, which no one in their right mind would have anything to do with, can become so delicious? How can they become addicting? How can we come to love the taste of alcohol and cocoa and chili pepper? Are these labels of some kind, whose value is acquired through experience? That is one possibility. Another possibility is that there is something about the learning or habituation process that we do not yet understand. Clearly, we need some new kind of learning process to explain the fact that a substance that is initially aversive becomes with experience not just neutral but positively addicting. We are not happy without it. The whole question of how food preferences change with experience remains a marvelous mystery.

Learning about Labels

We suspect that there are many bases for the learning about taste preferences. We know, for example, that if one is ill and consumes a distinctive food just prior to becoming suddenly well there is a "medicinal effect." The phenomenon has been convincingly demonstrated by Zahorik, Maier and Pies (1974), who showed that animals that were thiamin deficient and then given an arbitrary flavor, saccharin, just prior to being administered massive doses of thiamin subsequently showed increased preference for the saccharin flavor. The medicinal effect has also been demonstrated in animals recovering from the illness produced by apomorphine (Green & Garcia, 1971).[2] It would appear, then, that one kind of reinforcer that can serve to alter the hedonic value of tastes is recovery from illness. It is even more strongly documented that tastes that are associated with the onset of illness lose hedonic value. The Garcia effect is an important and powerful learning mechanism, which tends to follow Pavlovian conditioning principles (Domjan, 1980), and which can produce dramatic alteration in taste preferences. A substance which for rats is highly preferred, such as saccharin, becomes after just a couple of illness experiences highly aversive and distasteful. So we have both the poison effect and the medicinal effect as evidence of powerful mechanisms for altering food preferences.

There is, of course, a temptation to think of the marvelous, delicious tastes that were described above as signalling some sort of medicinal function. Perhaps alcohol, coffee, vanilla and chili pepper all come to be highly desired because they solve some nutritional problem. That is a most attractive hypothesis, but it does not square with the facts. First of all, some of our preferences are innate. Our preferences for sugar, salt and fat are no doubt all innate: there is not a scrap of evidence to suggest that we have to learn through medicinal benefits or nutritional benefits to prefer these labels. On the contrary, we appear to be born with a recognition and an appreciation of and a preference for certain tastes. The other part of the picture that is not convincing is that even with those agents that have pharmacological effects, alcohol, coffee and cocoa, it is yet to be demonstrated that they become preferred as a consequence of their pharmacological action. A physiologically inclined psychologist might make that assumption, but it remains only an assumption. There are as yet no data to indicate that we come to like alcohol because it makes us high, or that we come to like coffee because we learn that it picks us up.

Another possibility is that we come to like those substances with which we are familiar. Perhaps we like coffee just because we have drunk a lot of it. But, how did that happen? Why did we drink so much of it when it was initially aversive? Perhaps it was social pressure. After all, Daddy does it, big brother does it, all the kids do it, so I better do it too. So I suffer through it and eventually, because coffee becomes familiar and accustomed, and because there is continued social support, I come to like it and eventually to prefer it and crave it. This interpretation is also not entirely convincing. While it explains the habituation to the initial aversiveness, it does not explain the change in direction. It does not explain how a substance which is initially aversive ultimately becomes highly desired. Nor does it explain the virtual universality of some of these preferences. It is not just some people in some kinds of social settings. Virtually everyone who consumes substantial amounts of coffee or alcohol or pepper is affected in the same way.

The familiarity dimension is very puzzling. On the one hand, familiarity is important. If I drink my coffee black and you drink yours white, then we are basically incompatible. We are each no doubt reluctant either to experiment, or try the other concoction, or go for some kind of in-between grey coffee. We clearly do get stuck on what we are familiar with. People are notorious for preferring what they are used to. But if familiarity were sufficient to explain the phenomenon then it ought to be possible to put pepper consistently in the rat's diet and see the animal develop a preference for this new label for its food. Rozin, Gruss and Berk (1979) have reported in detail their failure to obtain such an effect. They report a scrap of evidence indicating that pepper may not be quite as aversive or painful as it was initially, but the experiments, which lasted almost a year, show no indication of the kind of affectionado feeling in rats that people develop toward peppery foods. They never do prefer it. However, an unpublished study by Purton and Booth suggests that if rats are continually hungry and receive pepper in a starch supplement to their meager diet, they will come to prefer pepper. But it is not clear why Purton and Booth obtained their results whereas Rozin's results were so negative. Perhaps the critical factor is that Booth's animals were hungry, and that it is only in the hungry organism, or one suffering some kind of nutritional deficit, that the familiarization effect obtains. Or perhaps the secret is that the pepper flavoring, initially aversive, was put in the animals caloric mainstay. Maybe it is the case that pepper becomes liked when it is established as an arbitrary label for calories, particularly when it labels the food providing the majority of calories that the organism takes in.

Such a concept is consistent with what we know about cultural differences in starch source preferences. It appears that every culture has its own favorite source of starches. There are potatoes, rice and corn, which are familiar in the temperate zones. In the middle part of the globe there are cassava, manioc, sweet potatoes and yams and tapioca, and so on. Each culture has a particular source of starch which it consumes in great quantity, and for which it acquires a strong taste preference. A remarkable instance of this phenomenon is the consumption of poi by Hawaiians. When you visit the Islands for the first time and encounter poi, you have no use for it. You cannot consume more than a spoonful. In texture, taste, and appearance it resembles nothing so much as stewed newspaper. It is not painful like pepper or aversive like coffee, it is just nothing; totally unappetizing, a non-food. But if you eat poi regularly for a few weeks, something amazing occurs. You sort of get to like it. If you eat it for years, then the transformation is complete; you love it. Your meal is not complete without it. A meal without poi is not a meal at all. Fish and fruit do not suffice, poi has to be there. The same phenomenon is seen across the two sides of the Pacific. Asians

eat rice primarily. They depend upon it and they like it.
They recognize many kinds of rice and they enjoy the different
varieties. Occidentals are stuck on potatoes, and we too
distinguish and enjoy different kinds. We relish baked ones
and fried ones and boiled ones, and little round ones and the
big long ones, and the red ones and the white ones. We become
potatoe connoisseurs, and we have to have the meally stuff with
our dinner or our dinner is not complete. The Westerner can
eat bowls of rice, but his meal is not satisfying without some
potato.

Why do people get so hung up on their own familiar source of
starch? Is it because starch provides for the majority of
humans the majority of their calories? Is it possible that in
each culture, as one becomes familiar with the prevailing source
of starch and its particular flavor and texture, these stimulus
properties become learned labels for calories? And if the
principle works for starches, could it not also work for other
macro-components of the diet. Does vanilla, for example, come
to be a delicious taste because it is associated with ice
cream and cookies, that is, sweets, that provide calories? Is
vanilla also a learned label for calories?

We have come back again to the question of the reinforcement
mechanism. What is the reinforcer that establishes the preference for a particular taste or the other stimulus properties
of a given food? What establishes particular stimuli as labels
for nutritional components, calories, protein, and so on? Is it
nutritional consequences? That seems terribly plausible (Booth,
1977). Perhaps the only way to like poi is to eat it when we
are hungry and derive calories from it. One ought to look first
for calories as a potential reinforcement mechanism because it
is, after all, calories that constitute the persistent problem
for most animals. There is generally enough protein and minerals in storage to tide the animal over for some time into the
future. But the overall energy balance is a persistent and
urgent problem for most animals. So if there are any learned
nutritional labels, labels whose value is acquired through
experience, then caloric labels ought to be foremost among
them. Such learning ought to be easy to demonstrate. But if
we look at the experimental literature, we find it strangely
devoid of good evidence for such a mechanism. There is only one
bright spot in terms of supporting data for such a concept, and
that is the work of Hogan (1973) on infant chicks. The learning
about caloric labels seems pretty clear for these animals.
Hogan's basic experimental procedure consisted of presenting
neonatal chicks, that had not ever eaten before, with two piles
of stuff. One pile was seeds, and the other was sand (birds
are supposed to eat a little sand or grit in order to make the

crop work properly, so there is nothing wrong in a young bird
ingesting a fair amount of sand). Hogan's little birds were
initially indiscriminate; they ate about 50% sand and 50% seeds
during their first exposure to the two alternatives. Hogan
found that if the animal was first presented just with sand, and
then an hour or so later was presented with the opportunity to
choose between seeds and sand, it was again discriminate,
going 50-50 to the sand. However, if the animals were offered
just seeds and then tested an hour or so later with both substances, they showed a strong preference for the seeds. The
seeds have a nutritional consequence that appears to change the
value of the label. We can suppose that the seeds come to look
like food because their ingestion is, in fact, followed by a
caloric gain. To clinch the argument, Hogan ran animals that
were allowed to consume just sand and were then intubed with a
high calorie load before being tested on the seed/sand choice
test. They showed a strong preference for the sand. It would
appear that labels for calories can be learned. They are
learned very quickly and at an early age, at least by chicks.

When we look at other species, however, the picture is not so
clear. There is a little bit of evidence by Booth, Lee and
McAleavey (1976) with human subjects indicating that a distinctive taste that is associated with a high calorie meal comes
to be preferred to an alternative taste associated with a lower
calorie meal. When we look at the rat, an omnivore that ought
to be capable of learning about calorie labels if any animal
is, we find just scraps of evidence. The fragmentary nature of
the evidence is surprising, particularly in view of the fact
that so many experimenters have no doubt sought to find it.
Mather, Nicolaidis and Booth (1978) have some indication that
calories can influence taste preference. The effects are hardly
more than suggestive, however.[3] Other evidence has been reported
by Holman (1975), but again the experimental evidence is much
less compelling than we would like to see. Holman found that if
an animal consumed a distinctive flavor, such as wintergreen and
then drank dextrose, there was a preference for wintergreen over
an alternative test flavor. Holman found that if there was a
delay of 30 min between the target taste and the ingestion of
dextrose, that preference conditioning was still possible.
However, if the reinforcer was saccharin, then it was not possible to bridge a 30-min gap. Saccharin was effective if it followed the test flavor immediately, but not if it was delayed.
Perhaps, then, it was the caloric consequences of sugar ingestion that made conditioning possible over the temporal gap.
The idea seems plausible. We would like to believe in it, but
the evidence, all things considered, is surprisingly meager.

If I had to design an omnivore, I would probably make it much
like a rat in most respects. Surely, I would equip it with
learning mechanisms, even if they had to have special properties, so that it could learn which foods were poisonous and
which ones were safe. At the same time, I would want to equip
it with other mechanisms so that it could learn what substances
were high in calories and what ones were poor in calories. I
would not want my omnivore to waste a lot of time eating vege-
table greens and watermelons. It should be able to learn about
grains and seeds and sweets, fatty foods, and other good calorie
sources. If I were designing an omnivore I certainly would
equip it with such learning mechanisms. Why then is it so difficult to find evidence of such mechanisms in our favorite long-
tailed, white-furred omnivore?

The answer may be that because the rat has innate recognition of
certain caloric labels, namely sugars and fats, it does not need
to learn about new caloric labels. If it innately recognized
and prefers fatty foods and sweet foods, and if its preference
for these substances increases with its need for calories, then
perhaps there is no need for it to learn about other caloric
foods. If calories occur sufficiently often in association with
sweets and fats, then perhaps that is all the rat needs to know.
Just as it does not need a variety of mineral detectors because
it can rely upon its salt detector and salt appetite to satisfy
all of its mineral needs, so it may be able to rely upon sugars
and fats as innate labels to solve all of its caloric problems.
And so we have a basic dichotomy of theoretical models. We
have the idea, which seems plausible enough, that omnivores
should be able to associate different tastes with what they
signal about caloric consequences. That is the learning model.
Then there is the innate model, the idea that no such learning
occurs because it is not necessary. The animal can solve its
caloric problems in the same way that it can solve its protein
problems by seeking fatty foods and its mineral problems by
seeking salt. There are bits of evidence, hardly more than
suggestive, to show the existence of both kinds of mechanisms.
Thus, there are innately recognized labels, and there is evidence for learning about calories, at least in baby chicks
(Hogan, 1973). Additionally, there is abundant evidence for
animals learning about the sickness and the medicinal consequences of different foods. So the innate mechanisms exist,
and the learning mechanisms exist. But the question remains,
what is the situation with real animals in the real world? Do
some animals lean toward the innate recognition strategy? One
might suppose that to be true of the koala bear. If Eucalyptus
tastes especially good to the koala bear, its feeding problem,
caloric problem and diet balancing problem are all solved at
once. Or if an animal eats nothing but meat all of the problems

get solved (the only difficulty for the meat eater is finding
enough meat). But what is the situation for the omnivore?

A Mixed Model

Perhaps the proper issue is not the dichotomy between innate
versus calorically learned labels. What I am going to propose,
as a sort of mix of these two concepts, is that taste labels
are learned but that the reinforcer is not the caloric value of
the food. Rather, the reinforcer is simply some previously preferred taste. A flavor that is associated with a sweet taste
becomes preferred itself, for example. The mixed model invites
us to look for conditioned preferences based on taste-taste
associations rather than taste-consequence associations (e.g.,
Booth, 1977).

There is recent evidence for taste-taste associations, as well
as support for the idea that a change in the value of one taste
will change the value of the other. The paradigm is the sensory
preconditioning experiment: Two stimuli are associated and then
when some signal value is established for one of them that new
value is conferred, at least in part, on the other. The sensory
preconditioning theme was used to demonstrate taste-taste
associations by Lavin (1976). One flavor, e.g. almond, was
paired with saccharin. Then in the second stage, the animals
were given saccharin and made ill with a lithium chloride injection. Later when the animals were tested for an almond preference, it was seen to be sharply reduced relative to controls.
The alteration in the hedonic value of saccharin was reflected
in a new value for almond. What we have, then, is a strong
suggestion that animals can learn a direct association between
two tastes, almond and saccharin.

The basic idea has been demonstrated in a most authorative
manner by Rescorla and Cunningham (1978). They used four
tastes, the gustatory primaries, salt, sweet, sour and bitter.
They presented them in pairs to various groups of animals in
the first stage of conditioning. In the second stage of conditioning one of the primary tastes was associated with sickness. Then subgroups were tested on the different tastes.
Sorting out the results from all the subgroups and the various
control groups, it is clear that just as illness-associated
tastes become aversive, so tastes associated with illness-
associated tastes can become aversive.

It is also possible to do the sensory preconditioning experiment the other way around, i.e., by increasing the value of
one of the associated tastes. The result was first reported
by Fudim (1978), who associated almond flavor with salt. That

was the preconditioning phase. Subsequently a salt need was
induced with formalin injections, and as we might expect, there
was a strong increase in preference for the almond flavor, the
flavor that had been associated with salt for which the animal
now had a specific hunger. So with either an upshift or a down-
shift in the value of one taste the sensory preconditioning
procedure demonstrates that tastes can be directly associated
with each other.

There is, however, a somewhat more direct, perhaps better, maybe
simpler methodology for getting at these taste-taste associations. It was first reported by Holman (1975). Holman
associated one flavor, almond, with a very sweet saccharin
solution and another flavor, vanilla, with a slightly sweet
saccharin solution. After the animals had some experience with
both of these combinations, a preference test was run for almond versus vanilla, and the animals showed a strong preference
for almond, the flavor that had been previously associated with
the sweeter substance. In effect, the animal tracks the label,
the associated flavor, in accordance with what the label used
to signal. It might be called a label tracing procedure. Perhaps the simplest and most direct experiment of this sort is
that reported by Fanselow and Birk (1982). For just three days
rats were given vanilla paired with saccharin and almond paired
with quinine. After exposure to these combinations, the animals were given the choice between vanilla and almond and plain
water. They opted for the vanilla, the label for sweet and
avoided the almond, the label for bitter. It appears that
tastes can be associated with each other.

The existence of such associations may be interpreted in terms
of the sensory preconditioning paradigm, as Lavin did, or they
may be regarded as within-compound associations, as Rescorla
and Cunningham did. But in whatever framework the phenomenon
is fit, it seems clear that animals can learn to associate
tastes, and that this learning occurs very rapidly, is robust,
and has considerable generality over experimental parameters.
The next consideration is whether the taste-taste association
reflects a mechanism that can replace conceptually the traditional caloric reinforcement mechanism to explain learned
taste preferences. Can the mixed model be made plausible and
workable? It remains to be determined if the concept of taste-
taste associations applied to the mixed model can begin to cope
with the great hord of mysteries and problems that have been
alluded to above. Let me recapitulate some of the problems.
Are sugars and fats innate labels for calories? Are fats innate
labels for protein? Is salt an innate label for all sorts of
minerals? Are our favorite pharmacological agents, those containing caffein and alcohol, preferred because they are

pharmacological agents or can the mixed model help explain them? Can the mixed model cope with the scrumptious flavors, vanilla, pepper, durian fruit, and the like? What about the pepper problem? Is it plausible to suppose that something as initially aversive as chili pepper comes to be desired simply because it is repeatedly associated with corn and beans? And if those foods are so good that they can produce such powerful learning, then why do people not eat them for their own deliciousness? What about the whole question of familiarity? How many trials does it take to prefer one brand of beer or one color of coffee? And what about the poi problem? Can we really dismiss the large contribution that starches make to one's caloric intake. There are a host of problems to be dealt with.

We have just begun working on some of these problems in our lab. In one recent experiment (Bolles, Hayward & Crandall, 1981), we took yet another look at the question of whether calories might be a reinforcer for taste preferences. We still find it hard not to believe in the plausibility of such a mechanism. In one experiment, we tested the possibility of labeling a meal that provided a characteristic number of calories with a distinctive taste. If one taste means a big meal and another taste means a smaller meal, will rats come to prefer the big-meal taste? The animals were given several days experience with almond flavored 6-calorie meals and vanilla flavored 12-calorie meals. We were anticipating that after a few days on this regimen the animals would prefer the vanilla over the almond, i.e., that they would track the taste that had signalled the larger meal. But when we tested for the tracking, the results were nill-no difference.

In another experiment, we correlated different tastes with different caloric densities of brief meals. For example, almond signalled meals that were diluted to 2 calories a gram, while vanilla labelled meals that were a full 4 calories per gram. The sizes of the meals were controlled so that the total caloric benefit was the same (6) for both kinds of meals. We found a large preference for the vanilla flavor, the taste that had been paired with the higher caloric density. So, the results are clear, but unfortunately, the conclusion is not. We do not really know whether our animals found the dilutant aversive, or whether the caloric enrichment (which was in the form of powdered starch) was initially preferred. We do not know whether rats can detect starch, or caloric density per se, or whether perhaps it is the speed of assimilation of a caloric load that constitutes reinforcement. In short, we found a powerful conditioner of taste preference, but we do not know what the US was. We do not really know if the results have anything to do with taste-taste associations. But we are pursuing the problem.

Footnotes

1. The question of fat brings us to a curious problem. we will
    be referring repeatedly to "taste" to designate different
    kinds of nutritional labels. The problem of fat is that
    fats do not have a distinctive taste per se. It is, rather,
    the consistency or the texture or some other oral charac-
    teristic of fats that distinguishes them and that actually
    defined the label. In what follows, then, we will take the
    liberty of using the word "taste" to refer to a great
    variety of oral and olfactory stimuli, and recognize that
    we are just postponing for the moment the interesting and
    important job of separating the different kinds of stimuli
    that distinguish different kinds of labels.

2. It should be noted that the medicinal effect is by no means
    as robust and easy to demonstrate as conditioned taste
    aversions. One is tempted to conclude that the hedonic
    values of tastes are readily moved down but are moved up
    only with difficulty. An interesting parallel is that
    frightened rats quickly learn about danger and only slowly
    learn about safety.

3. It should be observed, just to keep the record straight,
    that there are negative reports from similar investigations
    (e.g., Revusky, Smith & Chalmers, 1971). Moreover, even
    the positive reports of intragastric and post-ingestional
    reinforcing effects (e.g., Holman, 1969; Pureto, Deutsch,
    Molina & Roll, 1976) implicate other kinds of reinforcement
    mechanisms besides caloric benefit per se.

References

1 Bolles, R. C., Hayward, L. & Crandall, C. Conditioned
   taste preferences based on caloric density. Journal of
   Experimental Psychology: Animal Behavior Processes, 1981,
   7, 59-69.

2 Booth, D. A. Appetite and satiety as metabolic expectancies. In Y. Katsuki, M. Sato, S. F. Takagi & Y. Oomura
   (Eds.), Food intake and chemical senses. Tokyo: University of Tokyo Press, 1977.

3 Booth, D. A., Lee, M. & McAleavey, C. Acquired sensory
   control of satiation in man. British Journal of Psychology,
   67, 137-147.

4 Brandborg, S. M. Life history and management of the mountain goat in Idaho. Idaho Fish and Game Department Bulletin,
   1955, 2, 1-142.

5 Denton, D. A. The brain and sodium homeostasis. Conditioned Reflex, 1973, 8, 125-146.

6 Domjan, M. Ingestional aversion learning: Unique and
   general processes. In J. S. Rosenblatt, R. A. Hinde,
   C. Beer & M. Busnel (Eds), Advances in the study of behavior. Vol. 11. New York: Academic Press, 1980.

7 Fanselow, M. S. & Birk, J. Flavor-flavor associations induce hedonic shifts in taste preference. Animal Learning
   and Behavior, 1982.

8 Fudim, O. K. Sensory preconditioning of flavors with a
   formalin-induced sodium need. Journal of Experimental
   Psychology: Animal Behavior Processes, 1978, 4, 276-285.

9 Green, H. H. Perverted appetites. Physiological Reviews,
   1925, 5, 336-348.

10 Green, K. F. & Garcia, J. Recuperation from illness:
    Flavor enhancement for rats. Science, 1971, 173, 749-751.

11 Hogan, J. A. The development of food recognition in young
    chicks: II. Learned associations over long delays.
    Journal of Comparative and Physiological Psychology, 1973,
    83, 367-373.

12 Holman, E. W. Immediate and delayed reinforcers for
    flavor preferences in rats. Learning and Motivation, 1975,
    6, 91-100.

13 Holman, G. L.   Intragastric reinforcement effect.   Journal
    of Comparative and Physiological Psychology, 1969, 69,
    432-441.

14 Kreickhaus, E. E. & Wolf, G.   Acquisition of sodium by rats:
    Interaction of innate mechanisms and latent learning.
    Journal of Comparative and Physiological Psychology, 1968,
    64, 197-201.

15 Lavin, M. J.   The establishment of flavor-flavor associ-
    ations using a sensory preconditioning procedure.   Learn-
    ing and Motivation, 1976, 7, 173-183.

16 Puerto, A., Deutsch, J. A., Molinea, F. & Roll, P.   Rapid
    rewarding effects of intragastric injections.   Behavioral
    Biology, 1976, 18, 123-134.

17 Rescorla, R. A. & Cunningham, C. L.   Within-compound flavor
    associations.   Journal of Experimental Psychology: Animal
    Behavior Processes, 1978, 4, 267-275.

18 Revusky, S. H., Smith, M. H., Jr. & Chalmers, D. V.   Flavor
    preference: Effects of ingestion-contingent intravenous
    saline or glucose.   Physiology and Behavior, 1971, 6,
    341-343.

19 Rozin, P.   Psychobiological and cultural determinants of
    food choice.   In T. Silverstone (Ed.), Appetite and food
    intake.   Berlin: A. Verlags Gesellschaft, 1976.

20 Rozin, P., Gruss, L. & Berk, G.   Reversal of innate aver-
    sions: Attempts to induce a preference for chili pepper
    in rats.   Journal of Comparative and Physiological
    Psychology, 1979, 93, 1001-1014.

21 Rozin, P. & Schiller, D.   The nature and acquisition of a
    preference for chili pepper by humans.   Motivation and
    Emotion, 1980, 4, 77-101.

22 Rozin, P. & Kalat, J.   Specific hungers and poison avoidance
    as adaptive specializations of learning.   Psychological
    Review, 1971, 78, 459-486.

23 Steiner, J. E.   Facial expressions of the neonate infant
    indicating the hedonics of food-related chemical stimuli.
    In R. Weiffenbach (Ed.), Taste and development: The genesis
    of sweet preference. DHEW Publication, Bethesda: 1977.

24 Tannahill, R. Food in History. New York: Stein & Day, 1973.

25 Woods, S. C. & Weisinger, R. S. Pagophagia in the albino rat. Science, 1970, 169, 1334-1336.

26 Young, P. T. & Asdurian, D. The relative acceptability of sodium chloride and sucrose solutions. Journal of Comparative and Physiological Psychology, 1957, 50, 499-503.

27 Zahorik, D. M., Maier, S. F. & Pies, R. W. Preferences for tastes paired with recovery from thiamine deficiency in rats: Appetitive conditioning or learned safety? Journal of Comparative and Physiological Psychology, 1974, 87, 1083-1091.

ANIMAL COGNITION AND BEHAVIOR
Roger L. Mellgren, editor
© *North-Holland Publishing Company, 1983*

SPONTANEOUS BEHAVIOR:
INFERENCES FROM NEUROSCIENCE

Lynn D. Devenport

University of Oklahoma

For some reason laboratory animals exhibit behavior that is uncalled-for, unnecessary, superfluous, gratuitous. They do this in spite of obvious experimental rules that, if not proscribing deviations from goal-divided behavior, do not reward them. Dashiell's (1930) example still stands as one of the best instances: Having learned the shortest route from start to goal, rats proceed to trace-out virtually all of the many alternative paths available to them. Similarly, once having acquired an operant response, rats vary this response to the extent that the contingency will bear (Antonitis, 1951). Indeed, they frequently go beyond what the rules of reward will allow (Olton & Schlosberg, 1978).

This "misbehavior" of animals, especially their violation of principles of reinforcement, has, I suspect, a special poignancy for experimenters. Performing nearly flawlessly the correct, say, left turn, rats inexplicably turn right; or stop midway in the maze and sniff or refuse to run or refuse to eat. College student subjects, I am told (H. Williams, Personal Communication, 1981), exhibit similar incongruities.

These examples illustrate but, as will be seen, do not begin to exhaust the category of behavior that I am calling spontaneous. By spontaneous I do not mean to imply that the characteristic of behavior I am describing is without a cause; rather, that an external trigger is difficult to find. Although some attention will be given to the matter of neural sources of spontaneity, my principle aim is to address the issue of the nature of spontaneity, as this question can be answered with considerable confidence.

The question amounts to this: Is spontaneity error, the outcome of an imperfect nervous system, an inaccuracy; or is it in some sense deliberate, intentional? Is it the result of an occasionally malfunctioning nervous system or does a properly operating brain produce this sort of disorganization, and if so, why?

Plainly, I intend to argue on behalf of spontaneity; that it is the product of a properly operating--indeed advanced--nervous system and that it confers certain concrete advantages that are not to be had otherwise. The chief force of the argument depends upon the following premise. That the brain has evolved to perform its function--whatever this may be--in its intact undrugged state and that under normal circumstances the infliction of damage or the administration of potent drugs can only impair overall performance.

Perseveration

If this premise is accepted, and it hardly seems controversial, then it continues to remain a wonder that brain impairment so frequently improves learned performance. It might be countered that for every instance of performance facilitation, one can find a test that yields an impairment for a brain-damaged animal. True. But what I intend to establish is that for lesions of a number of telencephalic structures or following drug administration, decrements in learned performance are usually only obvious when established experimental rules are changed such that responses that were previously correct are now scored as errors.

In the sections that follow, I shall frequently refer to "telencephalic structures", not because the telencephalon is of a piece but because the evidence is not yet available to permit a more restrictive reference. Likewise it is not to be inferred that the function under consideration, spontaneity, is by any means the only one to be ascribed to the structures under discussion. Of the telencephalic structures most intensively investigated, probably none exceed the attention given the hippocampal formation, the septal nuclei, and the frontal cortex. And if one were to choose two drugs, a stimulant and a depressant, of similar experimental status they would probably have to be amphetamine and alcohol. I have restricted my discussion to these structures and drugs because of the wealth of information available and because of their special relevance to the topic.

A number of circumstances lead to DRL (differential-reinforcement-for-low-rates-of-responding) deficits. Among these are hippocampal (Clark & Isaacson, 1965), septal (Braggio & Ellen, 1976), and frontal (Numan, Seifert, & Lubar, 1975) lesions, amphetamine (Schuster & Zimmerman, 1961) and alcohol administration (Glowa & Barrett, 1976). It is exceptional to find an instance in which the DRL phase is not preceeded by continuous reinforcement training. Thus, at first the rate of reinforcement closely matches the rate of response, a condition quite unlike that of DRL. In the few instances when DRL training is carried out from the beginning, lesion deficits are not observed (Ellen, Aitken, & Walker, 1973). Extinction and passive avoidance training may be quite different procedurally. Nevertheless, each punishes a previously rewarded or strongly prepotent response, like moving into the darker of the two compartments. From this perspective, the retardation of extinction by alcohol (Barry, Wagner, & Miller, 1962), amphetamine (Heise, Laughlin, & Keller, 1970), or hippocampal (Kimble & Kimble, 1970), frontal (Kolb & Whishaw, 1981) and septal lesions (Kelsey & Grossman, 1971) might have been predicted from the frequently reported passive avoidance deficits for these drugs and lesions (Holloway & Wansley, 1973; Laties & Weiss, 1966; Riddell, 1968; Zielinski & Czarkowska, 1973; McCleary, 1961, respectively).

Perhaps one of the most obvious cases of this counter conditioning is reversal learning. If animals with CNS impairments have as one of their major deficits the persistance of first-learned responses or strategies, then one would expect deficiencies here. And this is usually the case. Hippocampal damage produces spatial (Kimble & Kimble, 1965) and nonspatial (Teitelbaum, 1964) reversal deficits as does frontal (reviewed by Fuster, 1980) and septal damage (Sikorszky, Donovick, Burright, & Chin, 1977), and alcohol injection (Devenport, Merriman, & Holloway, 1981; Devenport & Merriman, 1982). Amphetamine administration impairs nonspatial reversal (Carlton, 1961) while its effect on spatial reversal is not well studied.

Apart from some other specific deficits that seem to accompany selective telencephalic damage (e.g. frontal damage and temporal integration, reviewed by Fuster, 1980; hippocampal damage and spatial integration, reviewed by O'Keefe & Nadel, 1978) one of the commonest symptoms of limbic system damage is perseveration --a compulsion to repeat previously acquired strategies or responses. Thus it is not exactly accurate to characterize such animals as impaired learners (or performers) when one might argue with equal force that they have learned too well, too faithfully. And it should be remembered that original learning for brain-damaged animals is good and often superior (see Isaacson, 1981), provided the task does not incorporate the possibility of perseveration within trials or sessions such as multiple sites of reinforcement in a maze or alternation problems (e.g. Olton & Wertz, 1978). The other side of this coin is the implication that normal intact animals perform their learned responses less faithfully. They more readily abandon their rewarded experience when conditions change.

It is counterintuitive to entertain the notion that parts of the forebrain might have proliferated because of the advantage derived from a process or function that opposed learned performance. But it is ironic only to the extent that we think

of reinforcement-controlled behavior as the pinnacle of the
evolution of intelligence. Adaptation imples a preparedness to
abandon old responses for new, and it seems to be here that
animals which shun repetition--violate the principles of rein-
forcement--are so well fitted to adjust to change, whether
imposed by experimental design or the outcome of random natural
events. It seems likely then, since animals with telencephalic
damage are not (with exceptions noted earlier) simply poor
learners, the problem lies in a failure to afford themselves the
opportunity to learn. Without some method of sampling the outcome of alternatives, e.g. shifting behavior toward different
rates of response, different arms of the maze, etc., brain-
damaged or drugged animals may simply be depriving themselves
of new information. Here then is a function for spontaneity.
A small but consistent rate of alternative behaviors--however
experimentally perverse or "incorrect"--may be vital to the
welfare of animals generally, but especially for those that must
adjust to unpredictable shifts in weather, predation, prey avail-
abliity, plant growth, etc. And because many of these changes
are unpredictable, it probably behooves these animals to express
spontaneity somewhat unpredictably as well-or at least not according to a rigid method.

If these suppositions can be supported then the nature of spon-
taneity--the subject of this chapter--is clearer: It is behavioral variability, not cast in its customary role of "error",
but as a strategy for adaptation. The next section will be devoted to a more detailed examination of studies that were able
to assess directly the degree of variability contained in normal
behavior. It will, as before, contrast intact with CNS-impaired
animals: If spontaneity--behavioral variability (BV)--is a normal aspect of performance, then animals with CNS deficiencies
should display more structured, predictable performance. This
comparison constitutes a fairly severe test of the hypothesis,
as the predicted outcome is prima facie unlikely. One's best
guess about the brain might take this form: That the forebrain
has evolved in such a way that it delicately receives, structures, compares, integrates, and stores information and organ-
izes responses accordingly. Thus, is it plausible to expect
potent pharmacological or mechanical interference to result in
anything but a halting, awkward, more error-prone animal?

Behavioral Variability

Straight Alley Studies

Consider the effect of alcohol (incremental doses averaging
1 g/kg), extensive bilateral hippocampal lesions that destroyed
more than 85% of the structure, or both on acquisition in a

baited (2-45 mg. pellets) 1.8m runway. As seen in the upper
panel (A) of Figure 1, both drug and lesions dramatically improve one of the standard measures of alley performance, running time. Except for the combined lesion/drug condition which,
incidentally, significantly improves performance beyond the

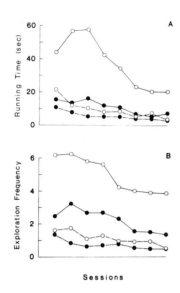

Figure 1.   (A) Mean running times and (B) exploration
   frequency across eight training sessions. Filled
   Circles, alcohol; open circles saline; broken lines,
   hippocampal lesions; solid lines, sham operates.
   ns = 6.

Note: From "Necessity of the Hippocampus for Alcohol's
   Indirect but not Direct Behavioral Action: by
   L. D. Devenport, J. A. Devenport, and F. A. Holloway,
   Behavioral and Neural Biology, 1981, 33, 476-487.
   Copyright 1981 by Academic Press, Inc. Reprinted
   by permission.

levels of either manipulation alone, findings like these have
been published before (e.g. Barry, Wagner, & Miller, 1962;
Wickelgren & Isaacson, 1963). What is new is contained in
panel B, but first some explanation. Although the literature
is generally silent on this point, rats do not simply run a
maze or alley. They advance, stop, rear, sniff; advance,
orient; advance, turn around, stop, sniff, bite, and so forth.

Finally, they make their way to the goal box. These behaviors
may be so extensive at first that the runway is not traversed
in the allotted time. It was our impression from earlier
studies that animals with lesions or injected with alcohol were
generally much more goal-directed. Which is to say, they
seemed to emit very few deviations from goal-directed activity,
viz., running. Counting the instances of these deviations per
trial, taking a mean of the 3 trials per session, and collect-
ing these data under the heading of "exploration", panel B of
Figure 1 was created.

These "exploration" data, which we now think of as representative of response variability, contain the same pattern of
significant alcohol and lesion effects that were found for running time. What they add to the standard measure is an explanation: Normal (in this case sham-operated) rats are slower
because they stop and engage in alternative activities. Another form of BV that parallels "exploration" is the variability of running times within a session (the standard deviation
of the mean for each session's 3 trials). This temporal variability is condensed in Table 1 for the first and second half
(4 sessions each) of the runway study. It is worth asking
whether or not the variability of running times might not be
closely related to incidences of exploration. We cannot answer

Table 1

Mean ($\pm$SE) Running Time Variability (Standard Deviation)
During Acquisition

Group	Sessions 1-4		Sessions 5-8	
	Alcohol	Saline	Alcohol	Saline
Hipps	5.10$\pm$1.10	12.47$\pm$2.62	1.96$\pm$0.68	4.66$\pm$1.63
Shams	13.17$\pm$3.62	38.34$\pm$3.51	5.21$\pm$1.32	18.83$\pm$2.98

Note. There was a significant effect of drug, $F(1,44)$
= 12.08, $p < .01$ and lesion, $F(1,44) = 13.10$ $p < .01$
during Sessions 1-4 but no interaction. During
Sessions 5-8, drug $F(1,44) = 10.50$, $p < .01$, and
lesion $F(1,44) = 11.88$, $p < .01$ had significant
effects and they interacted, $F(1,44) = 4.86$, $p < .05$.
n = 12/group.

this question, but can point out that the duration of each instance of exploration can be held constant or varied. Our finding for animals exhibiting the same number of stoppages, but different running times, on successive trials, indicates that they were varying their durations. Regardless of the origin of this variability, the tempo for lesion and drug groups was highly consistent from trial-to-trial but this was not the case for controls.

Virtually identical results were obtained by substituting d-amphetamine (1-2 mg/kg 30 minutes before testing) for alcohol (Devenport, unpublished). In this experiment sham-operated controls and rats with hippocampal damage (about 85% destroyed bilaterally) were all trained without drugs, receiving instead saline injections. Then, in a balanced design all animals received two d-amphetamine sessions or two baseline saline sessions followed by the opposite condition for two days. Also included in the design was the administration of haloperidol (.05-.07 mg/kg 40 minutes before testing). Because lesioned rats rapidly exhibit asymptotic performance levels, they were given fewer predrug training trials and the scores depicted in Figure 2 reflect this level of training. The relative effects of the drugs (2-day means are illustrated) are plain: Amphetamine significantly promoted running in the lesion condition and, as with alcohol in the previous study, did so by suppressing alternative responses ("stoppages" in Figure 2A and D). Accompanying the restriction of behavioral diversity was a significant homogenization of trial-to-trial variation in running time following amphetamine administration. Haloperidol, a dopamine antagonist (in contrast to d-amphetamine, an agonist) produced significant changes in the opposite direction, regardless of lesion type. The significance of dopaminergic mechanisms will be considered later.

The relatively simple act of running down an alley is in actuality a series of segment traversals punctuated by formally irrelevant responses. Running time, and especially its reciprocal, speed, is--if not a misnomer--at least misleading. Most of the time taken up in the alley is not owing to differences in speed of running; the time is given over to alternative behaviors and it is on this score that our groups differed. Performance is richly studded with these deviations. They are plainly susceptable to experimental manipulation and are certainly deserving of serious attention. There was a time when the hippocampus, septum, and frontal cortex were accorded an inhibitory function because of, among other things, the faster running times after the structure was damaged. Had the complexity of runway performance (and that of other tasks) been

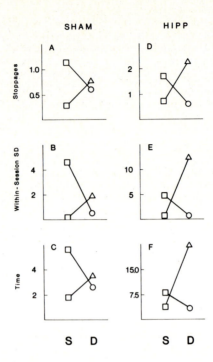

Figure 2. (A and D) Mean stoppages, (B and E) mean with-
in-session variability (between-trial standard deviations), (C and F) mean running time for sham-
operated (SHAM) rats in a straight alley, S = measurements taken after saline injection, D = measurements
taken after drug injection. Symbols: squares =
saline, circles = amphetamine 1-2 mg/kg, triangles=
haloperidol (.05-.07 mg/kg), ns = 6. From L. D. Devenport, unpublished data, 1980.

recognized, quite a different conclusion might have been drawn.
Rather than slowing the speed of animals, the hippocampus may
well promote behavioral diversity--perform an excitatory
function--which competes with running. Similar reformulations
might be drawn for theoretical acounts of other types of learned performance.

The exact nature of topographic deviation has been difficult
for us to describe and the evolution of our thinking can be
followed from the way our figure ordinates are labeled in this
and later sections. "Exploration," while describing the function we think deviations serve is inappropriate because some
of the relevant responses are orienting not exploratory re-

sponses and in any case exploration implies a gathering of information that we are not always able to demonstrate. "Stoppages" places too much emphasis on the termination of running when in fact behavior merely shifts from one topography to another. It is not in most cases being arrested. We have settled on the more descriptive terms, "deviation," or "topography variability" to convey more accurate and less speculatively our meaning.

Radial Maze Studies

Unlike other settings, mazes--particularly radial-arm mazes--emphasize the spatial behavior of animals and are ideally suited to examine variability along this dimension. The rat's spatial facility is justly celebrated and was recognized well before the radial-arm maze became popular. Of direct relevance were the observations that whether the arms of a Y- or T-maze are baited or not, normal rats exhibit a strong tendency to alternate the arms they visit. And this spontaneous alternation is abolished by septal (Dalland, 1970) or hippocampal (Roberts, Dember, & Brodwick, 1962) lesions, and by alcohol (Cox, 1970), or amphetamine administration (Bättig, 1963). Olton (e.g. Olton, 1979; Olton & Schlosberg, 1978) has been influential in drawing attention not only to the nature of the rat's memory processes in the radial-arm maze, but also its strong tendency to vary the places it visits in spite of reward contingencies; a disposition that according to Olton, psychologists ignore at their peril. Contrary to the prediction of most learning theories, rats rewarded for entering an arm do not typically return to that arm on the next trial. This tendency is expressed in Olton's notion of "Win-Shift" which describes a foraging strategy that stands in contrast to "Win-Stay" approaches, and is presumably suited to the habitat of various species. It is not this feature of behavior that most concerns us here. Rats alternate in two-choice situations and disperse themselves over as wide an area as is provided, whether they are rewarded or not. In fact, reward discourages dispersion to a significant extent (Gaffan & Davies, 1981). Spatial variability while no doubt of service in foraging, probably confers other advantages that have little to do with feeding. The dispute continues over the role of the hippocampus in spatial affairs. Whether, as O'Keefe and Nadel (1978) think, the hippocampus is the anatomical locus of spatial maps--cognitive representation of Euclidian space--or as Olton (e.g., Olton & Papas, 1979) thinks, it is the site of short-term "working" memory of where the animal has been, the final result of damage here is an animal that cannot vary its behavior across a purely spatial dimension. The hippocampal-lesioned rat's rigid, predictable patterns of arm visitations in the radial-arm maze are well known (e.g., Olton & Wertz, 1978) and will not be elaborated here. I will concentrate on the spatial behavior of rats impaired with varying doses of alcohol. Our analysis is not restricted to purely spatial matters; its usefulness is extended to include three independent measures of BV. Spatial variability is, of course, one of them and it refers to the degree of dispersion--number of different arms-that rats spontaneously select during trials. Sequential BV is another and refers to the extent to which successive arm choices are predictable or random. For example, given the selection of, say, arm 3, is the likelihood of a subsequent choice spread evenly over all maze arms or is a certain arm, say 6, favored? The descriptive statistic, $\underline{H_2}$ (Attneave, 1959; Frick & Miller, 1951) is employed to express this pairwise sequential independence. Finally, the "exploration" measure that was so sensitive in the straight alley is employed again except, as explained before, it now carries the more descriptive label, "deviations" (from goal-directed activity) and is taken as an index of the readiness of animals to divert their response topographies. Running times are also taken.

In one standard situation we have used (Devenport, Merriman, & Holloway, 1981), each of the eight arms is baited with 2-45mg pellets. Rats are placed in the central compartment and allowed to run from arm to arm until 8 rewards (16 pellets) are taken or until 10 minutes has elapsed. It should be appreciated that the rewards merely guarantee that the rats will continue to run. The replacement procedure employed does not oblige the subject to go to any arm in particular, rewards being replaced after they are taken and as the rat is returning to the central compartment. Animals are injected with High (2.0), Medium (1.5), Low (0.75 g/kg ethanol) alcohol doses or saline intraperitoneally 13 minutes before sessions (3 trials each).

Data were collected across 12 sessions. As criterion (taking all 8 rewards in the 10 minutes allotted) was not met by all animals until the fifth session, Figure 3 depicts results from the last 8 days only. Except for the lowest dose, which in no instance differed significantly from saline, alcohol reliably depressed every dimension of BV that was measured. Arm sequences were more predictable, choices were not widely dispersed, and deviations were reduced to about one-third that of saline and Low groups. The strength of these BV findings can be better appreciated in view of the failure of either conventional measure, running time (Figure 4) or sessions-to-criterion to be sensitive to the effects of alcohol. In fact,

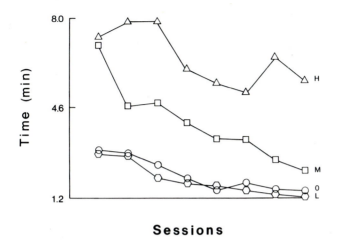

Figure 5. Mean running time for subjects to obtain
eight rewards in the eight-arm radial maze without
reward replacement. Abbreviations same as for
Figure 3. ns=6. From Devenport, Merriman, and
Holloway, 1981.

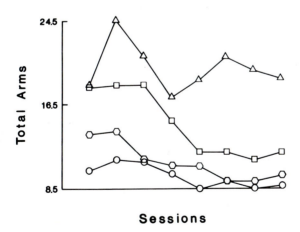

Figure 6. Mean number of total arms visited. Note that
a perfect score is eight; scores greater than this
reflect returns to previously visited arms. Abbreviations same as for Figure 3. ns = 6. From Devenport,
Merriman, and Holloway, 1981.

shows why. They visited an average of about 20 arms per trial
When perfect performance required only eight, return visits
making up the difference. In fact this figure fails to reflect the magnitude of the deficit as even during the last
session, rats in the High group were exceeding their 10 minute
time limit.

The Medium group displayed an interesting improvement of performance that generated a signific Dose x Session interaction.
Figure 6 indicates a progressive decrease in arms visited implying that the rats were solving the problem and not just
running faster. Accompanying this improvement in arm selection

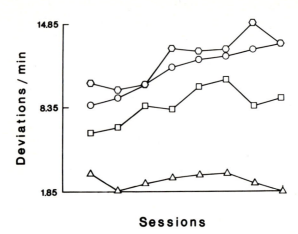

Figure 7. Mean deviation rate (deviations/running
   time). For details see Figure 6.

were parallel increases in topographic (expressed as deviations
per minute to permit group comparison, Figure 7) and sequential
BV (Figure 8). These findings are important and will be considered later.

One might well ask what were the relative contributions of the
ability to vary spatial behavior from the outset vs. the increasing tendency to repeat patterns and arm choices. The
question was answered by another study that was designed somewhat differently in order to provide some additional information.

Groups were administered the same alcohol doses as before and
were run in a nonreplacement task. However, in this case only
odd arms (1,3,5,7) were baited. Thus, we were able to assess

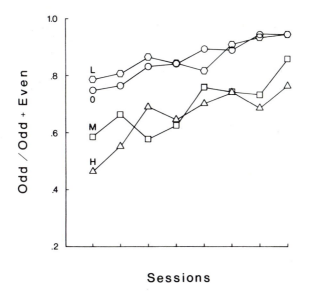

Figure 8. Mean sequential independence (H_2). For details see Figure 6.

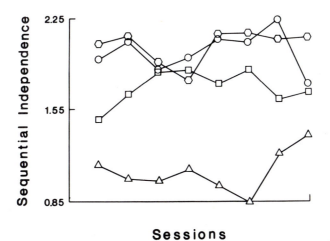

Figure 9. Mean proportion of odd (rewarded) to odd and even-numbered arms selected by rats during the last eight acquisition sessions. ns = 6. For abbreviations see Figure 3. From Devenport, Merriman, Devenport, 1982.

the extent to which subjects could find the correct set of arms as well as move economically among them (i.e., avoid previously entered arms). The results are shown in Figure 9. Expressed as a preference ratio for odd arms, this graph of the last 8 of 12 sessions indicates that while somewhat poorer, even animals with high doses of alcohol could locate the baited arms. These preference data, however, conceal a debilitating impairment that is unmasked by analysis of the number of odd-arms chosen (Figure 10). A perfectly efficient rat could obtain its

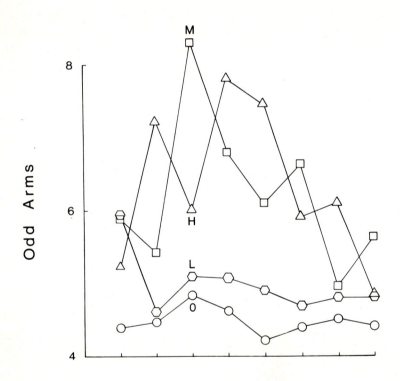

Figure 10. Total number of odd-numbered (baited) arms visited. Note that a perfect score is four. Scores greater than this reflect returns to previously visited arms. For details see Figure 9.

four rewards in four odd-arm visits. Owing to extensive with-in-trial perseveration, many of the Medium and High dose rats were unable to obtain their pellets in the allotted time. Consequently, there were wide group differences on the trials-to-criterion measure (Table 2).

Table 2

Number of Subjects Meeting Criterion During Acquisition of the Odd-Arm Nonreplacement Problem

Dose[a]	n						Sessions						
		1	2	3	4	5	6	7	8	9	10	11	12
sal[b]	12	4	6	7	6	8	8	9	11	10	11	10	11
0.75	12	5	6	9	12	12	12	12	12	12	12	12	12
1.5	12	0	0	2	2	3	4	6	7	9	10	9	9
2.0	12	0	0	2	2	3	3	3	4	4	3	4	4

Note. Criterion was defined as 2 successive sessions during which all four rewards were taken in the alloted
(10 min) time.
[a]In g/kg.
[b]Sal = saline injection.

This experiment can confirm that despite an ability to find their way around the maze, alcohol prompts rats to make repetitive arm choices within trials. Yet it also promotes perseveration across sessions. Phase 2 of the experiment found this to be the case. Here the nonreplacement regimen remained in force but the correct set of arms was displaced by 45°. With only even-arms baited now, how well could the groups find and restrict their movements to the new set of arms? Half of the rats trained to run to odd-arms were tested in this new problem, and the other half were exposed to different conditions (described below). As usual, Saline and Low dose rats adjusted apace while the rest never improved beyond chance proportions of odd to even choices (Figure 11). Few of the High dose animals ever met criterion (Table 3). Thus, even though their preferences for the odd-arms were not as strong as those of the other groups going into phase 2, Medium and

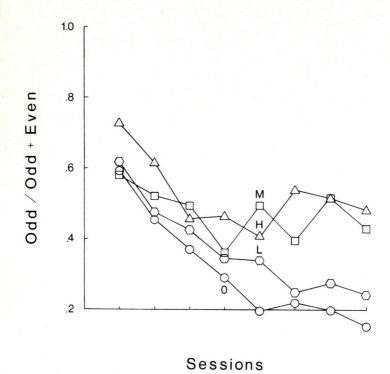

Figure 11.   Mean proportion of odd (not rewarded) to
   odd plus even (rewarded) -numbered arms selected.
   Note that the smaller the proportion, the more
   efficient the performance.   For details see Figure
   9.

High dose rats were nevertheless unable to refrain from running among the formerly correct set of arms.

In another part of this experiment, the rest of the animals
trained in phase 1 to run to odd-arms were now afforded an
opportunity to double their rewards.

Table 3

Number of Subjects Meeting Criterion in Phase 2
(Change from Odd- to Even-Arm Nonreplacement Problem)

		Sessions							
Dose[a]	n	1	2	3	4	5	6	7	8
Sal[b]	6	6	6	6	6	6	6	6	6
0.75	6	6	6	6	6	6	6	6	6
1.5	6	4	2	4	5	5	4	4	5
2.0	6	1	2	3	2	3	3	3	3

Note. Criterion was defined as 2 successive sessions during
which all four rewards were taken in the alloted (10 min)
time.
[a]In g/kg.
[b]Sal = saline injection.

All arms were baited in this phase. Higher dose groups were
unable to take advantage of the increased bounty (Figure 12
& Table 4) through their failure to sample new arms. As before, although we manipulated only spatial contingencies, the
other BV measures paralleled performance. Sequential and
topographic BV was gratuitously expressed at higher levels by
the flexible, rapidly adapting groups. And this is probably
to be expected as the subjects can never predict along what
dimension rules will change.

Foremost among our radial-arm maze findings is the conclusion
that normally functioning rats spontaneously interrupt their
goal-directed activities with alternative behaviors, diversify
the paths that they follow and as was known previously, spread
their visits over a wide area. In short, intact animals exhibit considerable superfluity of behavior. At least as far
as spatial BV is concerned, it acquits them well when change
occurs. It is reasonable to assume that new relationships
will not be discovered without at least some minimal rate
of sampling of new places, paths, and topographies. Variability in the radial-arm maze seemed to permit rapid adjustment to the displacement of reward, and it promoted serendi-
pity.

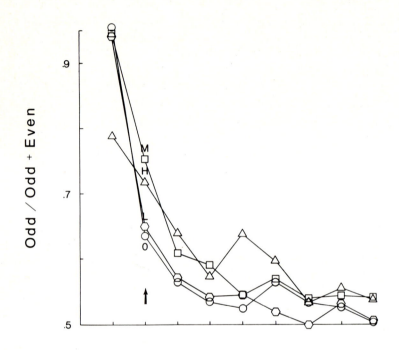

Figure 12. Mean proportion of odd (rewarded) to odd plus
even (also rewarded) -numbered arms selected. The
last training session is included for reference. The
new conditions begin at the arrow. Note that accurate
performance requires scores of about 0.5. $\underline{ns}$ = 6.
For abbreviations see Figure 3. From Devenport and
Merriman, 1982.

Table 4

Number of Subjects Meeting Criterion in Phase 2
(Change to All Arms Baited With Nonreplacement)

Dose[a]	n	Sessions							
		1	2	3	4	5	6	7	8
Sal[b]	6	6	6	6	6	5	5	6	6
0.75	6	6	6	6	6	6	6	6	6
1.5	6	3	4	3	6	5	6	6	6
2.0	6	0	0	1	1	1	1	1	2

Note. Criterion was defined as 2 successive sessions during which all four rewards were taken in the alloted
(10 min) time.
[a]In g/kg.
[b]Sal = saline injection

Operant Studies

While mazes tend to emphasize spatial aspects of performance,
these features are minimized in operant conditioning. This is
of importance for the present purpose as I am trying to establish BV as a fundamental dimension of behavior. Exhibiting its prevalence in dissimilar settings helps to demonstrate
the ubiquity of BV.

Besides their sometimes "rate-dependent" effects (e.g., Dews &
Wenger, 1977) both amphetamine (Weiss & Laties, 1964) and alcohol (Crow, McWilliams, & Ley, 1979) homogenize response
rates, the intervals separating successive operant responses
becoming more constant. Frontal ablations have the same effect (Crow & McWilliams, 1979) and so, apparently, do septal
lesions (Ellen, Wilson, & Powell, 1964), at least on interval
schedules of reinforcement: Similar findings have been obtained for the session-to-session stability of response rates
of rats with hippocampal lesions (Devenport & Holloway, 1980)
and for response durations in frontal-damaged animals (Crow
& McWilliams, 1979). Although only infrequently examined,
operant response topographies have been reported to be nar-
rowed by amphetamine administration (Nieto, Makhlouf, &

Rodriguez, 1979) and hippocampal damage (Devenport & Holloway, 1980). Finally, it has been demonstrated that hippocampal impairment (fornix damage; Osborne & Black, 1978) and alcohol injection (Crow, McWilliams, Ley, 1979) increase the predictability of response sequences in operant settings.

Although convincing, the studies I have cited are not as systematic as one might wish. With, I believe, the exception of Crow (e.g., 1977) no behavioral laboratory has undertaken a program of research into BV. This leaves us in the position of citing studies that were designed to answer rather specific questions (e.g., why do septal-lesional animals overrespond on interval schedules?), their chief aim not being an explication of BV or the establishment of a standard methodology. We have begun a major effort in this direction, examining alcohol-injected, frontal, and hippocampal-lesional animals in a setting designed to extract several indices of BV.

Operant chambers were modified in such a way that photobeams could detect visits to the recessed food cup or to the back of the chamber, away from the lever and source of reinforcement. These response sensors were continuously monitored by computer; and information about each response, its order with respect to other responses, the interval separating successive responses, etc., was stored on diskette. Manual shaping was carried out under the drug condition and schedule demands were progressively increased, at which point data collection began and continued for 16--30 minute sessions.

Because some of the findings from these studies are preliminary, I will restrict the discussion here to the effects of alcohol (saline 0.75, 1.0, 1.5 g/kg ethanol, i.p.) on random-interval 100 second schedules, which has been most thoroughly studied. Alcohol significantly reduced behavioral variability in a variety of ways. Topographic (or geographic) BV declined: Alcohol suppressed the tendency to withdraw from the site of the lever and food-cup (Figure 13). Saline and Low dose groups removed themselves frequently from the lever-pressing routine, and as Figure 14 indicates the visits to the rear of the cage were irregular in Saline animals and become more rhythmic as ethanol doses increased. Alcohol-injected rats also compared unfavorably on the variability of intervals separating successive lever presses (Figure 15). In this case all groups except the highest dose pressed at about the same mean rate yet the tempo was made more regular by the drug.

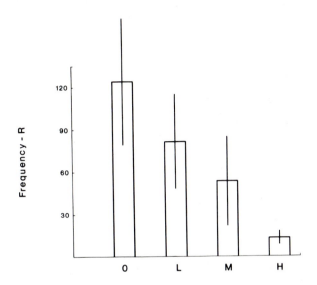

Figure 13. Mean ± SEM rear photocell interruptions during 16-30 minute $\overline{RI}60$-sec sessions for groups receiving
1.5(H), 1.0(M), 0.75(L), and 0.0 g/kg(0) ethanol doses.
From Devenport and Merriman, unpublished.

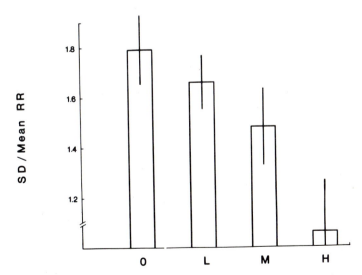

Figure 14. Mean variability (standard deviation in sec)
of successive rear photocell interruptions expressed as
a ratio of the mean interval (in sec). For details see
caption, Figure 13.

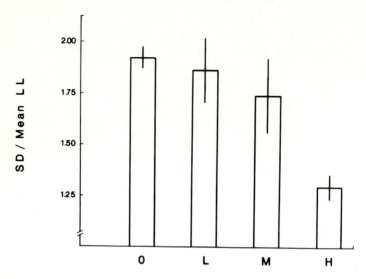

Figure 15. Mean variability (standard deviation in sec)
    of intervals separating successive lever presses ex-
    pressed as a ratio of the mean interval in sec. For
    details see caption, Figure 13.

In most respects, alcohol produced the sorts of changes characteristic of brain damage. With the narrowing of virtually all
aspects of BV, we guessed that superstitious-type performance
would be a likely outcome if the response-reinforcer contingency were broken. That is, if pellets continued to be delivered randomly, at the same rate but freely, intact animals
would soon extinguish but alcohol-injected animals' rates might
decline more slowly, as if the contingency remained. This
retarded extinction would not be attributable to perseveration
because--for reasons unknown--operant extinction is not influenced by alcohol (Devenport & Merriman, in preparation) nor
is it for that matter much affected, if at all, by brain (e.g.,
hippocampal: Schmaltz & Isaacson, 1967) damage. In any event,
the suspicion that superstition might be revealed was based on
the drugged rats' failure to sample diverse topographies,
rates, and places (withdrawing from the lever). Unless, this
sort of sampling is carried-out it seems unlikely that the
independence of pellet deliveries and, lever-pressing could
be discovered. This is the case with hippocampal (Devenport,

1980; Devenport & Holloway, 1980), septal (Devenport, 1979), and accumbens-lesioned (Devenport & Devenport, 1981) rats. False response-reinforcer associations are acquired in the absence of true contingency or, having broken a contingency, behavior--especially that of individual animals--continues at high levels for several sessions. Some brain-damaged animals eventually learn about the lack of correlation, others do not. Figure 16 illustrates in condensed form the shift from response-dependent (random-interval 60 seconds) to response-independent (random-time 60 seconds) conditions for animals receiving alcohol injections.

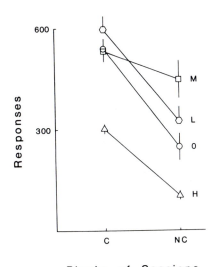

Blocks of Sessions

Figure 16. Mean lever presses per session collapsed across contingent (C) and noncontingent (NC) sessions for the various alcohol dose groups. For details see caption, Figure 13 and text.

Although rats with the highest dose of alcohol were in every case the most invariant, this group displayed no significant retardation of responding in the noncontingent phase of the experiment. Rather, it was the Medium group that failed to cease responding. In retrospect this might have been predicted. While all other groups pressed at about the same rate, High dose subjects spaced their responses widely, permitting quite fortuitously the opportunity to learn that pellet deliveries were now free. These animals did not need the services of a mechanism that forced them away from the bar and

away from homogeneous rates in order to discover the change.
Their long inter-response intervals provided ample time for
this. Accordingly, it was the Medium group that--of all those
pressing at high rates and of these being the most stereotyped-
-contined to superstitiously press the lever when it was no
longer reinforced. It is probably coincidental, but the relatively small decline in response rates (67%) following free
delivery for this group is exactly the extent to which animals
with hippocampal lesions decline (Devenport & Holloway, 1980).
It is worth observing that one of the ingredients of susceptibility to superstition generally is, besides invariance, the
tendency to respond at high rates. This was true of nearly all
the lesioned animals we have investigated and found superstitious.

Spontaneous vs. Induced BV

Taken together the alley, maze, and operant studies detailed
here, along with the considerable direct and indirect work
that has gone before, attests to the ubiquity of variability
and a beginning has been made toward explicating its function and how it may affect conventional measurements of performance. Far from "error" in the usual sense, variability can
be better thought of as a mechanism of discovery. We are accustomed to regarding exploration as the investigation of space
and its furnishings. This notion, I suggest, is in need of
liberalization: Rates, routes, sequences, times, and other
matters and presumably combinations of matters (relationships)
are explored. I can think of no other way in which environmental contingencies can be detected for the first time, how
their cessation can be disclosed, how adjustments can be made
to loss of necessities, or how more advantageous strategies,
locations, etc., can be found. The value of inconsistency
would seem to be well worth its cost: An occasional loss for
the potential opportunity that may be revealed. This, of
course, includes the confounding of predators and the discovery of potential routes of escape. Although this behavior
appears to be spontaneous, and despite its somewhat random
character, it is doubtless subject to at least some external
influences.

An interesting paper was published several years ago by
Wickelgren and Isaacson (1963), subsequently repeated by
Raphelson, Isaacson, and Douglas (1965) and others. Rats
with hippocampal damage together with control groups were
trained to run down an alley, after which a slight contextual
change was introduced (walls brightened or the floor given a
different texture). These changes occasioned some impressive

performance decrements in all groups except those that had
sustained hippocampectomy. This is particularly interesting in
view of the hippocampal-lesioned rats' normal or sometimes
exaggerated reaction to novel stimuli (Coover & Levine, 1972)
when not otherwise engaged in goal-directed activity (see
O'Keefe & Nadel, 1978 for a review of competitive vs. noncom-
petitive situations). These findings suggested to us that a
nonlocalizable contextual change might induce similar performance decrements. And as for the nature of these decrements,
we would consider them to be secondary to induced BV.

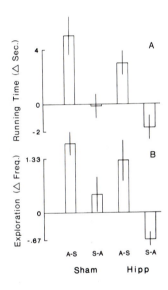

Figure 17. Mean ± SEM change in (A) running time, and (B)
    exploration frequency from the last training session for
    animals experiencing a change from the drug present dur-
    ing training. (A-S = alcohol in training, saline in
    testing, S-A = saline in training, alcohol in testing).
    From "Necessity of the Hippocampus for Alcohol's
    Indirect but not Direct Behavioral Action" by
    L. D. Devenport, J. A. Devenport, and F. A. Holloway,
    Behavioral and Neural Biology, 1981, 33, 476-487.
    Copyright 1981 by Academic Press, Inc. Reprinted
    by permission.

We (Devenport, Devenport & Holloway, 1981a) tested this possibility in a 2 x 2 x 2 design where some animals were trained
to run an alley under a comparatively low dose of alcohol while
others were trained under the saline condition. When these

drug conditions were suddenly switched the expected results
appeared (Figure 17). The internal stimulus or "state" change
induced considerable exploration (BV) and consequently longer
running times in sham-operated rats. This effect was mostly
restricted to the group shifting from alcohol to saline, a case
of the commonly observed "asymmetrical dissociation" characteristic of alcohol and other drugs (Overton, 1972). Rats with
hippocampal lesions displayed no overall effect of the change,
speeding up significantly when switched to alcohol, slowing
when switched to saline, indicating a sensitivity to the direct
effect of the drug on spontaneous BV (as discussed in the
section on BV in the runway) but no effect on induced BV. The
lesion effect we obtained was not owing to a decreased stimulus
value of alcohol vs. saline for hippocampal-lesioned animals
as indicated by their normal performance in an operant drug-
discrimination task (Modrow, Holloway, & Devenport, 1980).
Other groups underwent no drug change and as their performance
was unaffected we have not illustrated these results.

These findings suggest a further liberalization of the notion
of exploration; that changed items need not be localized to
induce exploration in normal animals. The drug state, or rather its change, could not possibly be explored in any customary
sense. Our tentative assumption is that for rats, context and
the stimuli it contains are relational and interdependent, one
change--even an internal one--prompts an episode of information
-gathering in the previously familiar apparatus. Again, a
forebrain structure was found to promote this sort of variability, suggesting its evolutionary utility as opposed to a
simple "disruption" of performance. Ordinarily, if one part
of a setting is changed, this may occasion the expectation that
other parts are changed as well.

Stereotypy

The evidence is strong that at least in part, the proliferation
of sizable amounts of telencephalic tissue provides for
spontaneous and induced BV. And these enforced encounters
with newness, while providing the raw material of learning,
quite plainly oppose previously acquired habits. In providing
information it disorganizes routine performance. So in a
sense, animals bearing lesions or administered drugs have
learned, or are performing, too well. Their behavior expresses
little of the tentativeness of intact animals for whom learning
seems more provisional. This explains why CNS-impaired animals

frequently learn and perform better. Providing the task does not require within-trial or between-session shifts, the performance of such subjects can be expected to contain more structure and less error. As the matter does seem to be one of performance, not learning (as was especially evident in the section on the radial-arm maze), and as reward is central to performance, a possible mode of action for the variability mechanism suggests itself.

No matter how it originates, learning as translated into performance is a process that diminishes BV. It is not merely, e.g., the high rate of lever-pressing or the making of right turns in a T-maze that distinguishes learning. It is that these responses are exhibited to the relative exclusion of other behaviors. This invariant repetition is what differentiates learned performance from activation. It is true that animals can be trained to shift performance away (in time, space, etc.) from that most recently rewarded. But this shifting pattern becomes progressively ingrained and a reversal of this rule offers some initial difficulty and could not be learned were it not for the normal animals' spontaneous tendency to sample occasional "incorrect" responses. At least in birds (Schwartz, 1982), variability itself cannot be established or maintained by reinforcing contingencies. And this (reinforcement) is probably the crux of the matter. Whatever the role of reward or reinforcement in learning, its effect on performance seems undeniable--it is an agent of stereotypy.

Although reinforcing stimuli are numerous and diverse, there is reason to believe that all natural reinforcers activate catecholamine (Poschel, 1968), most importantly dopamine (DA) circuits. Especially compelling are the observations that loci supporting the highest rates of self-administered brain stimulation are situated along catecholamine pathways (Crow, 1976) from which a measurable efflux of DA can be collected following natural (food or water) reward (Heffner, Hartman, & Seiden, 1980; Martin & Myers, 1976). Drugs that interfer with dopamine transmission reduce the efficacy of self-administered brain stimulation (e.g., Fouriezos, Hansson, & Wise, 1978) and produce an extinction-like effect on response output requiring higher current intensities (Esposite, Faulkner, & Kornetsky, 1979; Schafer & Michael, 1980) or multiple pulse (Franklin, 1978) in order to restore predrug response rates. And drugs that promote dopamine transmission (e.g., amphetamine, cocaine, apomorphine) serve as operant reinforcers (e.g., Yokel & Wise, 1978), except when animals are pretreated with DA receptor blockers.

Rewards or reinforcers, then, have much in common with drugs
that promote DA transmission and might themselves be considered to be functional DA agonists. If this is the case then
it should be worthwhile to review the behavioral effects of
pharmacological dopamine agonists in order to better understand
the nature of reward. Drugs can be administered in higher
"doses" than rewards, not being subject to the effects of satiety, sensory gating, and other limits of potency. In this way,
otherwise ambiguous aspects of reward action might stand in
bold relief.

Observed in birds (DeLaveralle & Youngren, 1978), rodents,
cats, and primates including man (Randrup & Munkvad, 1967),
stereotypy is the unequivocal result of DA agonist administration. The stereotypies seem to involve relatively primitive
movements that vary from species to species. The commonality
is not topography but invariance. The same response or sequence of responses are repeated over and over. An exception-
ally interesting report, (Ellinwood & Kilbey, 1975), along with
others of a similar vein, (Schallert, DeRyck & Teitelbaum,
1980, Stevens, Livermore, & Cronan, 1977) helped to establish
the idea that stereotypies were integrated environment-de-
pendent acts and not merely motor discharges. It was found
that in a less barren than usual testing environment, nearly
any class of behavior--species-specific acts, simple learned
responses, and more complex activity--is susceptible to drug-
induced stereotypy. According to these authors and in keeping with an earlier report (Teitelbaum & Derks, 1958) which
they cite, stereotypy can invade any ongoing behavior that
happens to be executed as amphetamine begins exerting its
central effects. A common and ironic case is that of exploration which itself can become absurdly repetitious.

In summary, DA agonists exact a stereotyping influence on ongoing behavior; they increase the likelihood of response repetition. With this we have come full circle. If DA causes
the repetition of whatever the animal is presently doing then
this "pharmacological" stereotypy does not differ from reinforced performance as commonly defined, except in intensity.
Perhaps more correctly stated, reinforced behavior is an instance of stereotypy.

Stereotypy seems fundamental and agents of reward seem to invoke it. Possibly, then, telencephalic mechanisms of BV have
evolved in such a way that they oppose the effects of reward
promoting BV by moderating reinforcement. If this is the
case then damage to brain structures involved in BV should un-
mask an unmodulated reward mechanism permitting a full-blown
pharmacological type of stereotypy. Stereotypy of this sort

has come to be assessed according to rating scales that attempt to determine the relative intensity of stereotyped acts. These usually include (in the rat) up-and-down and side-to-side head movements, sniffing, stereotyped locomotion, licking, and gnawing. Using such a scale, we (Devenport, Devenport, & Holloway, 1981b) have tested the idea that hippocampal lesions would release reward from neural opposition and permit it to induce uncommonly intense effects on behavior.

Rats with hippocampal lesions and their controls were blindly rated for stereotypy when free-feeding, deprived, or just before receiving an amount of reward (lab chow) sufficient to maintain their body weights at about 80% of free-feeding level. Gross locomotor activity was monitored automatically for the 1 hr period preceeding feeding (and the comparable interval for the previous two phases). Figures 18 and 19

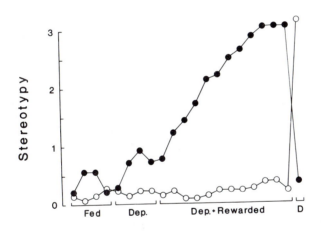

Sessions

Figure 18. Mean stereotypy scores of animals with hippocampal (filled circles) and neocortical damage (open circles) during free-fed (Fed), deprived (Dep.) during the signal for reward (Dep. + Reward) and drug(D) phases. Stereotypy scores: 0, asleep, or stationary; 1, active; 2, predominately active but with bursts of stereotyped rearing and sniffing; 3, constant stereotyped activity over a wide area; and 4, constant stereotyped sniffing or head-bobbing in one place. From "Reward-Induced Stereotypy:

Modulation by Hippocampus" by L. D. Devenport,
J. A. Devenport, and F. A. Holloway, Science, 1981,
212, 1288-1289. Copyright 1981 by the American
Association for the Advancement of Science. Reprinted by permission.

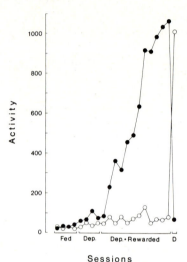

Figure 19. Mean locomotor activity scores of animals
    with hippocampal (filled circles) and neocortical
    (open circles) damage during the phases described
    in caption, Figure 18. From "Reward-Induced Stereo-
    type: Modulation by the Hippocampus" by L. D.
    Devenport, J. A. Devenport, and F. A. Holloway,
    Science, 1981, 212, 1288-1289. Copyright 1981 by
    the American Association for the Advancement of
    Science. Reprinted by permission.

illustrate a robust reward-induced stereotypy and locomotor
activation in hippocampal-lesioned, but not control rats. The
profile of this reward-induced effect in the hippocampal lesioned animals was virtually identical to that produced by
1.5 mg/kg d-amphetamine sulfate administered to the periodically fed controls. And the reward-induced effects were normal-
ized by DA antagonism (.06 mg/kg haloperidol).

These findings support the view that the evolution of the
hippocampus and probably other brain structures (e.g., see
Campbell, Ballantine, & Lynch, 1971) came to exert, among
other things, a functional opposition to stereotypy, quite
possibly by modulating the effects of reward. This idea is
supported by the recent finding (Goodman, Zacny, Osman, Azzaro,

& Donovan, 1981) that animals (pigeons) which lack well
developed frontal, septal and hippocampal formations are quite
susceptible to reward-induced stereotypies. In keeping with
the DA hypothesis, these stereotypies were found to be augment-
ed and suppressed by a dopamine agonist and antagonist respectively. This finding is also consistent with the pigeon's
tendency toward superstitious responding (e.g., Staddon &
Simmelhag, 1971) and other deficiencies relative to mammalian
behavior (e.g., DRL performance; Kramer & Rilling, 1970). This
is not to say that avians may not have developed other structures (e.g., the hyperstriatal complex) that may assist them
in varying behaviors that are appropriate for their niche.

Our results with reward-induced stereotypy and its possible
dopaminergic basis are suggestive. But they are only suggestive. Whether the hippocampus and possibly other structures enforce BV by modulating the effects of reward and whether this is accomplished by decreasing its intensity (peak DA
release?) or duration; by diminishing the efficacy of successive reward applications, or a host of other possibilities, we
can only guess. Moreover, the argument could just as well be
made that BV substrates oppose reward only in the sense that
they promote behaviors that compete with stereotypy. In the
absence of further research the only conclusion that can be
drawn at present is that stereotypy and its opponent process,
BV, are highly amenable to experimentation. Concrete predictions can be drawn and tested. And despite its somewhat
random character, BV will doubtless be found to be a lawful--
in a sense a predictable--feature of animal behavior.

Caveat and Summary

I have tried to place neuroscience at the service of psychology
for the purpose of drawing some broad inferences about animal
behavior. In doing this I might be accused of overlooking
important distinctions and glossing over exceptions and contra-
dictions; and so I have. For example, I am aware that septal
lesions affect spontaneous alternation in a way quite different from that of hippocampal lesions (Dalland, 1970), that
the projections of the mediodorsal thalamic nucleus (Leonard,
1969) more accurately define the cortical regions of interest
than my general inference to the "frontal area", etc. I have
not hesitated to omit discussions of drug action or brain
function that did not bear on the topic. This was done with
the understanding that this chapter is not a review of brain
function, rather a consideration of what neuroscience could
contribute in the way of understanding spontaneous behavior.
The septum, frontal areas and hippocampus each perform several
functions. And with respect to BV I can hardly help but think

that, even in this more restricted domain, specialization must prevail. Perhaps, the hippocampus contributes mostly to spatial BV, the frontal areas to temporal BV, and the septum to shifts in stimulus tracking. Finally, I have omitted much work that is relevant. Anticholinergic and opiate drugs might have replaced amphetamine and alcohol. And the proliferating literature on the limbic structures receiving extensive DA projections (e.g., N. Accumbens, Olfactory Tubercle) is becoming increasingly relevant to BV and stereotypy. However, the purpose was served without it.

I have tried to exhibit as plainly as possible the fact of CNS mediation of BV; that a brain-damaged animal is frequently a more structured, predictable, and--in unchanging conditions-- a more errorless performer than an intact subject. This permitted the adoption of the position that BV is not a "mistake" on the part of the animal. Nor is it necessarily a failing of theoretical prediction. Brain structures have evolved that dispense BV according to rules that are little understood, yet the result seems obvious. Animals are conducted away from routine and into the midst of novelty. This is not, I think, a difficult position to hold. It at once offers a legitimate explanation for the statistical variability that characterizes behavioral work and it explains how animals get themselves exposed to new information. Where some might part company is with the implication in these conclusions that the basic mechanism of performance may be quite primitive and that the strong influence of reward on behavior does not represent the pinnacle of mammalian evolutionary achievement. I think that this implication has some merit. The studies I have reviewed seem to represent animals with intact nervous systems as skeptics tentatively following rules that they can eagerly abandon, not inductivists holding that the future is a repetition of the past.

Footnote

Preparation of this chapter was supported by USPHS grant AA05699 to L.D.D.

References

1 Antonitis, J. J. Response variability in the white rat during conditioning, extinction, and reconditioning. Journal of Experimental Psychology, 1951, 42, 273-281.

2 Attneave, F. Application of information theory to psychology. New York: Holt, Rinehart and Winston, 1959.

3 Barry, H. III, Wagner, A. R. & Miller, N. E. Effects of alcohol and amobarbital on performance inhibited by experimental extinction. Journal of Comparative and Physiological Psychology, 1962, 55, 464-468.

4 Bättig, K. Differential psychopharmacological patterns of action in rats. In Z. Vatana (Ed.), Psychopharmacological methods. London: Pergamon, 1963.

5 Braggio, J. T. and Ellen, P. Cued DRL training: Effects on the performance of lesion-induced over-responding. Journal of Comparative and Physiological Psychology, 1976, 90, 694-703.

6 Campbell, B. A., Ballantine, P. & Lunch, G. Hippocampal control of behavioral arousal: Duration of lesion effects and possible interactions with recovery after frontal cortical damage. Experimental Neurology, 1971, 33, 159-170.

7 Carlton, P. Some effects of scopolamine, atropine and amphetamine in three behavioral situations. Pharmacologist, 1961, 3, 60.

8 Clark, C. V. H. & Isaacson, R. L. Effect of bilateral hippocampal ablation on DRL performance. Journal of Comparative and Physiological Psychology, 1965, 59, 137-140.

9 Coover, G. D. & Levine, S. Auditory startle response of hippocampectomized rats. Physiology and Behavior, 1972, 9, 75-77.

10 Cox, T. The effects of caffeine, alcohol, and previous exposure to the test situation on spontaneous alternation. Psychopharmacologia, 1970, 17, 83-88.

11 Crow, L. T. Is variability a unifying concept? Psychological Record, 1977, 27, 783-790.

12 Crow, L. T. & McWilliams, L. S. Relative stereotypy of
    water-ingestive behavior induced by frontal cortical
    lesions in the rat. Neuropsychologia, 1979, 17, 393-400.

13 Crow, L. T., McWilliams, L. S. & Ley, M. F. Relative
    stereotypy of water-ingestive behavior induced by chronic
    alcohol injections in the rat. Bulletin of the Psychonomic
    Society, 1979, 14, 278-280.

14 Crow, T. J. Specific monoamine systems as reward pathways:
    Evidence for the hypothesis that activation of the central
    mesencephalic dopamine neurons and norepinephrine neurons
    of the locus coeruleus complex will support self-stimula-
    tion responding. In A. Wauquier & E. T. Rolls (Eds.),
    Brain stimulation reward. Amesterdam: North Holland,
    1976.

15 Dalland, T. Response and stimulus perseveration in rats
    with septal and dorsal hippocampal lesions. Journal of
    Comparative and Physiological Psychology, 1970, 71, 114-
    118.

16 Dashiell, J. F. Direction orientation in maze running by
    the white rat. Comparative Psychology Monographs, 1930,
    7, No. 32.

17 DeLaveralle, N. D. & Youngren, O. M. Chick vocalization
    and emotional behavior influenced by apomorphine. Journal
    of Comparative and Physiological Psychology, 1978, 92,
    416-430.

18 Devenport, J. A. & Devenport, L. D. Mesolimbic mechanisms
    of behavioral superstition. Neuroscience Abstracts, 1981,
    7, 754.

19 Devenport, J. A., Merriman, V. J. & Devenport, L. D. Al-
    cohol mimics the effects of hippocampal lesions in the
    radial arm maze. Alcoholism: Clinical and Experimental
    Research, 1982, 6, 139.

20 Devenport, L. D. Inappropriate ingestive behaviors aris-
    ing from autoshaping procedures. Behavioral and Neural
    Biology, 1979, 27, 558-563.

21 Devenport, L. D. Response-reinforcer relations and the
    hippocampus. Behavioral and Neural Biology, 1980, 29,
    105-110.

22   Devenport, L. D., Devenport, J. A. & Holloway, F. A. Necessity of the hippocampus for alcohol's indirect but not
direct behavioral action. Behavioral and Neural Biology.
1981, 33, 476-487. (a).

23   Devenport, L. D., Devenport, J. A. & Holloway, F. A. Reward-induced stereotypy: modulation by the hippocampus.
Science, 1981, 212, 1288-1289. (b).

24   Devenport, L. D. & Holloway, F. A. The rat's resistance to
superstition: Role of the hippocampus. Journal of Comparative and Physiological Psychology, 1980, 94, 691-705.

25   Devenport, L. D. & Merriman, V. J. Alcohol and maze performance: Working memory or variability deficits? Al-
coholism: Clinical and Experimental Research, 1982, 6,
139.

26   Devenport, L. D., Merriman, V. J. & Holloway, F. A. Promotion of stereotypy by alcohol. Neuroscience Abstracts,
1981, 7, 159.

27   Dews, P. B. & Wenger, G. R. Rate-dependency of the be-
havioral effects of amphetamine. In T. Thompson &
P. B. Dews (Eds.), Advances in Behavioral Pharmacology.
New York: Academic Press, 1977.

28   Ellen, P., Aitken, W. C., Jr. & Walker, R. Pre-training
effects on performance of rats with hippocampal lesions.
Journal of Comparative and Physiological Psychology, 1973,
84, 622-628.

29   Ellen, P., Wilson, A. S. & Powell, E. W. Septal inhibition
and timing behavior in the rat. Experimental Neurology,
1964, 10, 120-132.

30   Ellinwood, E. H. & Kilbey, M. M. Amphetamine stereotypy:
The influence of environmental factors and prepotent behavior patterns on its topography and development. Biological Psychiatry, 1975, 10, 3-16.

31   Esposite, R. U., Faulkner, W. & Kornetsky, C. Specific
modulation of brain stimulation reward by haloperidol.
Pharmacology, Biochemistry and Behavior, 1979, 10, 937-940.

32   Fouriezos, G., Hausson, P. & Wise, R. A. Neuroleptic-in-
duced attenuation of brain stimulation reward in rats.
Journal of Comparative and Physiological Psychology, 1978,
92, 661-671.

33 Franklin, K. B. J. Catecholamines and self-stimulation: Reward and performance effects dissociated. Pharmacology, Biochemistry and Behavior, 1978, 9, 813-820.

34 Frick, F. C. & Miller, G. A. A statistical description of operant conditioning. American Journal of Psychology, 1951, 64, 20-36.

35 Fuster, J. M. The prefrontal cortex, New York: Raven, 1980.

36 Gaffan, E. A. & Davies, J. The role of exploration in win-shift and win-stay performance on a radial maze. Learning and Motivation, 1981, 12, 282-299.

37 Glowa, J. R. & Barrett, J. E. Effects of alcohol on punished and unpunished responding of squirrel monkeys. Pharmacology, Biochemistry and Behavior, 1976, 4, 169-173.

38 Goodman, I., Zacny, J., Osman, A., Azzaro, A. & Donovan, C. Dopaminergic nature of feeding induced behavioral stereotypies. Neuroscience Abstracts, 1981, 7, 43.

39 Heffner, T. G., Hartman, J. A. & Seiden, L. S. Feeding increases dopamine metabolism in the rat brain, Science, 1980, 208, 1168-1170.

40 Heise, G. A., Laughlin, N. & Keller, G. A. Behavioral analysis of reinforcement withdrawal. Psychopharmacologia, 1970, 16, 345-368.

41 Holloway, F. A. & Wansley, R. A. Factors governing the vulnerability of DRL operant performance to the effects of ethanol. Psychopharmacologia, 1973, 28, 351-362.

42 Isaacson, R. L. The limbic system (2nd Ed.). New York: Plenum, 1981.

43 Kelsey, J. E. & Grossman, S. P. Non perseverative disruption of behavioral inhibition following lesions in rats. Journal of Comparative and Physiological Psychology, 1971, 75, 302-311.

44 Kimble, D. P. & Kimble, R. J. Hippocamectomy and response perseveration in the rat. Journal of Comparative and Physiological Psychology, 1965, 60, 472-476.

45 Kimble, D. P. & Kimble, R. J. The effect of hippocampal lesions on extinction and "hypothesis" behavior in rats. Physiology and Behavior, 1970, 5, 735-738.

46 Kolb, B. & Whishaw, I. Q. Neonatal frontal lesions in the rat: Sparing of learned but not species-typical behavior in the presence of reduced brain weight and cortical thickness. Journal of Comparative and Physiological Psychology, 1981, 95, 863-879.

47 Kramer, T. S. & Rilling, M. Differential reinforcement of low rates: A selective critique. Psychological Bulletin, 1970, 74, 225-254.

48 Laties, V. G. & Weiss, B. Influence of drugs on behavior controlled by internal and external stimuli. Journal of Pharmacology and Experimental Therapeutics, 1966, 152, 388-396.

49 Leonard, C. M. The prefrontal cortex of the rat - I. Cortical projections of the mediodorsal nucleus. II. Efferent connections. Brain Research, 1969, 12, 321-343.

50 Martin, G. E. & Myers, R. D. Dopamine efflux from the brain stem of the rat during feeding, drinking and lever pressing for food. Pharmacology, Biochemistry and Behavior, 1976, 4, 551-560.

51 McCleary, R. A. Response specificity in the behavioral effects of limbic system lesions in the cat. Journal of Comparative and Physiological Psychology, 1961, 62, 263-269.

52 Modrow, H. E., Holloway, F. A. & Devenport, L. D. Effects of hippocampal lesions on alcohol discrimination in rats. Neuroscience Abstracts, 1980, 6, 313.

53 Nieto, J., Makhlouf, C. & Rodriguez, R. d-Amphetamine effects on behavior produced by periodic food deliveries in the rat. Pharmacology, Biochemistry and Behavior, 1979, 11, 423-430.

54 Numan, R., Seifert, A. R., & Lubar, J. F. Effects of medio-cortical frontal lesions on DRL performance in the rat. Physiological Psychology, 1975, 3, 390-394.

55 O'Keefe, J. & Nadel, L. The hippocampus as a cognitive map. Oxford: Claredon Press, 1978.

56   Olton, D. S.   Mazes, maps, and memory.   American Psycho-
     logist, 1979, 34, 583-596.

57   Olton, D. S. & Papas, B. C.   Spatial memory and hippocampal
     function.   Neuropsychologia, 1979, 17, 669-681.

58   Olton, D. S. & Scholsberg, P.   Food searching strategies in
     young rats: Win-shift predominates over win-stay.   Journal
     of Comparative and Physiological Psychology, 1978, 92,
     609-618.

59   Olton, D. S. & Wertz, M. A.   Hippocampal function and be-
     havior: Spacial discrimination and response inhibition.
     Physiology and Behavior, 1978, 20, 597-605.

60   Osborne, B. & Black, A. H.   A detailed analysis of behavior
     during the transition from acquisition to extinction in
     rats with fornix lesions.   Behavioral Biology, 1978, 23,
     271-290.

61   Overton, D. A.   State-dependent learning, produced by al-
     cohol and its relevance to alcoholism.   In B. Kissin &
     H. Begleiter (Eds.).   The biology of alcoholism, (Vol. 2):
     Physiology and behavior.   New York: Plenum, 1972.

62   Poschel, B. P. H.   Do biological reinforcers act via the
     self-stimulation pathways of the brain?   Physiology and
     Behavior, 1968, 3, 53-60.

63   Randrup, A. & Munkvad, I.   Stereotyped activities produced
     by amphetamine in several animal species and man.   Psycho-
     pharmacologia, 1967, 11, 300-310.

64   Raphelson, A. C., Isaacson, R. L. & Douglas, R. J.   The
     effect of distracting stimuli on the runway performance of
     limbic damaged rats.   Psychonomic Science, 1965, 3, 483-
     484.

65   Riddell, W. I.   An examination of the task and trial para-
     meters in passive avoidance learning by hippocampectomized
     rats.   Physiology and Behavior, 1968, 3, 883-886.

66   Roberts, W. W., Dember, W. N. & Brodwick, M.   Alternation
     and exploration in rats with hippocampal lesions.   Journal
     of Comparative and Physiological Psychology, 1962, 55,
     695-700.

67 Schaefer, G. J. & Michael, R. P. Acute effects of neuro-
    leptics on brain self-stimulation thresholds in rats.
    Psychopharmacology, 1980, 67, 9-15.

68 Schallert, T., DeRyck, M. & Teitelbaum, P. Atropine
    stereotypy as a behavioral trap: A movement subsystem and
    electroencephalographic analysis. Journal of Comparative
    and Physiological Psychology, 1980, 94, 1-24.

69 Schmaltz, L. W. & Isaacson, R. L. Effect of bilateral
    hippocampal destruction on the acquisition and extinction
    of an operant response. Physiology and Behavior, 1967,
    2, 291-298.

70 Schuster, C. R. & Zimmerman, J. Timing behavior during
    prolonged treatment with dl-amphetamine. Journal of the
    Experimental Analysis of Behavior, 1961, 4, 327-330.

71 Schwartz, B. Failure to produce response variability with
    reinforcement. Journal of the Experimental Analysis of
    Behavior, 1982, 37, 171-182.

72 Sikorszky, R. D., Donovick, P. J., Burright, R. G. & Chin,
    T. Experiential effects on acquisition and reversal of
    discrimination task by albino rats with septal lesions.
    Physiology and Behavior, 1977, 18, 231-236.

73 Staddon, J. E. R. & Simmelhag, V. L. The "superstition"
    experiment. Psychological Review, 1971, 78, 3-34.

74 Stevens, J., Livermore, A. & Cronan, J. Effects of deafen-
    ing and blindfolding on amphetamine induced stereotypy in
    the cat. Physiology and Behavior, 1977, 18, 809-812.

75 Teitelbaum, H. A comparison of effects of orbitofrontal
    and hippocampal lesions upon discrimination learning and
    reversal in the cat. Experimental Neurology, 1964, 10,
    452-462.

76 Teitelbaum, P. & Derks, P. The effect of amphetamine on
    forced drinking in the rat. Journal of Comparative and
    Physiological Psychology, 1958, 51, 801-810.

77 Weiss, B. & Laties, V. C. Effects of amphetamine,
    chlorpromazine, pentobarbital, and ethanol on operant
    response duration. Journal of Pharmacology and Experi-
    mental Therapeutics, 1964, 144, 17-23.

78 Wickelgren, W. O. & Isaacson, R. L.   Effect of the intro-
    duction of an irrelevant stimulus on runway performance of
    the hippocampectomized rat.   Nature, 1963, 200, 48-50.

79 Yokel, R. A. & Wise, R. A.   Amphetamine-type dopaminergic
    reinforcement.   Psychopharmacology, 1978, 58, 289-296.

80 Zielinski, K. & Czarkowska, J.   Go-No Go avoidance reflex
    differentiation and its retention after prefrontal lesions
    in cats.   Acta Neurobiologic Experimentalis, 1973, 33,
    467-490.

ANIMAL COGNITION AND BEHAVIOR
Roger L. Mellgren, editor
© *North-Holland Publishing Company, 1983*

SCHEDULE-INDUCED BEHAVIOUR

T.J. Roper

University of Sussex

Two or three decades ago psychologists interested in animal
learning or behaviour were mostly busy testing rats and pigeons
in Skinner boxes under various schedules of food reinforcement,
in an attempt to elucidate the processes of instrumental conditioning. But rats can drink as well as eat, so it is not surprising that sooner or later someone put a water bottle into a
Skinner box, as well as a food tray. Thus was discovered, quite
serendipitously, the phenomenon of schedule-induced polydipsia
or "SIP" -- when rats are allowed to obtain pellets of food
under certain intermittent schedules, and are given concurrent
free access to water, they develop the habit of taking a small
drink immediately after eating each pellet of food (Falk, 1961;
see Figure 1). The occurrence of drinking is not in itself surprising, because while the rat is waiting for food it must do
something, and that "something" might as well be drinking as any
other behaviour. Besides, it is well known that rats drink
prandially when food is freely available (e.g. Fitzsimon and
LeMagnen, 1969; Oatley, 1971). What is surprising, however,
is that given the opportunity the rat will go on drinking regularly after each pellet, for hours on end, with very little
sign of satiation (e.g. Keehn and Riusech, 1979). Consequently
the rat's cumulative water intake can be enormous by all conventional criteria. For example, Falk (1966a) reports a rat
consuming almost 100 ml of water per session under an FI 3-min
schedule of food delivery -- rather more water than a food-
deprived rat would normally drink in a week.

At the time of its discovery this anomalous drinking was of
interest for two major reasons. First, Skinnerian orthodoxy
held that the primary determinant of behaviour was the principle of response-reinforcement. According to this idea behaviour
was directly strengthened (i.e., the response in question increased in frequency) when a reinforcer was made contingent
upon it. In the case of SIP, however, enormous increases in
drinking were occurring despite the absence of any overt contingency between drinking and delivery of food (or of any other

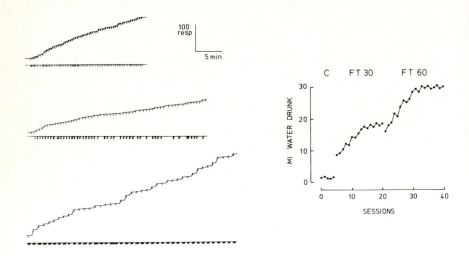

Figure 1.   Schedule-induced polydipsia in individual
    rats. Left-hand panel: Cumulative records of be-
    haviour during FI 20-s, FI 30-s and FI 60-s food
    reinforcement respectively, showing lever pressing
    and food delivery (upper trace) and drinking (lower
    trace). Note that bouts of drinking occur in the
    post-food period. Right-hand panel: Total amount
    of water drunk per session by a single rat when
    food was freely available (c), and when food was
    presented on FT 30-s and FT 60-s schedules. Note
    the gradual acquisition of polydipsia. Roper,
    unpublished results (1978).

tangible reinforcer). On the contrary, drinking was sometimes
so strong as to compete with operant responding, causing a
reduction in the frequency of food reinforcement. The existence
of SIP therefore challenged the supremacy of the principle of
response-reinforcement: it suggested that other factors were
exerting powerful control over behaviour in the Skinner box.

Second, in the early 1960's most people supposed that drinking
was essentially a homeostatically controlled activity, regu-
lated by negative feedback. According to this conception drinking was triggered in response to a real, physiological water
debt, and the amount drunk was directly related to the size of
the debt. Clearly, however, Falk's rats were not suffering
from a water debt equivalent to 100 ml -- they had not even been

deprived of water prior to the experiment. SIP was therefore
a glaring example of "non-homeostatic" or "secondary" drinking
-- i.e., drinking caused by factors other than a physiological
body water debt (e.g. Kissileff, 1973; Fitzsimons, 1971;
Toates, 1979).

As well as being of theoretical interest for these reasons,
SIP has proved to be an extraordinarily reliable phenomenon by
behavioural standards, technically very easy to demonstrate and
to measure in the laboratory. It is therefore easy to see why
SIP has continued to receive a great deal of experimental
attention. But despite the accumulation of much detailed information about this or that property of SIP, the phenomenon remains at bottom as much of a mystery as it was when it was first
discovered. Not only is there no coherent explanation of SIP:
there seems to be confusion about what precisely it is that
needs explaining (Roper, 1981).

To my mind there are two things about SIP that still require
an explanation. The first, and most striking, is the one that
caught Falk's attention, and I shall refer to it as the problem
of potentiation of drinking. It is simply the question of why
the rat drinks so much during certain schedules of food delivery; or, in other words, why intermittent delivery of food so
grossly enhances drinking relative to its normal level. The
second question, which I shall refer to as the problem of
selection of drinking, is why drinking occurs at all in the
intervals between eating. I have already said that the rat
might as well drink while awaiting the next food pellet; but by
the same token the rat might as well groom, run around the
chamber, rear, or perform any of the other activities available
to it. The fact that it does drink -- that drinking competes
successfully with all other available activities -- requires
explanation.

In what follows I shall attack these two problems in their re-
spective order. First, however, I shall briefly review what
seem to be the most important and/or least controversial properties of SIP. Detailed references for what follows are provided by Falk (1969; 1971; 1977); Segal (1972); Staddon (1977);
and Roper (1981).

Properties of SIP

Acquisition

SIP does not appear in fully developed form as soon as the rat
is introduced to an intermittent schedule of food delivery.
Rather, drinking gradually increases over several sessions

(see Figure 1). A possible explanation is that the rat takes several sessions to adapt to the food schedule, but Reynierse and Spanier (1968) found that prior experience of a VT schedule did not significantly alter the rate of acquisition of drinking. It seems reasonable to conclude from this that the drinking response itself takes time to develop.

Relatively little attention has been paid to the acquisition of SIP -- most of the research has concerned the effect of this or that variable of asymptotic level of drinking. This seems to me to be a pity, because if SIP is a learned habit then an understanding of the acquisition stage is crucial.

Type of Schedule

SIP occurs under fixed and variable ratio, interval and time schedules, and under differential-reinforcement-of-low-rate schedules. Direct comparisons between fixed versus variable schedules, and between response-dependent versus response-independent schedules, suggest that type of schedule is a relatively insignificant factor compared with overall food delivery rate (see below).

Food Delivery Rate

There have been many systematic studies of SIP under different food delivery rates, using various schedules (see review by Roper, 1980a). These unanimously show that the amount of water drunk per food pellet is maximal under intermediate rates of food delivery (about 0.5 to 2.0 pellets per min. depending on the study in question), and falls off at higher or lower rates (see Figure 2a). Rate of water ingested per unit time is also related to food rate by an inverted U-shaped curve, but with this measure of drinking the maximum intake occurs at somewhat higher food rates (Roper, 1980a; see Figure 2b).

Amount of Reinforcement

Amount of water drunk per food reinforcement varies approximately linearly with reinforcement size, in the range 1 to 20 45-mg food pellets per reinforcement (see review by Millenson, 1975).

Type of Food Reinforcement

There is some evidence that amount of water drunk per food pellet is inversely related to the sugar content of the pellets, but the effect seems to be relatively weak (e.g. Falk,

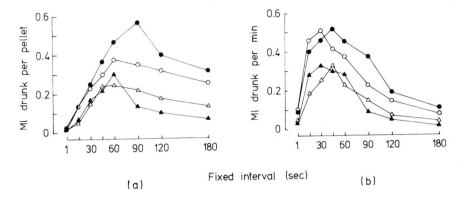

Figure 2. (a) Relationship between amount of water
drunk per food pellet and FI duration, in four
rats. (b) Relationship between rate of water intake per unit time and FI duration, in the same
four rats. Both graphs show a fall-off in water
intake when reinforcement rate is very high or
very low. From Roper (1980a).

1967; Christian and Schaeffer, 1975; Colotla and Keehn, 1975).
SIP has also been reported with certain types of liquid or
semi-liquid food reinforcement, but not with others (e.g.
Stricker and Adair, 1966; Falk, 1967).

One or two investigators have succeeded in obtaining SIP using
brain stimulation, rather than food, as reinforcer; but rather
more have failed (see review by Roper, 1981).

Level of Food Deprivation

Several studies have shown an inverse relationship between
amount of water drunk and level of food deprivation -- i.e.
the hungrier the rat, the more it drinks (see review by
Roper and Nieto, 1979).

Omission of Food

Several studies of SIP have involved the use of second-order
schedules, in which one or more components of the schedule
terminate in presentation of a discriminative stimulus rather

than in presentation of food. These studies show that drinking is less likely to follow a non-food discriminative stimulus
than it is to follow food, though there is disagreement concerning the precise extent to which drinking is reduced by
omission of food (for reviews see Corfield-Sumner, Blackman and
Stainer, 1977; Roper, Edwards and Crossland, reference note 1).

Level of Water Deprivation

Depriving the rat of water as well as of food increases the
asymptotic level of SIP, but not its rate of acquisition
(Brush and Schaeffer, 1974; Roper and Posadas-Andrews, 1981).

Conversely, SIP is reduced by preloading the rat with liquid
prior to testing, but these results are difficult to interpret
because preloading techniques inevitably introduce confounding
variables (e.g. Cope, Sanger and Blackman, 1976; Corfield-
Sumner and Bond, 1978).

Type of Fluid

Besides water, rats can be induced to drink relatively large
amounts of alcohol, saccharine, saline and quinine solutions
under intermittent food schedules. They will also avidly
"drink" compressed air (e.g. Mendelson and Chillag, 1970). The
use of palatable solutions enhances SIP and the use of unpalat-
able solutions reduces it (e.g. Keehn, Colotla and Beaton,
1970; Riley, Lotter and Kulkosky, 1979; Roper and Posadas-
Andrews, 1981); but in general rats seem to be less finicky
about the taste of the solution when drinking is induced by
food schedules than when it is induced by water deprivation or
by intracranial stimulation (see review by Roper, 1980b).

Rate of Water Availability

When water is presented in drinking tubes with different-sized
internal diameters, rats tend to drink a constant volume of
water per inter-food interval rather than to drink for a constant amount of time (Freed, Mendelson and Bramble, 1976;
Freed and Mendelson, 1977). This suggests that SIP is not
completely immune from regulatory influences, and that individual bouts of drinking are terminated by a short-term feedback
process (Roper et al., reference note 1).

Contrast-like Effects

If drinking is allowed to develop under both components of a
multiple schedule, and is then prevented during one component,

the amount drunk during the other component increases (Allen and Porter, 1975). Thus, SIP is subject to contrast-like effects.

Temporal Characteristics of Drinking

One of the most firmly established properties of SIP is that bouts of drinking tend to occur immediately after eating -- i.e., drinking occupies the post-reinforcement part of the interreinforcement interval (see Figure 3). Because of this Staddon (1977) terms drinking an "interim" activity, to distinguish it from "terminal" activities, such as lever-pressing, which increase in frequency towards the end of the interreinforcement interval (see also Staddon and Simmelhag, 1971; Staddon and Ayres, 1975).

This property of drinking is not however immutable, because drinking will occur in other parts of the interreinforcement interval if access to water is restricted to those parts (e.g. Flory and O'Boyle, 1972; Gilbert, 1974; Daniel and King, 1975). Drinking will also extend into the middle and terminal parts of the interval if air or saccharine solution are provided instead of water (Mendelson and Chillag, 1970; Keehn et al., 1970; Roper and Posadas-Andrews, 1981).

Morphology of Drinking

In all obvious respects schedule-induced drinking is morphologically similar to ad libitum or deprivation-induced drinking: the rat adopts a typical posture at the water spout and licks at a typical rate. There can be no doubt that SIP involves consumption and not spillage of water: one can observe water being swallowed, and the accompanying diuresis is usually all too evident. Rats do not usually become physiologically over-hydrated as a consequence of SIP, but this is only because of the efficiency of their kidneys as eliminators of excess fluid (e.g. Stricker and Adair, 1966).

Reinforcing Value of Drinking

The usual procedure in studies of SIP is to make water freely available in the operant chamber. However Falk (1966b) and Stricker and Adair (1966) have shown that rats will perform an operant response (lever pressing) to obtain water when food is intermittently scheduled. In one of Falk's rats, drinking supported a fixed-ratio requirement of up to 50 lever-press responses per 0.1-ml portion of water. These studies show that when food is intermittently scheduled, water acts not only as an eliciting stimulus for drinking but also as a reinforcer.

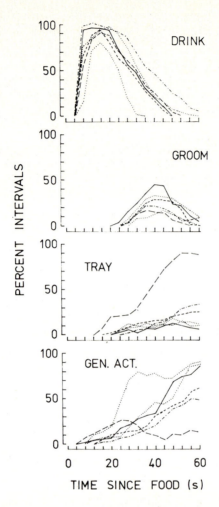

Figure 3. Temporal distributions of four different behaviours within the inter-food interval, in six individual rats, under an FT 60-s schedule of food delivery. Behaviours are drinking, grooming, scrabbling or nosing in the food tray, and general activity. Drinking is a post-reinforcement or "interim" behaviour, grooming is a "facultative" behaviour, and food-tray responses and general activity are "terminal" behaviours (see also Figure 5). From Roper and Nieto (1979).

Species Differences

Besides being a highly reliable phenomenon in laboratory rats
of various strains, SIP of comparable magnitude occurs in at
least some species of non-human primate (e.g. Allen and
Kenshalo, 1976, 1978; Barrett, Stanley and Weinberg, 1978),
and in some strains of mouse (Palfai, Kutscher and Symons,
1971; Symons and Sprott, 1977).

On the other hand, SIP seems to be at best a weak and variable
phenomenon in pigeons, hamsters, gerbils, Guinea pigs, cotton
rats and wild-caught rats (see review by Roper, 1981).

Conclusions

Two main conclusions emerge from this brief survey. First,
SIP is affected both by food-related factors (rate of food
delivery, reinforcement size, reinforcement quality, level of
food deprivation) and by water-related factors (type of fluid
available for drinking, level of water deprivation, rate of
water availability). Second, although SIP occurs under a
fairly wide range of experimental conditions there are definite
limitations -- e.g. SIP does not occur when food rate is very
low or very high; although its occurrence is not restricted to
the laboratory rat it is not universal in other species; and
there is little evidence that it occurs with reinforcers other
than food.

The Problem of Potentiation of Drinking

As already noted, the most striking feature of SIP is the sheer
volume of water that rats will consume, despite the absence of
any apparent physiological or psychological incentive to drink.
This is the problem of potentiation of drinking. There have
been several attempts to explain this aspect of SIP, but first
it is worth examining the question of whether drinking really
is "potentiated" in the first place.

Does SIP Constitute Excessive Drinking?

There is no doubt that within a certain range of food delivery
rates the rat's total intake of water is greatly enhanced
(see Figure 2a). Clearly, however, total intake is a function
of number of opportunities to drink, which is itself a function
of session duration. Given the natural tendency of rats to
drink prandially, the excess drinking could perhaps be caused
by prolongation of the rat's feeding behaviour, rather than by
the intermittency of the food schedule per se. In other words,
we might be able to account for SIP by saying that the rat

tends to drink periodically for as long as it continues to eat, regardless of the actual rate of food intake. This may seem far-fetched, but consider the case, already cited, of a rat that drank about 100 ml of water during a single session of FI 3-min food reinforcement (Falk, 1966a). There is no denying that this is a prodigious quantity of water: on the other hand, the demands of the schedule forced the rat to spread its eating out over a very long period of time (about 9 h). It is not inconceivable that if a rat were made to continue eating for 9 h by some other means (e.g. by applying brain stimulation) its prandial intake of water would amount to 100 ml.

Clearly, what we need to know to assess the validity of this idea is the rate of water intake rather than the total amount of water consumed per session. Only if rate of intake increases as a function of food schedule can drinking be described as "schedule induced" in a meaningful sense (Roper, 1980a). In fact there is evidence that rate of water intake is enhanced by intermittent presentation of food, within the range 0.5 to 4.0 pellets per min (see Figure 2b). This increase in rate of drinking makes it impossible to claim that SIP is nothing more than normal prandial drinking continued for an abnormal length of time.

The Hypothesis of Adventitious Reinforcement

Skinner (1948) observed that when pigeons were given food according to a fixed-time 15-s schedule (i.e., food was presented every 15 s regardless of the birds' behaviour) they developed stereotyped responses such as circling around or pecking at the chamber wall. Skinner termed this "superstitious" behaviour, and attributed its acquisition to chance correlations between the response in question and the delivery of food. Given the paramount importance attached to the principle of response-reinforcement at the time (see above, p. 127), and given the similarity between Skinner's procedure and that used to generate SIP, it is not surprising that some investigators were quick to interpret SIP as an example of superstitious behaviour (e.g. Clark, 1962).

This interpretation of SIP has since been discredited for a number of reasons. First, SIP typically consists of post-food bouts of drinking, whereas reinforced responses increase in frequency towards the end of the inter-food interval (cf. Figure 3). Second, several studies have shown that acquisition of drinking is not prevented by imposing a changeover delay between drinking and food delivery, such that accidental pairing of drinking with reinforcement is prevented (e.g. Falk, 1964; Segal and Oden, 1969a; Flory and Lickfett, 1974; Iversen,

1975). Third, if drinking is explicitly reinforced with food
it tends to differ morphologically from both schedule-induced
and normal drinking -- the lick rate is slow, and rats often
paw and gnaw at the spout instead of licking it (e.g. Segal
and Deadwyler, 1964). Finally, and perhaps most importantly,
a variety of evidence now indicates that the "superstitious"
responses originally observed by Skinner owe their existence
more to a process of classical conditioning than to the operation of covert instrumental conditioning (e.g. Staddon and
Simmelhag, 1971; Hearst and Jenkins, 1974; Staddon, 1977;
Schwartz and Gamzu, 1977). Thus the hypothesis of adventitious
reinforcement, besides being unsatisfactory as an account of
SIP, is unsatisfactory as an account of the other types of behaviour that it was originally formulated to explain.

If "superstitious" behavior is a product of classical conditioning, might not the same be true of SIP? This question has received relatively little attention. Once again, a major reason
for doubting the validity of a classical conditioning interpretation is the temporal location of drinking within the interreinforcement interval: drinking seems to be elicited by food
rather than to occur in anticipation of food. Staddon's (1977)
distinction between "terminal" (i.e., pre-reinforcement)
responses such as classically conditioned pecking in the pigeon,
and "interim" (i.e., post-reinforcement) responses such as
schedule-induced drinking in the rat, is intended to emphasize
this theoretical distinction between behaviour that can
plausibly to attributed to known conditioning processes and
behaviour that cannot.

To summarize the argument so far: SIP cannot be dismissed as a
special case of prandial drinking, nor can it be attributed to
instrumental or classical conditioning. We need to look elsewhere for an explanation of the phenomenon of response potentiation.

"Thirst" Explanations

One approach to SIP is to look for physiological rather than
psychological determinants: in other words, to suggest that
intermittent presentation of food somehow makes the rat thirsty.
As already noted, the sheer amount of water consumed rules out
an explanation in terms of a bodily water debt of equivalent
magnitude, but other types of "thirst" explanation might be
possible.

An early suggestion was that drinking occurred in response to
peripheral rather than central dehydration -- specifically, in
response to a dry mouth (e.g. Stein, 1964). This idea has been

dismissed on the grounds that SIP can develop when the reinforcer is liquid rather than solid food (e.g. Falk, 1971), but
in point of fact the relevant evidence is inconclusive: SIP
has been observed with liquid monkey diet as reinforcer, but
not with liquid "Metrecal", milk, sucrose solution, or oil
(Stein, 1964; Stricker and Adair, 1966; Falk, 1967). In any
case the "dry mouth" idea need not be interpreted absolutely
literally -- for example the oral receptor might be sensitive
to the salt content of the diet rather than to its water content, which might explain why less drinking occurs when dry
sugar pellets are used as reinforcer (e.g. Falk, 1967). Another complicating factor is that different diets might support different levels of SIP because of differences in their
reinforcing value: for example, milk as reinforcer does not
support SIP, but then it does not support lever pressing very
strongly either (Stein, 1964).

For other reasons which will appear below I do not believe that
any variant of the "dry mouth" idea can completely account
for schedule induction. On the other hand the idea that some
kind of oral "thirst" factor contributes to the appearance of
SIP has perhaps been dismissed too lightly, and may yet provide a key to understanding the puzzling and contradictory
literature on the efficacy of different types of food reinforcers as supporters of SIP.

Subsequent "thirst" hypotheses have been more ingenious, but
are to my mind less plausible. Carlisle (1971) suggests that
intermittent delivery of food causes hypothalamic overheating,
and that the rat drinks to cool down its central temperature
receptors. However there is little evidence of overheating at
a central level, and variables such as ambient temperature and
water temperature do not affect SIP in the manner predicted by
the hypothesis (Carlisle, 1971, 1973; Carlisle and Laudenslager,
1976; Carlisle, Shanab and Simpson, 1972). Freed, Zec and
Mendelson (1977), taking another tack altogether, suggest that
SIP is caused by an insulin-mediated fall in blood glucose
level -- but again a variety of physiological manipulations
have failed to provide confirmation (Berrios et al., 1979).

As I have pointed out elsewhere (Roper, 1981; Roper and
Posadas-Andrews, 1981), "thirst" hypotheses, by their very
nature, tend to focus more or less specifically on the interaction between eating and drinking when searching for the
cause of SIP. Consequently, this type of hypothesis predicts
that drinking will be more or less unique as a schedule-induced
activity with food reinforcement -- in other words, the phenomenon of schedule induction should not extend to activities
other than drinking, or to reinforcers other than food. Some

"thirst" hypotheses might even expect SIP to be restricted to species whose mechanisms of water regulation resemble those of the rat. It therefore becomes pertinent when judging the plausibility of the "thirst" approach to ask whether schedule induction is a specific or a general phenomenon. This is the subject of the next section.

How General is Schedule Induction?

Many investigators have asserted unequivocally that the phenomenon of schedule induction extends across a variety of activities, reinforcers and species (e.g. Falk, 1971, 1977; Segal, 1972; Wayner, 1974; Wallace and Singer, 1976; Porter, Hastings and Pagels, 1980). In a recent review, however, I suggested that the evidence is less conclusive than might appear (Roper, 1981). The essence of my argument is that the term "schedule induced" implies potentiation of the activity in question by the schedule: in other words, it implies that the activity in question occurs to a greater extent when the schedule is present than when it is absent. It follows that a proper demonstration of schedule induction requires some kind of non-schedule baseline condition for purposes of comparison -- yet very few of the relevant studies include appropriate baseline data. Consequently, although it is clear that many different kinds of behaviour can occur in conjunction with intermittent schedules of reinforcement, it is not usually known whether they are potentiated in the way that is characteristic of induced drinking.

In my review I concluded that aggression in pigeons, drug self-injection in rats, and drinking in certain species in addition to the rat, probably qualify as schedule-induced activities. Subsequently fairly good evidence has been obtained of schedule-induced wood-chewing in rats (Roper and Crossland, 1982a: see Figure 4), and of induced wheel-running in gerbils (Edwards and Roper, 1982). In none of these cases, however, is schedule induction anything like as pronounced or as reliable as with SIP in rats.

In addition to these positive instances of schedule induction there have been a significant number of reports of failure to induce certain kinds of behaviour, the best documented examples being wheel-running and grooming in rats with food reinforcement, eating in rats with water reinforcement, aggression in rats with food reinforcement, and drinking in certain species other than the rat (for references see Roper, 1981). In these latter cases the activity in question may occur during a significant proportion of interreinforcement intervals, but it does not do so with greater than baseline frequency. Staddon

(1977) has termed such activities "facultative" behaviour, to distinguish them from schedule-induced behaviour. The relevance of this distinction is that for facultative activities the "problem of potentiation" simply does not arise, because the activities are not potentiated by the schedule.

The only reasonable conclusion from this mixed bag of evidence seems to be that schedule induction is neither a completely general nor a completely specific phenomenon: with any one reinforcer in any one species there is a limited range of activities susceptible to schedule induction. The implication of this is that any very specific "thirst" account of SIP is unlikely to constitute a complete explanation -- for example, it is difficult to see why schedule-induced wood-chewing should occur in response to a dry mouth. On the other hand "thirst" could still act as a contributing factor, and this might explain why other types of schedule-induced behaviour are, by comparison with SIP, so weak and variable in their occurrence.

"Low-Probability-of-Reinforcement" Hypotheses

"Thirst" explanations imply that drinking is induced because some consequence of food ingestion acts as an elicitor of drinking. In the "dry mouth" hypothesis, for example, eating a pellet of solid food is supposed to make the rat's mouth dry, which in turn constitutes an eliciting stimulus for drinking. "Low-probability-of-reinforcement" hypotheses, by contrast, ascribe a different role to eating: they suggest that eating induces drinking because eating is a discriminative stimulus signalling a period of low reinforcer probability. According to this type of hypothesis the ingestion of food per se, and the various hedonic and physiological consequences thereof, are irrelevant.

Historically speaking, "low-probability-of-reinforcement" hypotheses arose in order to account for the striking temporal distribution of drinking within the inter-food interval: drinking typically occurs early in the interval, when the probability of reinforcement is indeed at a minimum (e.g. Falk, 1971 -- cf. Figure 3). Furthermore the same is true of schedule-induced aggression in pigeons while, conversely, wheel-running in rats, which does not seem susceptible to schedule induction, tends to occupy the middle region of the inter-food interval (see review by Staddon, 1977). On the basis of this evidence Staddon (op. cit.) proposes that during the period immediately following reinforcement the animal enters a special "motivational state" conducive to the appearance of schedule-induced behaviour. Later in the interval a different state, not conducive to schedule induction, takes over (see Figure 5).

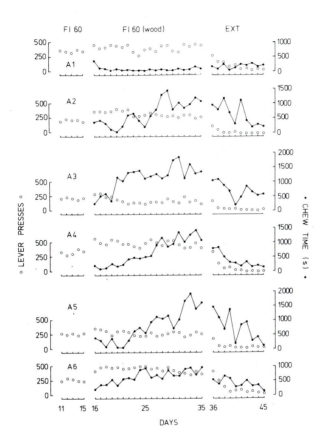

Figure 4. Acquisition of schedule-induced wood-chewing
in six individual rats under an FI 60-s schedule of
food reinforcement (days 16-35), followed by extinction
(days 36-45). During the first five sessions of FI 60
(days 11-15) no wood was available. Note the relatively
slow acquisition of induced chewing, the large day-
to-day variability, and the absence of induced chewing
in one rat. From Roper and Crossland (1982a).

I believe that Staddon is wrong, and that schedule-induced behaviour is not necessarily confined to the early part of the
interreinforcement interval. The arguments are presented in
detail elsewhere (Roper et al., reference note 1), but a
summary follows.

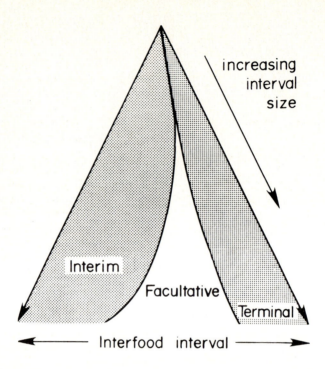

Figure 5. Schematic representation of the proportion
of the inter-food interval devoted to interim,
facultative and terminal behaviour, as interval
duration increases. Short inter-food intervals contain only interim (post-food) and terminal (pre-
food behaviours, while longer intervals also contain facultative behaviour. From Staddon (1977).

First, as already noted, induced drinking is able to occur anywhere in the interreinforcement interval provided access to
water is appropriately restricted. In addition drinking can
extend right through the early and middle parts of the interval
if air or a palatable (e.g. saccharine or saline) solution
is provided instead of water, or if rate of water availability
is reduced by providing a narrow drinking tube. This suggests
that drinking is normally confined to the early part of the
interval not because this corresponds to a period of especially
low reinforcer probability, but because individual bouts of
drinking are terminated by a short-term satiatory process (i.e.,
a process having a time-course measured in seconds rather than
in minutes or hours). (Other evidence for short-term satiation
of drinking in the rat is presented by Blass and Hall, 1976;
and by Roper and Crossland, 1982b).

Second, it now appears that wood-chewing in rats and wheel-running in gerbils are susceptible to schedule induction, despite the fact that both these activities tend to occupy the middle region of the inter-food interval (Roper and Crossland, 1982a; Edwards and Roper, 1982 -- see Figure 6). Thus there is no longer reason to suppose that the temporal location of an activity within the inter-food interval is a perfect predictor of its schedule inducibility. (I shall return later to the question of why different activities do occupy different post-food times).

Although these findings contradict Staddon's view they do not necessarily mean that any kind of "low-probability-of-reinforcer" approach is wrong: there may be other evidence favouring this type of hypothesis. In fact there have been various attempts to test directly the predictions of the "low-probability-of-reinforcer" idea, and it is to these that I shall now turn. The relevant studies fall into two categories: those involving random interval schedules and those involving second-order schedules.

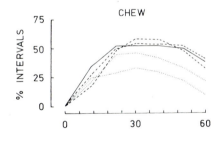 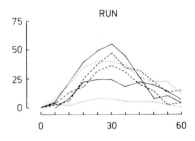

TIME SINCE FOOD (s)

Figure 6a. Temporal distribution of schedule-induced wood-chewing within the inter-food interval (results from five individual rats), under an FI 60-s schedule of food reinforcement. From Roper and Crossland (1982a),

Figure 6b. Temporal distribution of schedule-induced wheel-running within the inter-food interval (data from six individual gerbils), under an FT 60-s schedule of food delivery. From Edwards and Roper (1982).

In a random interval (RI) schedule the probability of a food
pellet being delivered is no lower immediately after food than
at any other post-food time. Accordingly in an RI schedule the
delivery of food does not constitute a discriminative stimulus
signalling low probability of reinforcement, and so post-food
SIP should not develop. Two studies have been carried out to
test this prediction but both have yielded equivocal results:
SIP seems to occur less reliably under RI than under VI
schedules, but it is not abolished altogether (Millenson,
Allen and Pinker, 1977; Keehn and Burton, 1978).

The logic of tests involving second-order schedules is that if
drinking follows eating because eating signals a period of low
reinforcer probability, then drinking should follow any other
non-food discriminative stimulus that has equal signal value.
A second-order schedule, in which certain schedule components
terminate in presentation of a non-food discriminative stimulus,
can be used to test this prediction. Once again, however, the
relevant experiments have yielded inconclusive results: drinking is substantially less likely to occur after a non-food
stimulus than after food, but it is usually not abolished altogether (Rosenblith, 1970; Allen, Porter and Arazie, 1975;
Porter et al., 1975; Corfield-Sumner, Blackman and Stainer,
1977). More recently we have conducted a similar type of
experiment using schedule-induced wood-chewing in rats (Roper
et al., reference note 1). In contrast to induced drinking,
induced wood-chewing continues almost as strongly after a non-
food discriminative stimulus as after food (see Figure 7).

In conclusion, the evidence for a "low-probability-of-reinforcement" approach is tantalisingly mixed. There may be
something in this type of explanation of SIP, but it is unlikely to provide a full account. But why should drinking be
induced by a period of low probability of reinforcement in the
first place? In the next section I consider various possible
answers to this question.

"Disinhibition", "Escape" and "Frustration" Hypotheses

Disinhibition. The simplest type of "low-probability-of-reinforcement" hypothesis states that intermittent schedules create
periods of spare time into which, because reinforcement is not
available, the animal can fit other activities. This is
tantamount to saying that the rat drinks because it has nothing
better to do, its highest priority activity (eating) being
temporarily impossible. I have called this the "disinhibition"
hypothesis because it has much in common with the ethologists'
explanation of displacement activities (e.g. Andrews, 1956;
Zeigler, 1964; McFarland, 1966). Indeed, it has sometimes been

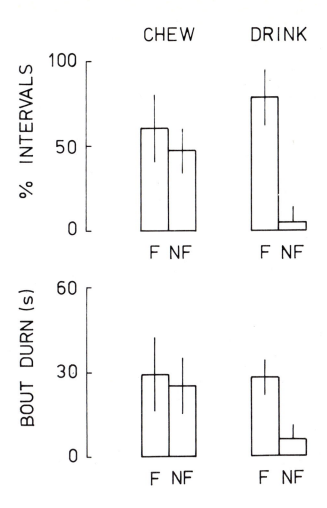

Figure 7. Effect of reinforcement omission on the frequency and bout duration of schedule-induced wood-chewing and schedule-induced drinking, under an FT 60-s schedule in which food delivery was occasionally omitted. F=intervals beginning with food, NF=intervals beginning with a non-food discriminative stimulus. Wood-chewing is somewhat reduced by omission of food, whereas drinking is virtually eliminated. Unpublished results: Roper, Edwards and Crossland (reference note 1).

suggested that SIP is a kind of displacement activity (Falk, 1971, 1977; McFarland, 1970; see also Roper and Posadas-Andrews, 1981).

One problem with this idea is that no answer is offered to the problem of selection of drinking: if all that is required of the behaviour is that it occupies spare time, why is it specifically drinking that occurs most readily? But more important from the point of view of the present discussion is that the dis-inhibition idea fails to explain how a schedule-induced activity comes to be enhanced beyond its baseline rate of occurrence (the problem of potentiation). For example, why does drinking not occur to an equal extent when food is absent altogether, since the spare time available would be just as great, or greater?

In order to explain the potentiation of schedule-induced be-haviour it is not enough to point to the absence of any more strongly determined competing behaviour -- we need to postulate some kind of active, potentiating factor. The other hypotheses considered in this section attempt to meet this requirement.

Escape. The essence of the "escape" idea is that a period of low reinforcer probability is aversive, and that schedule-induced behaviour occurs because it offers an escape from this aversive situation (e.g. Staddon, 1977, p. 139). The first part of this hypothesis has empirical support: there is a variety of evidence suggesting that stimuli associated with non-reinforcement have aversive properties (e.g. Ferster, 1958; Azrin, 1961; Thompson, 1964, 1965; Rilling et al., 1969; Brown and Flory, 1972). However there is very little direct evidence for the second and more crucial part of the hypothesis -- the idea that engaging in schedule-induced behaviour reduces the aversive aspects of an intermittent schedule of reinforcement. Attempts have been made to test this idea by using tranquil-izing drugs (e.g. Sanger and Blackman, 1977) or by making the test situation still more aversive by the addition of electric shock (e.g. Segal and Oden, 1969b; Hymowitz and Freed, 1974; King, 1975), but the results have not been clear-cut -- not surprisingly in view of the host of poorly understood variables which such procedures introduce. One possible prediction that might be more open to test is that if the drinking tube were located a considerable distance away from the manipulandum, so that the animal would no longer be in contact with the supposedly aversive food-related stimuli, SIP should be reduced. There is already evidence that proximity to the food site does enhance SIP (Clark, 1972; Staddon and Ayres, 1975), but the effect has not been investigated in detail.

Apart from this lack of direct empirical evidence the "escape" idea suffers from a number of other shortcomings. First, it is not intuitively obvious why engaging in an activity such as drinking should reduce the aversiveness of non-reinforcement. McFarland (1965) claims that it makes functional sense for an animal to "switch attention" from a fruitless activity (trying to get food when none is available) to a fruitful one, but by what sane criterion can the consumption of an enormous excess of water be regarded as fruitful? Second, like the "dis-inhibition" idea the "escape" hypothesis fails to answer the problem of selection of drinking: i.e., it fails to explain why drinking offers a more effective escape than alternative types of behaviour.

Frustration. Schedule-induced behaviour has often been ascribed to some kind of hypothetical internal activating state such as "frustration" (e.g. Thomka and Rosellini, 1975), "arousal" (e.g. Killeen, 1975), "stress" (e.g. Wallace and Singer, 1976) or "general motor excitability" (Wayner, 1970, 1974). There are two ways in which such a hypothetical state might potenti-ate schedule-induced behaviour: it might elicit the behaviour directly; or the state itself might be aversive, and its intensity be reduced by engaging in schedule-induced behaviour. "General motor excitability", as postulated by Wayner, seems to be a direct activating process, acting more or less indis-criminately on the animal's behavioural output like the gain control on an amplifier. On the other hand "arousal", "frustration" and "stress" are usually (not always) seen as aversive states from which schedule-induced behaviour offers an escape (cf. "escape" theories as discussed in the previous section).

An elementary criticism of this type of approach is that unless the hypothetical internal state can be verified independently of the phenomena it is invoked to explain, then the explanation is obviously circular. A hungry rat when it first experiences intermittent delivery of food behaves in an extremely active manner: it rushes about the test chamber, rears, chews the floor grid, chases its tail, scrabbles and bites at the food tray and so on. When we are observing such behaviour terms such as "arousal" leap unavoidably to mind because the be-haviour of the rat reminds us of subjective states that we our-selves have experienced. However, as Staddon notes (1977, p. 138): "Unfortunately a feeling of familiarity is not the same thing as exact knowledge, and may even hinder the search for it." Thus the question to ask of terms such as "arousal" or "frustration" is: "do they have any scientific respectability, or do they just explain schedule-induced behaviour away by re-minding us of subjectively familiar moods?" There are two ways

in which such terminology might be independently validated: by
testing predictions about behaviour, and by establishing
physiological correlates.

As regards behaviour the main feature shared by this class of
explanations, as distinct from "thirst" explanations, is that
it predicts that schedule induction will be a more or less
general phenomenon (cf. Roper, 1981). This implication is
clear, for example, in the term "general motor excitability"
(Wayner, 1974). But as we have already seen, schedule induction is neither a completely specific nor a completely
general phenomenon; and explanations couched in terms of
"frustration" and the like are no better than any others in
account for this.

Because of the motivational connotations of terms such as
"frustration" and "arousal", this type of hypothesis is consistent with the fact that SIP is affected by various motivational variables. For example, intuition suggests that the
frustrating and arousing effects of a food schedule, and hence
the frequency of SIP, should be greater the hungrier the animal
or the larger the reinforcement size; and these expectations
are confirmed. This type of hypothesis also predicts that different food reinforcers should support SIP in proportion to
their incentive value, rather than in relation to their food
content per se -- a prediction that is readily open to test.
Similarly, a "frustration" or "arousal" hypothesis would
account for the decline in SIP at low reinforcement rates
(cf. Figure 2) by saying that when the inter-food interval is
too long the food loses its incentive value.

Finally, this type of account does make genuine predictions
about the properties of other types of schedule-induced behaviour. Specifically, it suggests that variables such as
deprivation level, amount of reinforcement and rate of reinforcement should affect other induced activities in the way
that they affect SIP. So far there is little evidence on this
point, but what there is is encouraging: both schedule-induced
aggression in pigeons and induced wood-chewing in rats are
related to reinforcement rate by an inverted U-shaped function
(see Figure 8), and induced chewing in rats is directly related to deprivation level (Figure 9).

As regards physiological knowledge of the supposed internal
states little progress has been made. The idea of "general
motor excitability" was explicitly linked to the hypothalamus as the supposed controlling centre for schedule-induced behaviour (Wayner, 1970, 1974), but there is no evidence
favouring a special role for this area of the brain (e.g.

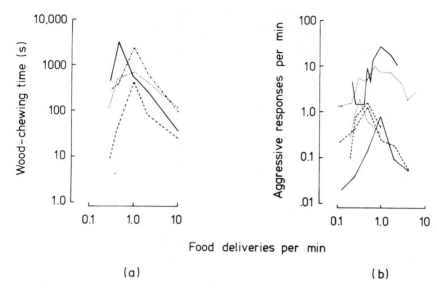

Figure 8a. Time spent wood-chewing per session in four
   individual rats, as a function of food delivery rate.
   Food was available according to FI schedules. From
   Roper, Edwards & Crossland (reference note 1).
Figure 8b. Number of aggressive responses (pecks at a
   target bird) per minute in six individual pigeons, as
   a function of food delivery rate. From Staddon (1977).

Singer, Armstrong and Wayner, 1975; Wayner, Loullis and Barone,
1977). The concepts of "arousal" and "stress" have often been
linked with the pituitary-adrenal system, and Brett and Levine
(1979) have recently suggested that this system is implicated
in the causation of SIP. Specifically, they report that while
intermittent delivery of food increases pituitary-adrenal
activity in the rat (see also Goldman, Coover and Levine, 1973),
indulgence in SIP tends to restore it to normal. They therefore interpret SIP as a "coping response" which prevents
normonally-mediated over-arousal (cf. Delius, 1967, for a similar suggestion regarding displacement activities). This attractive hypothesis deserves to be followed up, though later
evidence already suggests that it may be oversimplified (e.g.
Devenport, 1978, reports that rats which develop SIP have
larger adrenal glands than those which do not). Finally it
has been suggested that schedule-induced behaviour is related
to activation of the nigro-striatal dopamine system (Roper,
1980b), but this idea is nothing more than an interesting speculation.

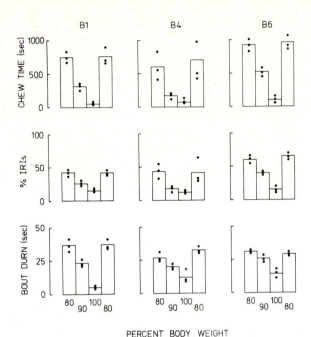

Figure 9. Time spent wood-chewing per session, percent-
    age of interreinforcement intervals (IRI's) contain-
    ing chewing, and bout duration of chewing, in three
    individual rats at different body weights. All three
    measures of chewing decline as body weight increases.
    From Roper and Crossland (1982a).

In conclusion, accounts of schedule induction in terms of "frustration", "arousal" or similar hypothetical states are quali-
tatively consistent with some of the known properties of schedule-induced behaviour, and do yield some rough behavioural
predictions. However until the behavioural and/or physiological
correlates of these hypothetical states are known in substantially more detail their explanatory value will remain more
apparent than real.

Conclusions and Future Directions

Clearly we are still a long way short of a satisfactory, coherent explanation of the problem of potentiation of schedule-
induced behaviour. How might we get closer to this objective?

First and foremost we need to go a lot further towards achieving
what Staddon (1977, p. 128) terms a "natural history" of

schedule-induced behaviour. What this means is (a) knowing
more about which activities are, or are not, susceptible to
schedule induction with different reinforcers and in different
species, and (b) knowing how induced activities other than
drinking are affected by variables such as deprivation level,
and size, rate and quality of reinforcement. Only when we have
this sort of information will we know which of the properties
of SIP are traceable to specific and relatively trival features
of the interaction between eating and drinking in the rat, and
which are relevant to the general problem of schedule induction.
(For example, a great deal of time and elaborate theorizing
would have been saved had it been known earlier that schedule-
induced behaviour need not necessarily occupy the post-reinforcement part of the interreinforcement interval -- cf. Figure
6.)

Second, I believe that the time has now come to abandon the
search for a unitary explanation of schedule induction (be it
in terms of "thirst", "frustration", "disinhibition", or whatever) and to adopt a more flexible, multi-dimensional approach.
My reasons for suggesting this are simply that no single explanation that has so far been put forward comes anywhere near
to explaining all the available data; and that whenever direct
tests of competing hypotheses have been attempted they have
yielded inconclusive results. For example, schedule induction
is neither a very general phenomenon as predicted by an
"arousal" hypothesis, nor is it absolutely restricted to drinking as predicted by a "thirst" hypothesis; when a non-food discriminative stimulus is substituted for food presentation, SIP
is neither abolished altogether as predicted by a "thirst"
hypothesis, nor does it continue at its normal level as predicted by a "low-probability-of-reinforcement" hypothesis; SIP
is affected by various food-related variables as predicted by
a "frustration" account, but it is also affected by water-related variables as predicted by a "thirst" account; SIP is
abolished with some liquid reinforcers as predicted by a "dry
mouth" hypothesis, but with other liquid reinforcers strong
induced drinking persists; and so on. Considering the robust-
ness of the basic phenomenon of SIP the literature concerning
its finer properties seems to give nothing but complex answers
to even the simplest of questions. This seems to me to indicate
that SIP is itself a complex phenomenon, in the sense that a
variety of factors contribute to its appearance. In particular,
it seems to me that any satisfactory account of the phenomenon
will have to postulate both a specific eliciting effect of food
consumption and some kind of general, response-facilitating
process.

The Problem of Selection of Drinking

When food is intermittently presented rats almost always develop SIP, even if opportunities to engage in other activities
are also present (though there is some evidence that the
ability of other activities to compete with drinking depends
on reinforcement rate -- see Roper, 1978). Why is it that
drinking is so invariably selected as the post-food behaviour,
and why is it that other activities such as grooming, running
and wood-chewing each have their own preferred temporal distributions within the inter-food interval (cf. Figures 3 and 6)?
This is the problem of selection. Note that unlike the problem
of potentiation, the problem of selection applies to non-induced
("facultative") as well as to induced activities -- it applies
to any behaviour that occurs within the inter-food interval.
Compared with the problem of potentiation the problem of
selection of interreinforcement behaviour has received relatively little experimental or theoretical attention, perhaps because of our general ignorance about the processes whereby
response sequences are organized.

All attempts to deal with the problem of selection have used as
their starting point the observation that normal, unconstrained
feeding in the rat tends to be accompanied by drinking (e.g.
Fitzsimons and LeMagnen, 1969; Oatley, 1971). More specifically, a rat that is given free access to food switches more
frequently from eating to drinking than from eating to any
other activity such as grooming or exploring (e.g. Roper and
Crossland, 1982b). This led me to suggest (Roper, 1980c) that
one could predict the sequence of activities that would occur
during interreinforcement intervals with a particular reinforcer by observing the sequence of activities occurring during self-initiated pauses in normal, unconstrained consummatory
behaviour directed towards the same reinforcer. In other words,
interreinforcement intervals merely preserve the animal's
natural tendency to sequence particular activities in particular ways.

In an experiment intended to test this idea with interreinforcement activities other than drinking, I observed the behaviour
of rats during self-imposed pauses in either drinking or feeding (Roper, 1980c). The results showed that during pauses in
drinking rats tended to groom first, then explore the chamber,
whereas during pauses in eating they tended to explore first,
then groom (see Figure 10a). When water and food were subsequently made intermittently available in the same apparatus,
intervals between successive water presentations tended to
contain grooming followed by exploration, whereas intervals
between food presentations tended to contain exploration

followed by grooming (Figure 10b). For some reason, then,
certain pairs of activities are especially strongly linked in
the animal's normal ongoing behaviour (e.g. eating with drinking or exploration; drinking with grooming), and these same
pairs of activities preserve their preferential linkage during
scheduled interreinforcement intervals.

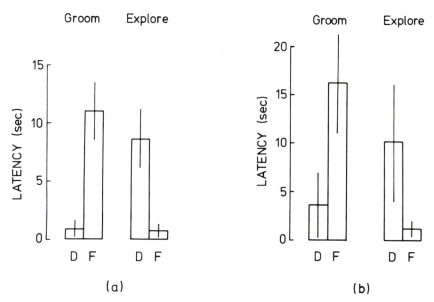

Figure 10a. Latency to groom and to explore during
    self-initiated pauses in drinking (D) or feeding
    (F), when water or food were freely available to
    thirsty or hungry rats. Pauses in drinking tended
    to start with grooming, whereas pauses in feeding
    tended to start with exploration.
Figure 10b. Latency to groom and to explore following
    drinking (D) or feeding (F), when food or water were
    presented on an FT 60-s schedule. Inter-drink
    intervals tended to start with grooming whereas
    inter-food intervals tended to start with exploration.
    From Roper (1980c).

This approach to explaining the characteristic sequencing of
activities that one sees in interreinforcement intervals is
attractive, and it accounts for observations concerning quite
a number of different behaviours. For example, freely feeding
rats never switch straight from eating to wheel-running or from
eating to wood-chewing, even when water is not available -- they
always move away from the food tray and wander around for a few
seconds before approaching the wheel or wood (unpublished
observations). Correspondingly, when food is intermittently
presented running and wood-chewing occupy the middle, rather

than the post-food, part of the interval (e.g. Roper, 1978).
However this hypothesis merely puts back by one stage the
question of why activities are sequenced in particular ways
to start with: for example, why does the rat normally interrupt
bouts of eating in order to drink? Ethologists interested in
the general question of behaviour sequencing have developed
various models which may go some way towards answering this
question (e.g. McFarland, 1969, 1974; Ludlow, 1976, 1980;
Slater, 1978), but thinking in this area is still at a very
rudimentary state of development.

The idea that interreinforcement behaviour merely reflects the
activities that normally occur in association with that particular consummatory activity also suffers from another difficulty, which is that sometimes interreinforcement activities
develop despite having a baseline frequency of virtually zero.
Wood-chewing is a case in point: it is rarely seen when food is
freely available, yet it develops when food is intermittently
scheduled (e.g. Roper, 1978). This suggests that intermittent
schedules may, after all, select particular types of behaviour
for special reasons. For example there has traditionally been
thought to be a link between frustration and aggressive behaviour, so one might expect that intermittent schedules of
reinforcement, insofar as they are frustrating, would tend to
select aggressive activities. Here the problem of selection
begins to merge with the problem of potentiation: in considering
why a schedule-induced activity is selected one may need to
consider the function of schedule-induced behaviour.

Reference Notes

1. Roper, T. J., Edwards, L and Crossland, G. Schedule-
    induced wood-chewing in rats and its relevance to
    theories of schedule induction. Submitted for publi-
    cation, 1982.

References

1 Allen, J. D. & Kenshalo, D. R. Schedule-induced drinking as a function of interreinforcement interval in the rhesus monkey. Journal of the Experimental Analysis of Behavior, 976, 26, 257-267.

2 Allen, J. D. & Kenshalo, D. R. Schedule-induced drinking as a function if interpellet interval and draught size in the Java macaque. Journal of the Experimental Analysis of Behavior, 1978, 30, 139-151.

3 Allen, J. D. & Porter, J. H. Demonstration of behavioral contrast with adjunctive drinking. Physiology and Behavior, 1975, 15, 511-515.

4 Allen, J..D., Porter, J. H. & Arazie, R. Schedule-induced drinking as a function of percentage reinforcement. Journal of the Experimental Analysis of Behavior, 1975, 23, 223-232.

5 Andrew, R. J. Some remarks on behaviour in conflict situations with special reference to Emberiza spp.. British Journal of Animal Behaviour, 1956, 4, 41-45.

6 Azrin, N. H. Time-out from positive reinforcement. Science, 1961, 133, 382-383.

7 Barrett, J. E., Stanley, J. A. & Weinberg, E. S. Schedule-induced water and ethanol consumption as a function of interreinforcement interval in the squirrel monkey. Physiology and Behavior, 1978, 21, 453-455.

8 Berrios, N., Carlson, N. R., Sawchenki, P. E., Gold, R. M. & Mui, A. H. Insulinogenci mediation of schedule induced polydipsia? Physiology and Behavior, 1979, 23, 237-240.

9 Blass, E. M. & Hall, W. G. Drinking termination: inter-actions among hydrational, orogastric, and behavioral controls in rats. Psychological Review, 1976, 83, 356-374.

10 Brett, L. D. & Levine, S. Schedule-induced polydipsia suppresses pituitary-adrenal activity in rats. Journal of Comparative and Physiological Psychology, 1979, 93, 946-956.

11 Brown, T. G. & Flory, R. K. Schedule-induced escape from fixed-interval reinforcement. Journal of the Experimental Analysis of Behavior, 1972, 17, 395-403.

12   Brush, M. E. & Schaeffer, R. W.   Effects of water depri-
     vation on schedule-induced polydipsia.   Bulletin of the
     Psychonomic Society, 1974, 4, 69-72.

13   Carlisle, H. J.   Fixed-ratio polydipsia: thermal effect of
     drinking, pausing and responding.   Journal of Comparative
     and Physiological Psychology, 1971, 75, 10-22.

14   Carlisle, H. J.   Schedule-induced polydipsia: blockade by
     intrahypothalamic atropine.   Physiology and Behavior,
     1973, 11, 139-143.

15   Carlisle, H. J. & Laudenslager, M. L.   Separation of water
     and ambient temperature effects on polydipsia.   Physiology
     and Behavior, 1976, 16, 121-124.

16   Carlisle, H. J., Shanab, M. E. & Simpson, C. W.   Schedule-
     induced behaviors: Effect of intermittent water reinforce-
     ment on food intake and body temperature.   Psychonomic
     Science, 1972, 26, 35-36.

17   Christian, W. P. & Schaeffer, R. W.   Motivational properties
     of fixed-interval reinforcement: a preliminary investi-
     gation.   Bulletin of the Psychonomic Society, 1975, 5, 143-
     145.

18   Clark, F. C.   Some observations on the adventitious rein-
     forcement of drinking under food reinforcement.   Journal
     of the Experimental Analysis of Behavior, 1962, 5, 61-63.

19   Colotla, V. A. & Keehn, J. D.   Effects of reinforcer-
     pellet composition on schedule-induced polydipsia with
     alcohol, water and saccharin.   Psychological Record, 1975,
     25, 91-98.

20   Cope, C. L., Sanger, D. J. & Blackman, D. E.   Intragastric
     water and the acquisition of schedule-induced drinking.
     Behavioral Biology, 1976, 17, 267-270.

21   Corfield-Sumner, P. K., Blackman, D. E. & Stainer, G.
     Polydipsia induced in rats by second-order schedules of
     reinforcement.   Journal of the Experimental Analysis
     of Behavior, 1977, 27, 265- 273.

22   Corfield-Sumner, P. K. & Bond, N. W.   Effects of preloading
     on the acquisition and maintenance of schedule-induced
     polydipsia.   Behavioral Biology, 1978, 23, 238-242.

23 Daniel, W. & King, G. D. The consequences of restricted
    water accessibility on schedule-induced polydipsia.
    Bulletin of the Psychonomic Society, 1975, 5, 297-299.

24 Delius, J. D. Displacement activities and arousal. Nature,
    1967, 214, 1259-1260.

25 Devenport, L. D. Schedule-induced polydipsia in rats:
    adrenocortical and hippocampal modulation. Journal of
    Comparative and Physiological Psychology, 1978, 92, 651-
    660.

26 Edwards, L. & Roper, T. J. Schedule-induced running in
    gerbils. Behaviour Analysis Letters, 1982, in press.

27 Falk, J. L. Production of polydipsia in normal rats by an
    intermittent food schedule. Science, 1961, 133, 195-196.

28 Falk, J. L. Studies on schedule-induced polydipsia. In:
    M. J. Wayner (ed.), Thirst: First international symposium
    on thirst in the regulation of body water. New York:
    Pergamon Press, 1964, 95-116.

29 Falk, J. L. Schedule-induced polydipsia as a function of
    fixed interval length. Journal of the Experimental
    Analysis of Behavior, 1966, 9, 37-39 (a).

30 Falk, J. L. The motivational properties of schedule-induced
    polydipsia. Journal of the Experimental Analysis of Be-
    havior, 1966, 9, 19-25 (b).

31 Falk, J. L. Control of schedule-induced polydipsia: type,
    size and spacing of meals. Journal of the Experimental
    Analysis of Behavior, 1967, 10, 199-206.

32 Falk, J. L. Conditions producing psychogenic polydipsia in
    animals. Annals of the New York Academy of Science,
    1969, 157, 569-593.

33 Falk, J. L. The nature and determinants of adjunctive be-
    havior. Physiology and Behavior, 1971, 6, 577-588.

34 Falk, J. L. The origin and functions of adjunctive behavior.
    Animal Learning and Behavior, 1977, 5, 325-335.

35 Ferster, C. B. Control of behavior in champanzees and
    pigeons by time out from positive reinforcement. Psycho-
    logical Monographs, 1958, 72 (Whole No. 461), 1-38.

36 Fitzsimons, J. T. The physiology of thirst: a review of the extra-neural aspects of the mechanisms of drinking. In: E. Stellar & J. M. Sprague (Eds), Progress in Physiological Psychology, Vol. 4, pp. 119-201. New York: Academic Press, 1971.

37 Fitzsimons, T. J. Y LeMagnen, J. Eating as a regulatory control of drinking in the rat. Journal of Comparative and Physiological Psychology, 1969, 67, 273-283.

38 Flory, R. K. & Lickfett, G. G. Effects of lick-contingent time-out on schedule-induced polydipsia. Journal of the Experimental Analysis of Behavior, 1974, 21, 45-55.

39 Flory, R. K. & O'Boyle, M. K. The effect of limited water availability on schedule-induced polydipsia. Physiology and Behavior, 1972, 8, 147-149.

40 Freed, W. J. & Mendelson, J. Water-intake volume regulation in the rat: schedule-induced drinking compared with water-deprivation-induced drinking. Journal of Comparative and Physiological Psychology, 1977, 91, 564-573.

41 Freed, W. J., Mendelson, J. & Bramble, J. M. Intake-volume regulation during schedule-induced polydipsia in rats. Behavioral Biology, 1976, 16, 245-250.

42 Freed, W. J., Zec, R. F. & Mendelson, J. Schedule-induced polydipsia: the role of orolingual factors and a new hypothesis. In: J. A. W. M. Weijnen & J. Mendelson (Eds.), Drinking Behavior: Oral Stimulation, Reinforcement and Preference. New York: Plenum, 1977.

43 Gilbert, R. M. Ubiquity of schedule-induced polydipsia. Journal of the Experimental Analysis of Behavior, 1974, 21, 277-284.

44 Goldman, L., Coover, G. D. & Levine, S. Bidirectional effects of reinforcement shifts on pituitary adrenal activity. Physiology and Behavior, 1973, 10, 209-214.

45 Hearst, E. & Jenkins, H. M. Sign-tracking: the stimulus-reinforcer relation and directed action. Psychonomic Society Monograph, 1974.

46 Hymowitz, N. & Freed, E. X. Effects of response-dependent and independent electric shock on schedule-induced polydipsia. Journal of the Experimental Analysis of Behavior, 1974, 22, 207-213.

47   Iversen, I. H.   Interactions between lever pressing and collateral drinking during VI with limited hold.   Psychological Record, 1975, 25, 47-50.

48   Keehn, J. D. & Burton, M.   Schedule-induced drinking: entrainment by fixed- and random-interval schedule-controlled feeding.   T.-I.-T. Journal of Life Sciences, 1978, 8, 93-97.

49   Keehn, J. D., Colotla, V. A. & Beaton, J. M.   Palatability as a factor in the duration and pattern of schedule-induced drinking.   Psychological Record, 1970, 20, 433-442.

50   Keehn, J. D. & Riusech, R.   Schedule-induced drinking facilitates schedule-controlled feeding.   Animal Learning and Behavior, 1979, 7, 41-44.

51   Killeen, P.   On the temporal control of behavior.   Psychological Review, 1975, 82, 89-115.

52   King, G. D.   The enhancement of schedule-induced polydipsia by FR-20 and FR-80 lick-contingent shock.   Bulletin of the Psychonomic Society, 1975, 6, 542-544.

53   Kissileff, H. R.   Nonhomeostatic controls of drinking.   In: A. N. Epstein, H. R. Kissileff & E. Stellar (Eds.),   The Neuropsychology of Thirst, pp. 163-198.   New York: Wiley, 1973.

54   Ludlow, A. R.   The behaviour of a model animal.   Behaviour, 1976, 58, 131-172.

55   Ludlow, A. R.   The evolution and simulation of a decision maker.   In: F. M. Toates & T. R. Halliday (Eds.), Analysis of Motivational Processes, pp. 273-296.   London: Academic Press, 1980.

56   McFarland, D. J.   The role of attention in the disinhibition of displacement activities.   Quarterly Journal of Experimental Psychology, 1965, 18, 19-30.

57   McFarland, D. J.   On the causal and functional significance of displacement activities.   Zeitschrift fur Tierpsychologie, 1966, 23, 217-235.

58   McFarland, D. J.   Mechanisms of behavioural disinhibition.   Animal Behaviour, 1969, 17, 238-242.

59 McFarland, D. J. Adjunctive behaviour in feeding and drinking situations. Revue de Comportement Animal, 1970, 4, 64-73.

60 McFarland, D. J. Time-sharing as a behavioural phenomenon. In: D. S. Lehrman, J. S. Rosenblatt, R. A. Hinde & E. Shaw (Eds.), Advances in the Study of Behavior, Vol. 5, pp. 201-225. New York: Academic Press, 1974.

61 Mendelson, J. & Chillag, D. Schedule-induced air licking in rats. Physiology and Behavior, 1970, 5, 535-537.

62 Millenson, J. R. The facts of schedule-induced polydipsia. Behavior Research Methods and Instrumentation, 1975, 7, 257-259.

63 Milleson, J. R., Allen, R. B. & Pinker, S. Adjunctive drinking during variable and random-interval schedules. Animal Learning and Behavior, 1977, 5, 285-290.

64 Oatley, K. Dissociation of the circadian drinking pattern from eating. Nature, 1971, 229, 494-496.

65 Palfai, T., Kutscher, C. L. & Symons, J. P. Schedule-induced polydipsia in the mouse. Physiology and Behavior, 1971, 6, 461-462.

66 Porter, J. H., Arazie, R., Holbrook, J. W., Cheeck, M. S. & Allen, J. D. Effects of variable and fixed second-order schedules on schedule-induced polydipsia in the rat. Physiology and Behavior, 1975, 14, 143-149.

67 Porter, J. H., Hastings, M. T. & Pagels, J. F. Schedule-induced polydipsia in the cotton rat (Sigmondon Hispidus). Bulletin of the Psychonomic Society, 1980, 16, 15-18.

68 Reynierse, J. H. & Spanier, D. Excessive drinking in rats' adaptation to the schedule of feeding. Psychonomic Science, 1968, 10, 95-96.

69 Riley, A. L., Lotter, E. C. & Kulkosky, P. J. The effects of conditioned taste aversions on the acquisition and maintenance of schedule-induced polydipsia. Animal Learning and Behavior, 1979, 7, 3-12.

70 Rilling, M., Askew, H. R., Ahlskog, J. E. & Kramer, T. J. Aversive properties of the negative stimulus in a successive discrimination. Journal of the Experimental Analysis of Behavior, 1969, 12, 917-932.

71   Roper, T. J.   Diversity and substitutability of adjunctive
     activities under fixed-interval schedules of food rein-
     forcement.   Journal of the Experimental Analysis of
     Behavior, 1978, 30, 83-96.

72   Roper, T. J.   Changes in rate of schedule-induced be-
     haviour in rats as a function of fixed-interval schedule.
     Quarterly Journal of Experimental Psychology, 1980, 32,
     159-170 (a).

73   Roper, T. J.   "Induced" behaviour as evidence of nonspecific
     motivational effects.   In: F. M. Toates & T. R. Halliday
     (Eds.), Analysis of Motivational Processes, pp. 221-242.
     London: Academic Press, 1980 (b).

74   Roper, T. J.   Behaviour of rats during self initiated
     pauses in feeding and drinking, and during periodic re-
     sponse-independent delivery of food and water.   Quarterly
     Journal of Experimental Psychology, 1980, 32, 459-472 (c).

75   Roper, T. J.   What is meant by the term "schedule-induced",
     and how general is schedule induction?   Animal Learning and
     Behavior, 1981, 9, 433-440.

76   Roper, T. J. & Crossland, G.   Schedule-induced wood-chewing
     in rats and its dependence on body weight.   Animal Learning
     and Behavior, 1982, in press (a).

77   Roper, T. J. & Crossland, G.   Mechanisms underlying eating-
     drinking transitions in rats.   Animal Behaviour, 1982, 30,
     602-614 (b).

78   Roper, T. J. & Nieto, J.   Schedule-induced drinking and
     other behavior in the rat, as a function of body-weight
     deficit.   Physiology and Behavior, 1979, 673-678.

79   Roper, T. J. & Posadas-Andrews, A.   Are schedule-induced
     drinking and displacement activities causally related?
     Quarterly Journal of Experimental Psychology, 1981, 33B,
     181-193.

80   Rosenblith, J. Z.   Polydipsia induced in the rat by a
     second-order schedule.   Journal of the Experimental
     Analysis of Behavior, 1970, 14, 139-144.

81   Sanger, D. J. & Blackman, D. E.   The effects of drugs on
     adjunctive behavior.   In: D. E. Blackman & D. J. Sanger
     (Eds.), Contemporary Research in Behavioral Pharmacology.
     New York: Plenum Press, 1977.

82   Schwartz, B. & Gamzu, E.   Pavlovian control of operant be-
     havior. In: W. K. Honig & J. E. R. Staddon (Eds.),
     Handbook of Operant Behavior. Englewood-Cliffs, N. J.:
     Prentice-Hall, 1977.

83   Segal, E. F.   Induction and the provenance of operants.
     In R. M. Gilbert & J. R. Milleson (Eds.), Reinforcement:
     behavioral analyses. New York: Academic Press, 1972.

84   Segal, E. F. & Deadwyler, S. A.   Water drinking patterns
     under several dry food reinforcement schedules. Psychonomic
     Science, 1964, 1, 271-272.

85   Segal, E. F. & Oden, D. L.   Schedule-induced polydipsia:
     effects of providing an alternate reinforced response and
     of introducing a lick-contingent delay in food delivery.
     Psychonomic Science, 1969, 15, 153-154 (a).

86   Segal, E. F. & Oden, D. L.   Effects of drinkometer current
     and of foot shock on psychogenic polydipsia. Psychonomic
     Science, 1969, 14, 13-15 (b).

87   Singer, G., Armstrong, S. & Wayner, M. J.   Effects of
     norepinephrine applied to the lateral hypothalamus on
     schedule induced polydipsia. Pharmacology, Biochemistry
     and Behavior, 1975, 3, 869-872.

88   Skinner, B. F.   "Superstition" in the pigeon.   Journal of
     Experimental Psychology, 1948, 38, 168-172.

89   Slater, P. J. B.   A simple model for competition between
     behaviour patterns. Behaviour, 1978, 67, 236-258.

90   Staddon, J. E. R.   Schedule-induced behavior.   In
     W. K. Honig & J. E. R. Staddon (Eds.), Handbook of operant
     behavior. New York: Prentice-Hall, 1977.

91   Staddon, J. E. R. & Ayres, S. L.   Sequential and temporal
     properties of behavior induced by a schedule of periodic
     food delivery. Behaviour, 1975, 54, 26-49.

92   Staddon, J. E. R. & Simmelhag, V. L.   The "superstition"
     experiment: a re-examination of its implications for the
     study of adaptive behaviour. Psychological Review, 1971,
     78, 3-43.

93   Stein, L.   Excessive drinking in the rat: superstition or
     thirst? Journal of Comparative and Physiological Psycho-
     logy, 1964, 58, 237-242.

94 Stricker, E. M. & Adair, E. R.   Body fluid balance, taste and post-prandial factors in schedule-induced polydipsia. Journal of Comparative and Physiological Psychology, 1966, 62, 449-454.

95 Symons, J. P. & Sprott, R. L.   Genetic analysis of schedule induced polydipsia.   Physiology and Behavior, 1977, 17, 837-839.

96 Thomka, M. L. & Rosellini, R. A.   Frustration and the production of schedule-induced polydipsia.   Animal Learning and Behavior, 1975, 3, 380-384.

97 Thompson, D. M.   Escape from S^D associated with fixed-ratio reinforcement.   Journal of the Experimental Analysis of Behavior, 1964, 7, 1-8.

98 Thompson, D. M.   Punishment by S^D associated with fixed-ratio reinforcement.   Journal of the Experimental Analysis of Behavior, 1965, 8, 189-194.

99 Toates, F. M.   Homeostasis and drinking.   Behavioral and Brain Sciences, 1979, 2, 95-139.

100 Wallace, M. & Singer, G.   Schedule-induced behavior: a review of its generality, determinants and pharmacological data.   Pharmacology, Biochemistry and Behavior, 1976, 5, 483-490.

101 Wayner, M. J.   Motor control functions of the lateral hypothalamus and adjunctive behavior.   Physiology and Behavior, 1970, 5, 1319-1325.

102 Wayner, M. J.   Specificity of behavioral regulation. Physiology and Behavior, 1974, 12, 851-869.

103 Wayner, M. J., Loullis, C. C. & Barone, F. C.   Effects of lateral hypothalamic lesions on schedule dependent and schedule induced behavior.   Physiology and Behavior, 1977, 18, 503-511.

104 Zeigler, H. P.   Displacement activity and motivational theory: a case study in the history of ethology. Psychological Bulletin, 1964, 61, 363-376.

ANIMAL COGNITION AND BEHAVIOR
Roger L. Mellgren, editor
© *North-Holland Publishing Company, 1983*

APPETITIVE STRUCTURE
AND STRAIGHT ALLEY RUNNING

William Timberlake

Indiana University

Traditional research in learning has treated animals as bundles of reflexes and random behaviors that are selectively strengthened in the presence of particular stimuli by the appropriate presentation of events called reinforcers. Lloyd Morgan (1896, p. 23) set the stage for this view in his claim, "Just as a sculptor carves a statue out of a block of marble, so does acquisition carve an activity out of a mass of random movements." Behaviorist researchers continue this emphasis in remarks such as: "Learned behavior is constructed by a continual process of differential reinforcement from undifferentiated behavior, just as the sculptor shapes his figure from a lump of clay" (Skinner, 1953, p. 92); and "An ideally adequate theory even of so-called purposive behavior ought, therefore, to begin with colorless movement and mere receptor impulses as such, and from these build up step by step both adaptive and maladaptive behavior" (Hull, 1943, p. 25). The result of this view of learning has been to focus attention on the carefully controlled manipulations of the experimenter and to de-emphasize the contribution of the subject.

In actuality, good researchers are extremely aware of the subject's contribution in terms of the sensory-motor organization and limitations it brings to the experiment. But this awareness is expressed in the design of the apparatus and procedures, rather than acknowledged in theoretical concepts. Ethologists have often disparaged laboratory research because of its artificiality, but good experimenters clearly create spaces, manipulanda, and procedures in which the organism is at home. In an odd twist, the fiercest allegiance to the concept of a randomly organized, malleable organism is often coupled with a remarkable sense of an animal's capabilities and how adjustments and contingencies can be arranged to produce a particular behavioral outcome. In short, though most experimenters have been aware of the contribution of the animal to the learning

situation, their theories read as though reinforcement were a
completely general process set in motion and its outcome
determined solely by their operations.

The resultant conception of an animal as a disorganized re-
cipient of the response-strengthening value of reinforcers
stands in stark contrast with the complex and sophisticated
organization apparent in the behavior of even the simplest
animal in less constrained and more "natural" learning environments (e.g., Jennings, 1906). An early attempt to reconcile
this gulf and make the animal a more active participant in the
learning process was made by cognitive psychologists such as
Tolman (1932). However, for many, Tolman's solution came
uncomfortably close to creating little humans inhabiting animals at the choice points of experimental apparatus, complete
with demands, cognitive maps, and decision making mechanisms.
Reinforcement theorists had much the advantage in the ensuing
debate. Not only were they able to shake their heads over the
anthropomophic tinges which muddied the cognitive view, their
own approach was more firmly grounded in observable behavior
and had the advantage of an apparently more clearly defined
learning mechanism, namely reinforcement.

Unfortunately the relatively easy critique of much of early
cognitive theory appeared to confirm the importance of a
general reinforcement mechanism at the expense of concern with
the contribution of the subject. The study of learning became
isolated from the actual phenomena of learning in the animal's
selection environment (Johnston, 1981; Timberlake, in press).
The possibility that reinforcement could be invoked to explain
all behavior change led to a narrowing of experimentation to a
few paradigms which produced consistent results under manipulation of a few variables. Reinforcement became an inadequately defined but powerful conceptual force that could ex-
plan any modification in behavior after the fact, but actually
predicted very little in advance in novel situations
(Timberlake, 1980). The unbridled enthusiasm for reinforcement
also led to downplaying apparent limitations on its effects
(Bolles, 1970; Shettleworth, 1972).

In brief, researchers in animal learning have attributed
learning at the conceptual level to the presumed effects of
the reinforcers they manipulate, and have incorporated the
contribution of the animal into the design of their presumably
arbitrary procedures and apparatus. The result has been a
schizophrenic concern with arbitrary general process learning
theories on one hand, and experimental lore on the other. One
way to heal this split would be to study learning only in "real
world" settings, thereby forcing explicit attention to the
animal's contribution. Though I think there is much to be

learned from the examination of learning in natural settings,
I am convinced that the laboratory remains an ideal place to
analyze the animal's contribution to learning. We have only
to use experimental manipulations to investigate the nature of
the organism, rather than taking advantage of our intuitive
assessment of the organism to investigate the presumably general effects of experimental manipulations.

The Ecological Approach:
Behavior Systems and Appetitive Structure

Recently agreement has begun to develop around an ecologically
based alternative to a general reinforcement theory (e.g.,
Johnston, 1981; Rozin and Kalat, 1971; Timberlake, in press).
This ecological view is based on the assumption that learning
is an evolved capacity that occurs in the service of functions
promoting the survival and reproduction of the organism.
Learning is seen as something the animal is organized to allow
along certain lines, rather than as something the experimenter
produces by critical manipulations of independent variables.
Though an ecological approach does not necessarily rule out a
general reinforcement mechanism, there are important differences in how learning is viewed: (1) Learning occurs against
a backdrop of the complex and sophisticated organization of
response elements, stimulus sensitivities and motivational
hierarchies which underlie an animal's behavior. (2) Any
learning that occurs is not automatic but is dependent on and
mediated by the organization underlying the animal's behavior.
(3) Learning is not limited to circumstances in which a reinforcer closely follows a particular response or stimulus.
Learning occurs whenever and however there has been selection
for efficient local adaptations of behavior to deal with predictable variation in the environmental.

A key element of the ecological approach is that the motor and
sensory potential of an organism is structured prior to entering a learning situation. A simplified way of viewing this
structure is in terms of behavior systems (Baerends, 1976;
Timberlake, in press; Tinbergen, 1951). At the molar level,
these systems consist of sets of response patterns and stimulus
sensitivities that are potentially related to obtaining a
particular terminal stimulus (or performing a particular
terminal response). For example, there are behavior systems
associated with feeding, drinking, mating, parenting, body
care, and thermoregulation. Each system is made up of more or
less independent groups of functional patterns and stimulus
sensitivities I will call perceptual-motor modules. Each
module consists of responses that show a moderate to high level
of sequential and/or temporal relatedness, and can be elicited,

controlled, and terminated by stimuli with particular characteristics. These modules may vary considerably in degree of
typical completeness, integration, fineness of perceptual
tuning, and degree of independence of a particular system.

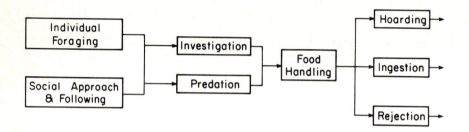

Figure 1. A schematic of the feeding behavior system
in rats. The modules correspond roughly to a search
phase (individual foraging and movement; social
approach), a more specific capture phase (investigation; predatory behavior), and food handling and
disposal (hoarding; ingestion; rejection). Though
the arrows from each block point to the right, it
is certainly possible for an animal to interrupt
this flow and stop, or return to the same or earlier
blocks, depending on the stimuls conditions.

In most behavior systems modules are loosely organized in one
or more sequences that lead to the terminal event. Figure 1
illustrates a simplified view of the feeding system of the
rat. The diagram is derived from reports by Barnett, 1975;

Ewer, 1971; Galef and Clark, 1972; Steininger, 1966; Telle, 1950; and from personal observation. These modules are organized in a left to right order that corresponds roughly to typical sequences of behavior in the feeding system, though modules may be skipped or repeated in any actual sequence.

A second key assumption is that the learning capacities of an organism evolved within particular functional systems of behavior to modify the expression and elicitation of behavior to meet predictable variation in the environment. In other words, the animal comes equipped in varying degrees with appetitive structures of stimulus sensitivities and response elements that predispose it to learn certain appetitive behaviors in the presence of particular classes of stimuli.

The third assumption is related to the second and argues that learning is not limited to changes in responding which accompany repeated temporal contiguity (or contingency) between a response or stimulus and a reinforcer. In the present view learning may occur in the form of changes in the frequency, order, timing, integration and elicitation of responses, modules, and systems. Different aspects of a system may be sensitive to modification by stimulus pairing, response feedback, and/or simple response repetition and stimulus exposure, all dependent on the circumstances within which the learning ability evolved.

Previous work specifically testing the present behavior system analysis has determined the relation of the appetitive structure of feeding in rats to the effects of stimulus pairing and response feedback (Timberlake, in press; Timberlake, Wahl and King, 1982). This work examined unusual and highly organized appetitive behaviors within the traditional paradigms of Pavlovian and operant conditioning. The present research reverses this pattern by closely analyzing the organization, elicitation, and support of a traditional appetitive behavior (maze running in rats) in the unusual context of the absence of reward in the goal box. The intent of this research was to examine how the unconditional organization (appetitive structure) related to maze running contributes to its acquisition and maintenance. Such an analysis of the determinants of maze running is important not only because it provides further characterization of the rats appetitive structure, but because many researchers have assumed that maze performance is based primarily on the frequency and intermittent nature of reward. A clearer conception of the nature of unrewarded maze running might contribute considerably to a clarification of the role of reward in learning.

In short, in the ecological view learning is presumed to occur within a functional appetitive structure which the animal brings to the experimental situation. Previous studies have shown that this structure can be of great importance in determining the results of stimulus pairings and response contingencies (see Timberlake, Wahl and King, 1982, and Timberlake, in press, for reviews). The present experiments attempt to analyze and illustrate the effects of the appetitive structure underlying a traditional instrumental behavior, namely, straight alley running in rats.

The Problem

Small (1900) introduced the maze as a means of studying learning as it related to the animal's natural ecology. He claimed that maze running was one of the psychobiological attributes of the rat, and that reasonable results in studies of learning would not be obtained unless these attributes were considered. Studies of rats' ecology support Small's presumption of the ecological relevance of maze learning. Rats follow trails in the wild (Steininger, 1966; Telle, 1950; Calhoun, 1962) learning them sufficiently well to traverse them even in deep snow over sleet ice (Calhoun, 1962). The laboratory has provided considerable evidence from both the latent learning literature and the radial arm maze literature that rats are unusually able at memorizing complex spatial mazes (e.g., Blodgett, 1929; Olton, 1978). All of these studies point to the relevance of studying the elicitation and support of maze running to provide a picture of the appetitive structure the rat brings with it into the learning situation.

Unfortunately, the majority of maze studies following Small's (1900) original work have been predicated on the position that maze running is organized and produced by the action of reward on random movements by the rat. Thus, few studies have asked about what elicits and controls maze running independent of reward. There was a flurry of interest in the 1930's in concepts such as centrifugal swing (cf., Munn, 1950), and a more abiding interest in latent learning (e.g., Thistlewaite, 1951), but these were largely subservient to the interest in the presumed effects of reward.

In actual fact, there are a considerable number of studies that show acquisition of maze running in the absence of overt reward. Simmons (1924) found that rats run without reward in a Hampton Court maze significantly decreased their run times and number of errors over trials. Tolman and Honzik (1930) and Anderson (1941), both using a multiple T-maze, showed similar improvement in the run times and errors of rats run over repeated trials without reward. Ligon (1929) found that rats

run in a multi-chambered runway decreased run times over trials.
Brant and Kavanau (1965) showed that mice living in a large
free environment containing several very complex mazes markedly
decreased their run times over successive self-administered
trials.   In simpler apparatus, many experimenters have shown
that rats receiving no overt rewards clearly improve performance over trials in the T-maze (Clark and Miller, 1966;
Glanzer, 1961), and the straight alley (Chapman and Levy,
1957; Daly, 1969; Fowler, 1963, 1967; King and Appelbaum, 1973;
Logan, 1960; Mote and Finger, 1942; Paul, 1969; Paul and
Calabrese, 1973; Spear and Spitzner, 1967; Weinstock, 1958;
Wong, 1971).

The Experiments

The purpose of these experiments was to examine the appetitive
structure of straight alley running in rats.   The fundamental
question was not how does reward determine maze running, but
what is the nature of the appetitive structure that the rat
brings to the maze environment, and how does this structure
interact with the eliciting and supporting qualities of the
maze environment to produce improved performance over trials?
The first set of experiments attempted to establish the phenomenon of unrewarded maze running more firmly and evaluated some
subtle and vague reinforcement explanations for its occurrence.
The second set of experiments further analyzed the stimuli
and procedures that appeared to control the development and
expression of straight alley running.   The final set of
experiments briefly explored the contribution of unrewarded
straight alley running to rewarded running.

Two things must be made clear at this point.   The first is
that anyone with an investment in reinforcement explanations
may not be convinced that the present experiments considered
and eliminated all possible sources of reinforcement for unrewarded running.   With a concept as ubiquitous and nonspecific
as reinforcement has become, there will always be a way to
invoke it to explain any particular outcome.   However, it is
important to note that reinforcement explains much more in
retrospect than prospect, and that its speculative use tends
to hinder rather than facilitate the kind of analysis reported
here.   For these reasons, I think the burden of proof that
reinforcement occurs should be shifted to those invoking it.
If it is presumed that reinforcement must have occurred because
behavior changed, the reinforcement concept serves only to
block further analysis of maze running.

Second, I chose to study straight alley running because it is
the simplest and least "cognitive" of various forms of maze
running.   The simplicity of the straight alley facilitated

performing the experiments and should provide the groundwork
for any further analysis of maze running. It also decreased
the likelihood of invoking presumed cognitions with implied
motivational force (the rat knows the maze, therefore it runs)
as explanations of behavior. The straight alley has the disadvantage that sources of reinforcement such as exploration
or conditioned general activity are much more difficult to
eliminate or control for than in complex choice mazes. In the
straight alley the structure of the apparatus guides the
direction of activity whereas in choice apparatus the animal
must provide its own direction of running. It is for these
reasons that considerations of the effects of deprivation and
exploration occupy a large portion of the initial experiments.

General Method

With a few exceptions that will be noted, the subjects, apparatus, and procedures described below were "standard" across
all experiments.

Subjects. The subjects were male Wistar albino rats, 90-150
days of age at the beginning of the experiment. They were
housed one-per-cage under a 12:12 hr light-dark cycle. The
rats were run during the light part of the cycle. Deprived
rats were maintained between 80-85% of their body weight by
providing each with a restricted amount of food following
running. Water was always available. All animals were
handled for a minimum of three days, a minimum of 5 min each
day before exposure to the apparatus or food restrictions
began.

Apparatus. The apparatus was a 2.3 m black straight alley,
10 cm wide, 30 cm high, and divided into 10 23-cm units by
small pieces of tape on the outer surface of the Plexiglas
wall. A 30.3 cm long start box was separated from the alley
by a guillotine door. In most experiments there was no
seperate goal-box, just the end of the alley. One side of
the alley and the start box was made of Plexiglas to facilitate
observation. The apparatus was located on a 30-in high cabinet-
top and oriented so that it was directly under a fluorescent
fixture in the ceiling that ran the length of the alley. The
observer sat approximately 1.5 m from the apparatus and approximately at the mid point.

Start latencies were measured from the opening of the start
box door until the animal broke a photo-beam at the end of the
first 23 cm unit. Run times were recorded from this point
until the rat entered the last 23 cm unit of the apparatus.
Total activity was measured by the number of times a rat

placed all four feet into a new 23-cm unit of the alley, or
re-entered the start box.

Procedure. The subjects were run unrewarded one trial per day
for 18 or 20 days. The relatively small number of total trials
was chosen for convenience given that initial pilot work
showed little difference between 18 or 60 trials in terms of
stable levels of performance. Within each group, and when
possible between groups, the order of running was changed
daily. If animals were on a deprivation schedule, none were
fed until all were returned to the colony room to eliminate
excitation due to other rats being fed (Wong and Traupman,
1971). Thus, deprived rats were fed an average of 1.5 hours
after running. On each daily trial the experimenter placed
the rat in the start box, tail first, through a rear door.
After 15 sec the start box door was raised and left open for
the remainder of the trial. The subject was removed from the
apparatus 2 min after it completely left the start box, and
the start and run times, and total units entered were recorded.
The latter provided a measure of activity in the apparatus.

A time of 2 min to move about the apparatus was chosen when
pilot work revealed no apparent differences in start or run
times given 2 or 4 min exposure to the apparatus. If the subject did not leave the start box within 2 min after the door
was raised, or did not complete the alley within 2 min after
starting, it was assigned a time of 2 min for the appropriate
measure(s) and removed from the apparatus. Between individual
trials, the apparatus was wiped clean with a deodorizing
solution made of 2 ml of a commercial organic acid detergent
obtained from a dairy in 1-1 of water. A natural logarithm
transformation was performed on the start and run times to
normalize their distributions before analyses of variance. In
most cases, the start times were a more variable version of the
run times and, thus, were not reported. To control for experimenter bias the initial studies were run by naive experimenters
who were told the experiments concerned the effects of deprivation on exploration. Subsequent studies were performed by
naive experimenters who had no idea of the number of previous
experiments or what the data looked like.

Section 1:   Reinforcement Explanations of Unrewarded Running

Old ways of thinking die hard. We should not expect it to be
otherwise. Old ways have been successful in analyzing new
effects and summarizing old ones, in eliminating erroneous
alternative explanations, and in clarifying our thinking
about complex phenomena. Presented with the fact that rats
improve their performance in both simple and complex mazes
in the absence of overt food reward, the most tempting

explanation is that running is being reinforced, though in
some covert manner. A number of alternative sources of reinforcement and activation will readily occur to the reader:
(1) Conditioned reward from association of the subjects arrival
in and removal from the goal box with subsequent feeding. (2)
The development of a conditioned activity state based on the
anticipation of subsequent feeding and the expression of that
state in the alley. (3) Food odors or particles in the alley
and goal box left by rewarded rats run in the same or a prior
experiment. (4) An excitatory effect produced by feeding nearby rats in the housing environment (Wong and Traupman, 1971).
(5) Unconditioned reward associated with handling and removal
from the goal box (Sperling and Valle, 1964). (6) Preservation
of recent stimulus-response associations (entering the goal
box from the alley) by removal from the goal box. (7) Unconditioned reward based on play or exploration related to the
alley.

Nearly all of these explanations are ruled out by one or more
of the studies reviewed in the introduction. The demonstration
of improvement in choice mazes rules out conditioned activity,
activation states, and exploration as the fundamental determinants of maze acquisition. None of these sources of increased activity has a differential directional component.
The demonstration that ad lib rats acquire alley-running rules
out the effects of any but the most indirect and ephermeral
sources of conditioned reward, and allowing rats to shuttle
back-and-forth in a runway (Payl and Calabrese, 1973) rules
out the importance of immediate handling and the preservation
of recent stimulus-response associations in determining improved maze performance. Thus, no single explanation can
account for all the cases of unrewarded acquisition in the
literature.

However, each single study reviewed has obvious alternative
explanations that were not specifically considered. Studies
that controlled for the possible effects of deprivation typically ignored the potential effects of handling, and those
that controlled for the effects of handling (e.g., Brandt and
Kavanau, 1974) did not consider the potential effects of
deprivation or feeding following completion of the maze. Even
the careful study of Paul and Calabrese (1973) using a shuttle
maze to avoid handling and ad lib animals to avoid the effects
of deprivation did not consider the possibility of food particles in the runway and the potential effect of consistent removal from one of the end boxes. Thus, the possibility remains
that unrewarded maze running, though not controlled by a single
source of reinforcement, is nonetheless fundamentally determined by a number of different sources of reinforcement acting
through a general mechanism.

The purpose of the studies in this section was threefold: first, to attempt in a single experiment to evaluate most of the possible sources of reinforcement for alley running; second, to determine the nature of any effect of deprivation on unrewarded alley running; third, to examine as carefully as possible how exploration and play might contribute to unrewarded running.

Experiment 1

The primary purpose of this experiment was to eliminate, control for, and/or test the majority of explanations of unrewarded alley running proposed in the introduction. Any potential effects of conditioned reward or food deprivation were minimized by using naive ad lib animals that had no history of experimenter-imposed deprivation. The possibility that food particles or odors were present in the apparatus to reward running was eliminated by running only unrewarded rats in a newly constructed alley that had never been used for reward studies. On the chance that an ad lib rat might bring in some food on its paws, the apparatus was washed between subjects with an organic acid detergent used by dairy farmers to clean milk machines.

Table 1

Number of Subjects Removed From
Each Quarter of the Alley at Asymptote (Trials 15-18).[a]

Trial	SB-1	2-4	5-7	8-10	x^2 Values
15	11	5	3	8	5.7
16	10	4	4	7	3.5
17	10	6	7	4	2.8
18	7	5	4	8	1.7

[a]Only those subjects which left the start box on a
given trial are counted in this table.

The typical association between the goal box and the handling and removal of the animal was broken by allowing the animals to move freely around the alley, removing them from wherever they were 2-min after they left the start box (c.f., Kimball, Kimball and Weaver, 1953). Table 1 shows the point of removal (in fourths of the alley) for all animals that left the start box on Trials 15-18. Chi-squares calculated for each trial were not significant, indicating that the distribution of removal points did not differ significantly from chance.

Further, examination of the data of individual animals showed
no differential tendency to return to the location from which
they were removed on previous trials. For example, only one
animal was removed from the same location for Trials 15-18, a
number expected by chance. Thus, though no effects were noted,
any potential effects of handling and removal were distributed
randomly across the apparatus and, so, could not be a determinant of running to the goal box. Removal from a point
earlier in the alley should at best interfere with rather than
facilitate running to the goal box.

The second intent of this experiment was to determine the relation of unrewarded alley performance to deprivation. Deprivation has been an important determinant of performance in
rewarded alley running (e.g., Mackintosh, 1974), but several
theorists have assumed that its effect is indirect, being
mediated through changes in the incentive value of the reward
(Bolles, 1967, Ch 12). Unless there is clear evidence of
an incentive in the goal box, any deprivation effect in the
present study will be considered a direct effect on performance. Deprivation effects were assessed by comparing the
performance of the ad-lib group (Group NS--not scheduled)
with that of a group placed on a feeding schedule beginning
18 days before the first trial (Group PS--pre-scheduled) and
that of a group placed on a feeding schedule beginning the
first day of the experiment (Group S--scheduled). Any effect
of deprivation on alley performance should be apparent from
the start for Group PS and should develop over trials for
Group S.

Last, the experiment allowed partial assessment of the contribution of body weight loss and conditioned activity state to
alley performance. If asymptotic running is related to a
conditioned activity state, run times and activity after completing the alley (subtracting out the units contributed by
the initial trip down the alley) should show a significant
positive correlation at asymptote. If asymptotic running
is related to body-weight loss, then there should be a high
positive correlation between percent body weight and performance at asymptote.

Figure 2 shows that the animals decreased run times and increased activity over trials, $Fs(17,459)= 25.3$ and 25.5, both
$p<.01$. As indicated in the method section, start times were
not analyzed because they showed effects similar to, but more
variable than the run time effects. Analyses of the simple
main effects of trials for each group showed that the decrease
in run times and the increase in activity were significant for
each group.

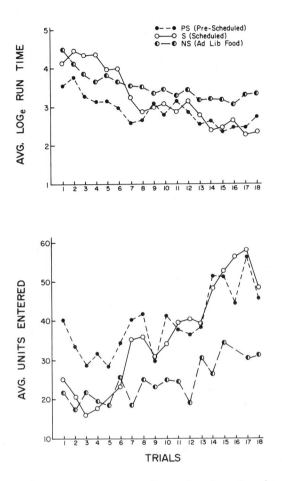

Figure 2. Average $\log_e$ run times (top) and units
entered (bottom) over trials for Group NS (ad lib.),
Group S (schedule beginning with Trial 1), and
Group PS (scheduled beginning 18 days prior to
Trial 1 (Experiment 1).

Deprivation appeared to facilitate run times and activity, an
effect supported by a significant deprivation by trials interaction for both run times and activity, $F_s(34,459) = 3.18$ and
4.19, both $p < .01$. However, separate analyses of variance run
over the last four trials failed to show a significant group
effect. The reason for this anomaly was that deprivation
had its primary effect on performance by increasing the probability that an animal would complete the alley (see Table 2),

Table 2

The number of animals in Experiment 1 completing the alley on each trial for Group NS (ad lib. food), Group S (on deprivation schedule beginning with trial 1), and Group PS (on deprivation schedule beginning 18 days before trial 1). The number of subjects per group was 10.

Trials

Group	1	2	3	4	5	6	7	8	9	10	11	12	13	14	15	16	17	18
NS	3	5	5	5	5	5	5	5	5	5	5	5	6	6	6	6	5	6
S	4	4	3	3	4	4	7	7	7	7	7	7	8	9	8	9	9	8
PS	8	7	7	8	8	8	8	8	7	8	8	8	8	9	9	8	9	9

Table 3

Rank order correlations at asymptote between run time and activity (measured as total units, and as rate of activity after completing the alley), and between run time and percent body weight for Groups NS (ad lib.), S (scheduled beginning with Trial 1), and PS (scheduled beginning 18 days before the first trial). (Experiment 1).

Run Time by Activity

Group	Run Time by Total Units	Run Time by Rate Post Run	Run Time by % Body Weight
NS	-.87**	.14	-.65*
S	-.86**	-.26	.23
PS	-.46	-.18	.22
ALL	-.75**	.06	.12

*significant at the .05 level
** significant at the .01 level

not by decreasing the run times of the animals that completed the alley. A comparison of the average total number of trials on which the alley was completed shown in Table 2 revealed a significantly larger number of completions for the deprived groups (PS and S) than for the ad lib. group (NS), $t(28) = 2.19$, $P<.05$.

Because Figure 2 shows an inverse relation between run times and activity for each group, it might be concluded that part of the improvement in the performance of the deprived animals was due to some form of conditioned activity based on anticipation of feeding or unconditioned activity due to body weight loss. This relation was supported by the large rank order correlations between run time and activity over the last four trials, both within and between groups, as shown in column 1 of Table 3. However, the relationship between activity and run times drops nearly to zero if one substracts the 10 units of activity contributed by completing the alley, and eliminates the cases of animals that did not leave the start box (if the animal did not start it received a score of zero activity and an arbitrary long run time, producing a spurious anchor for a high negative relation between activity and run time). Thus, the critical contributors to the initial correlation were tendencies to complete the alley and a tendency not to leave the start box.

The third column of correlations in Table 3 shows the relation between percent body weight (with respect to ad lib weight) and run times averaged over the last four trials both within and between groups. Since general activity levels are usually highly correlated with percent body weight (Collier, 1969; Finger, Reid and Weasner, 1960), if run times are determined by the same factors as activity, there should be a large correlation between run times and percent body weight. The only significant correlation occurred in the ad lib group.

In brief, the first experiment strongly discouraged explanations of unrewarded straight alley performance in terms of conditioned and unconditioned reinforcement due to handling or goal box removal, activity states due to conditioned anticipation of feeding or body weight loss, presence of food particles or odor in the maze, or preservation of stimulus-response relations through removal from the goal box. Food deprivation had a significant effect on run times and activity, primarily expressed as an increase in the probability that a deprived animal would complete the alley.

Experiment 2

The only hypothesis not explicitly considered in the first experiment was the possibility that straight alley running is reinforced by exploration. Because exploration can be such a vague concept, I tried to treat it more carefully in separate experiments. It should be clear that exploration alone cannot readily account for unrewarded performance in complex mazes; in the absence of distinctive novel cues at particular locations in the environment, there is no directional component to

simple exploration. However, the straight alley has no choice
points so a general tendency to explore might determine
straight alley running. The primary purpose of the next four
experiments was to test as specifically as possible various
interpretations of how exploration might contribute to unrewarded straight alley running.

A secondary purpose of the present experiment was to clarify
whether the effects of deprivation shown in the previous experiment were due to conditioned anticipation of feeding, or
loss in body weight. In the previous study these two factors
were procedurally confounded. The present study sought to
separate them by comparing the performance of a group of rats
fed two hours before (Group FB-Fed Before) with that of a
group fed one hour after running (Group FA--Fed After). Food
intake for these groups was adjusted to make their percent
body weights approximately equal when they were in the alley.
It should be clear that Group FB was not incapacitated by being fed two hours prior to running. Rats are perfectly capable
of eating every two hours.

The notion that exploration or play is a reinforcer has a long
experimental and conceptual history (e.g., Berlyne, 1960).
Unfortunately, this history includes a good deal of ambiguity
about just how exploration serves to modify behavior. In the
present case there are a variety of ways in which exploration
could account for the results. First, the goal box may represent a unique exploratory incentive which served to reinforce running the alley (e.g., Berlyne, 1960). Second, maze
running itself may be produced by a series of exploratory
responses linked together by a tendency to leave a less interesting place and enter the next more interesting part of the
alley (Glanzer, 1961). Third, maze running may be produced by
a motivation to play (Paul and Calabrese, 1973). Fourth,
exploration may not be a reinforcer at all, but rather may represent a competitor with maze running for time and energy.
On this account run times should decrease over trials as competing tendencies to explore dropped out. The present experiment focused on the first three explanations, leaving the
fourth hypothesis to subsequent experiments.

I will focus first on the second possibility. Glanzer (1961),
in studying the behavior of rats in a complex maze, proposed
a model in which the probability (and by extension, the speed)
of moving from one location to the next was a product of the
proportion of familiar stimulus elements at the first location
multiplied by the proportion of novel stimulus elements at the
second location. Essentially Glanzer's model is a drive times

incentive formulation in which the drive to leave a location
is related to the number of familiar elements and the incentive for approaching a new location is related to the number of novel elements.

This model has two interesting predictions, the first is that
animals should progress further and more rapidly down the alley
as they become adapted to the apparatus. The second prediction
is based on the assumption that in the reasonably homogenous
alley environment, the proportions of novel and familiar
elements in any two adjacent locations sum to approximately 1
(the proportion of novel elements equals one minus the proportion of familiar elements). In such a case the relation
between latency of movement and trials should be bitonic because as the proportion of novel elements drops toward 50%,
movement will increase, but as the proportion of novel elements
falls below 50%, movement will decrease again. Neither the
results of the first experiment, those from a pilot experiment
in which rats ran for 60 days, or those from Experiment 10
in which animals ran 32 unrewarded trials, showed this expected
U shaped relation between performance and trials. Instead, run
times and activity appeared reasonably stable once asymptote
had been reached. An explanation for the stable run times
that is compatible with Glanzer's model is that the procedure
of running one trial per day allowed an identical recovery of
novelty each day to a reliable level of at least 50% novel
stimulus elements.

Such an account seems unlikely because the requirement of an
identical recovery in novelty each day for each rat. If the
recovery were not identical each day, run times and activity
for individuals would either continue to increase with subsequent trials or begin to decrease. However, since the identical recovery of novelty each day is (marginally) a possibility,
the present study tested this interpretation by using very old
rats (380 days) as subjects. Several studies have shown that
rats decrease exploration as a function of age, and most importantly, old rats show more decrement in exploration with
repeated tests (Furchtgott, Wechkin and Dees, 1961; Goodrick,
1966; Valle, 1971). Thus, it appears that novelty does not
recover as rapidly or completely for older rats as for younger
rats. In the present situation, if the asymptotic performance
of young animals represented identical marked recovery in
apparatus novelty between trials, then older rats run in the
same situation should show a much less complete recovery that
deteriorates over exposures. Thus, older rats would be expected to show a bitonic relation between trials and performance.

The use of older rats also should shed light on the hypothesis
that running is reinforced by exploration at the end of the
alley or by the opportunity to engage in "self reinforcing"
play responses (Paul, 1969; Paul and Calabrese, 1979). If
novel stimuli in the goal are the critical reinforcer, then
older rats again should show a bitonic relation between trials
and run times as the goal box novelty lessens. In the case of
play Muller-Schwarze (1971) and Welker (1962) have reported
that the frequency of play in rats approaches zero around 70
days of age. If play is responsible for unrewarded alley running, one would expect no improvement in the performance of the
380 day old rats in this study. It should be noted that this
experiment does not provide a direct comparison of the performance of old and young rats. However, the prediction of a
bitonic function makes a direct comparison unnecessary.

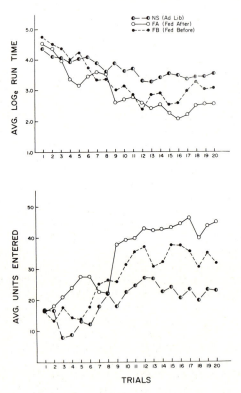

Figure 3. Average $\log_e$ run times (top) and units entered
(bottom) over trials for Group NS (ad lib.), Group FB
(scheduled and fed before running), and Group FA
(scheduled and fed after running) (Experiment 2).

Table 4

The number of animals in Experiment 2 completing the alley on
each trial for Group NS (ad lib. food), Group FA (on deprivation schedule, fed 1 hr after running), and Group FB (on
deprivation schedule, fed 2 hr before running). The number
of subjects per group was 10.

Trials

Group	1	2	3	4	5	6	7	8	9	10	11	12	13	14	15	16	17	18	19	20
NS	4	4	3	3	3	3	5	6	6	7	7	7	7	7	7	6	7	7	7	7
FB	3	2	3	4	5	6	7	7	7	7	9	9	8	7	9	9	9	8	8	8
FA	3	3	5	6	6	7	6	5	9	9	10	10	10	9	9	10	10	10	10	10

Except as noted standard procedures were used in this experiment for a total of 20 trials. Figure 3 shows that all groups
decreased run times and increased activity as a function of
trials, $Fs(19,513) = 15.1$ and 14.7, $p<.01$. The results are
similar to those of Experiment 1 and clearly suggest that
neither play nor exploration, as they are usually defined, provides a critical basis for the reinforcement of unrewarded alley running. There is a slight hint of a bitonic function relating activity and trials in the ad lib group, and between
run times and activity in the group fed before running. The
latter effect was largely due to a decrease in the number of
animals completing the alley, a result not affecting the run
times of the other subjects. The former effect, though quite
marginal, was more consistent and may reflect the predicted
effect of age on exploration (see also Valle, 1971). For
those who are interested, a comparison of the performance of
the old animals in Figure 3 with the young animals in Figure 2
suggested no significant differences, a conclusion confirmed
by analyses of variance.

The apparent effect of deprivation on run times and activity
in Figure 3 was not confirmed by overall tests, $Fs(2,27) =$
2.24 and 2.83, $p>.10$ and $.10 >p>.05$. However, individual tests
of each deprivation groups versus the control group showed a
significant effect of deprivation on the run times and activity
of Group FA, $ts(27) = 2.24$ and 2.37, both $p<.05$; but no significant effect on the scores of Group FB, $ts(27) = 1.15$ and 1.23,
both $p>.10$. Table 4 shows a similar effect in that deprivation
increased the probability of completing the alley for both
groups, but only Group FA differed significantly from the ad
lib group, $t(19) = 1.73$, $p<.05$.

In short, given the level of performance of the old animals,
it does not appear that exploration and play responses can
account for stable unrewarded running in straight alleys,
though age may affect activity. If the young animals of
Experiment 1 were just able to recover the novelty or the play
eliciting value of the alley during the day long inter-trial-
interval, then the old animals should not have been able to
recover the alley novelty given the same inter-trial-interval.
It might be argued that the session time or number of trials
was insufficient to produce an adequate reduction in novelty
even for the old animals. But this is argued against by the
simplicity of the alley and the finding of reduced exploration
by old rats in open fields with comparable exposure periods
and fewer trials. Thus, it seems inescapable that the old
animals should have performed much worse than the young animals as the trials increased. The results do not support this
view. In addition, the present results supported a role for
conditioned activity in increasing the probability of completing the alley, but did not rule out a role for body weight
loss. On all measures the performance of the group fed before running fell between that of the group fed after running
and the ad lib group.

Experiment 3

Because an exploration account of unrewarded alley running
seems so attractive, the intent of this study was to test
further the relation of alley running to exploration. The present experiment contrasted the view that alley running is
based fundamentally on exploration (Glanzer, 1961) with the
possibility that reliable running was related to a decrease in
exploration and running produced by repeated exposure to the
alley. The experimental procedure involved two groups of 10
rats, both run under standard methods except that Group Two-
Trial received two trials per day, one run approximately 30
min after the other.

If the stable straight alley performance in Experiments 1 and
2 was based on continued expression of exploration and recovery of novelty over the 24 hr interval between each exposure, then asymptotic performance of the animals in Group
Two-Trial should be considerably less on Trial 2 than on Trial
1, and performance on both trials should be lower than that of
the Control Group. The first prediction follows because 30
minutes should be markedly insufficient to recover the level
of novelty found with a 24 hr inter-trial-interval. The
second prediction follows because Group Two-Trial received
twice the amount of exposure to the apparatus per day and,
thus, had a lower level of novelty from which to recover.

In short, at asymptote the run times and activity of the two groups should be ordered with the control group best, followed by Trial 1 of Group Two-Trial, and then by Trial 2.

On the other hand, if acquisition of maze performance is related to a decreased tendency to cautiously explore the alley, then performance should not differ at asymptote because by that point the tendency to explore should have been reduced approximately equally in all cases. A common prediction of the two approaches is that during early acquisition the performance of Group Two-Trial should be better than the control, and better on Trial 2 than on Trial 1. This prediction follows from the competing exploration view because the performance of Group Two-Trial, on Trial 2 in particular, should reflect a greater adaptation of exploratory tendencies. The prediction follows from Glanzer's (1961) model because movement toward 50 percent novel stimulus elements (producing maximum performance) should occur at a faster rate for Group Two-Trial, and at a slightly faster rate in Trial 2 than in Trial 1.

Figure 4 shows clear acquisition of straight alley running and activity in all cases. Individual analyses of variance comparing the control group with each trial separately for Group Two-Trial showed significant trials effects for both run times and activity, all $ps < .01$. However, there was no effect of groups at asymptote (the last two-day blocks). Tests comparing performance on Trials 1 and 2 with that of the control did not approach significance, all $ts(18) < 1$.

In contrast to the absence of an effect on asymptotic performance, the two-trial procedure approximately doubled the rate of approach to asymptote. Tests of the differences between the average run time and activity of Group Two-Trial and the control group showed no significant difference on the first trial block, both $ts(18) < 1$; but on the next three trial-blocks Group Two-Trial showed significantly shorter run times, $ts(18) = 2.15, 2.28,$ and 2.17, all $p < .05$, and greater activity, $ts(18) = 2.57, 3.00,$ and 2.07, all $p < .05$. Further, on the first two blocks of trials, performance on Trial 2 was better than performance on Trial 1, though this difference reversed on subsequent trials.

In short, asymptotic unrewarded alley running did not appear to be based in a fundamental way on a general tendency to explore the alley. Instead, the results supported the view that running was related to the adaptation of exploratory tendencies as a function of repeated exposure to the alley. The acquisition data supported the common predictions of both approaches in that Group Two-Trial acquired alley running more rapidly than the control group, and over the first four trials,

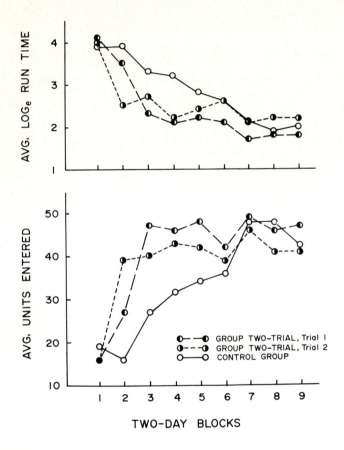

Figure 4. Average $\log_e$ run times (top) and units entered (bottom) over two day blocks for Group Two-Trial (Trial 1 and Trial 2) and the control group (Experiment 3).

performance on Trial 2 was better than on Trial 1. Further, some effect of the exploratory attractiveness of the alley on performance was indicated by the fact that the start times at asymptote were significantly longer for Trial 2 than for Trial 1, $\underline{t}(8) = 1.96$, $\underline{p}<.05$, one-tailed test.

It seems reasonable that the exploratory attractiveness of the alley may exert some effect on alley running if only to increase persistence of running under large numbers of massed trials. Mote and Finger (1942) found that animals run under successive massed trials on an elevated maze eventually stopped running. However, this effect occurred also with rewarded animals, so it is not clear whether this effect should be related only to exploration or unrewarded running, or to some general inhibition of movement. A more complete account of alley running should consider the contribution of exploratory variables to persistence of running. However, the important point of the last two experiments is not that asymptotic performance was completely invariant across all manipulations, but that it did not reliably vary across a range of manipulations that should affect performance if it were based solely on a general tendency to explore the novel stimuli of the alley.

Experiment 4

The previous results are compatible with the view that improvement in straight alley performance is related to the adaptation of exploration tendencies through exposure to the apparatus. The present study attempted to specify more precisely the nature of this adaptation by coding the exploratory behavior of subjects in the alley into the categories of: Specific Sniff--rhythmically oscillate the nose and whiskers within 1 cm of the floor or wall of the alley. Rear--lift both front feet off ground, leaving weight on hind feet. Point--tilt the head above the horizontal position and sniff. By examing the duration of these exploratory behaviors we hoped to determine the relation between overt exploratory responses and running. A major question was whether overt exploratory responses actually competed directly for time with running behavior, or whether the novelty of the alley produced a state of caution or fear which directly inhibited running and provided the opportunity for cautious exploration. If peripheral competition is the key, exploratory response during the animals run times should decrease in duration over trials. If a more central inhibition is important, the animal may simply run faster over trials, showing little or no change in exploratory behavior during run times.

Data were acquired using a 16 key keyboard connected to a small minicomputer which recorded the duration of holding down a particular key to the nearest half second. Intercoder reliability checks produced tetrachoric correlations of .94 or better for presence versus absence of a code for the two coders. Circumstances beyond our control prevented examining more than 10 days of acquisition, but the first ten trials contained sufficient data to suggest an answer to the question posed. Ten rats were run in each of two groups; one group was deprived of food and run at 85% body weight, the other was run <u>ad lib</u>. The deprivation manipulation was used because of the possibility that differences in exploratory behavior were related to the effects of deprivation on performance.

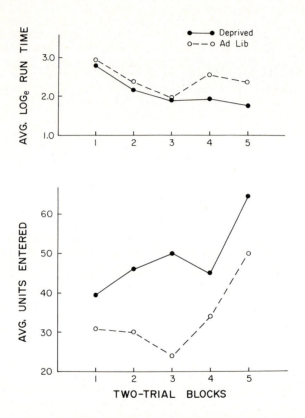

Figure 5. Average $\log_e$ run times (top) and units entered (bottom) over two trial blocks for deprived and undeprived (<u>ad lib</u>) animals (Experiment 4).

Figure 5 shows a small but significant improvement in run times and activity over 5 two-trial blocks, $\underline{F}s(4,72) = 8.4$ and 18.1, $\underline{p}<.01$. There was no significant effect of deprivation over this number of trials. The top of Figure 6 shows the duration of the exploratory behaviors during the run times over two trial blocks. Separate analyses of variance for each category showed no significant changes due to deprivation or trials. The bottom of Figure 6 shows the duration of exploratory behaviors by two trial blocks over the entire

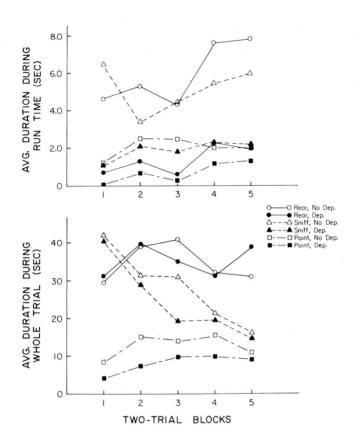

Figure 6. Average duration of time spent in Specific
Sniffing, Rearing, and Pointing over two trial blocks
during the run time (top) and during the entire
trial (bottom) for deprived and undeprived (ad lib)
rats (Experiment 4).

time the animal was in the apparatus. Here, all the categories
showed significant changes across trials: Specific Sniff
markedly decreased, $F(4,72) = 40.4$, $p<.10$, Rear and Point
slightly increased, $\overline{F}s(4,72) = 2.93$ and 4.32, $\underline{p}s<.05$ and .01.
In addition there was a nonsignificant trend toward less
Specific Sniff and Point in the deprived groups, $.10>\underline{p}>.05$.

The results clearly show that run times did not decrease as a
function of the dropping out of the specific exploratory
responses we measured. Instead, the animal simply moved
faster down the alley. In conjunction with the marked drop

in Specific-Sniff over the entire trial, these data suggest
the presence of an exploration/caution state in the early
trials which inhibited running, and elicited or allowed
exploration. As trials in the alley increased, the caution
state waned, leading to more rapid running, and less exploration. However, that some forms of specific exploration slightly increased over trials suggests that there may be some role
of directed exploration in maze performance, at least in
determining total activity in the alley. This also indicates
that overt exploration does not reflect a single internal
state.

Though these data appear to support the role of a relatively
central state in inhibiting running and encouraging exploration, it could be argued that the animal initially runs more
slowly because it is exploring while it moves, a clearly peripheral competition. It did not appear to us that such peripheral competition was the key to slower running. The animal
simply appeared more hesitant and deliberate in its actions
in initial trials, both in its overt exploratory responses
and in its movement. However, whether interference with running is primarily central or peripheral, the evidence strongly
supports the notion that increased performance is related to a
reduction in some forms of exploration and the expression of
a cautious "attitude".

Experiment 5

Taken together the previous studies suggest that acquisition
of straight alley running is related to a reduction in strength
of an exploration/caution state through repeated exposure to
the alley. However, rats do not always procede cautiously
when introduced into an unfamiliar environment. Barnett (1975)
reported that rats introduced into a strange territory explore
almost compulsively despite drawing the hostile attention of
the residents. Male rats in the presence of estrus females
show similar directed exploration. Further, several pieces
of data suggest a role of directed exploration in at least the
total activity of the animals in the maze. The present experiment tested whether the presence of a novel visible stimulus
at the end of the alley would be sufficient to engage a directed exploratory state that would supercede the exploration/
fear state and increase speed of acquisition of alley running
(cf., Schneider and Gross, 1965).

The novel object was an unlit light bulb mounted base-down in
a small porcelain fixture fixed to the floor of the alley.
This object had previously been shown to elicit considerable
exploration in other studies and to be visible to animals in

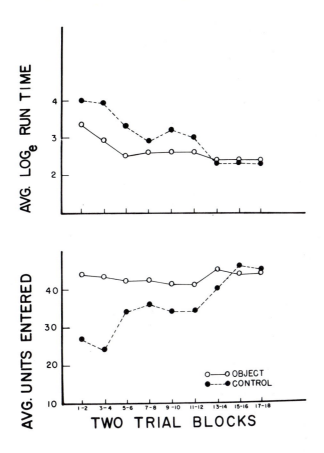

Figure 7. Average $\log_e$ run times (top) and units entered
(bottom) over two-trial blocks for the group with an
exploratory incentive at the end of the alley (Object Group) and a control group (Experiment 5).

the start box against the black end of the alley. Figure 7
shows that the presence of the light bulb did facilitate both
decreases in run times and increases in units entered. Analyses of variance applied to the run times and activity showed
no significant over-all effect of the object, $F_s(1,19)$ =
1.21 and 1.42, $p > .10$, but a significant effect of trials,
$F_s(8,152)$ = 8.31 and 3.84, $p < .01$. Separate t-tests run on the
first two trial blocks showed significant differences between
the groups on run times and activity, Block 1--$t_s(19)$ = 1.84
and 2.78; Block 2--$t_s(19)$ = 2.09 and 2.58, all at least $p < .05$.

The initial effect on activity was not entirely expected because exploring the light bulb competed with moving about the
alley. However, apparently the shift to a directed exploration state produced by the presence of the light bulb was
sufficiently general to facilitate moving about in the period
after completing the alley. The results demonstrate that
visible novel stimulus in the goal area can facilitate acquisition of straight alley performance, apparently by inducing a
state of directed exploration. The possibility of such a
state is supported by the fact that rats will choose a compartment associated with novel stimuli (Berlyne, 1960), change
start/run times as a function of shifts in novelty (Chapman
and Levy, 1957), and even explore a novel area regardless of
aggressive interaction with territory residents (Barnett,
1975).

Discussion

In conjunction with previous data, the results of the first
five experiments argue against a reinforcement account of the
acquisition and maintenance of rewarded straight alley performance. Improvement in alley running occurred in the absence
of food deprivation, conditioned excitatory states, food cues
in the alley, and differential association of handling and
removal with the goal box. Further, arguments that alley running is intrinsically reinforced by exploration or play were
weakened by the demonstration of clear acquisition and maintenance of straight alley running by 1 year old rats, and by
the failure of two massed trials to decrease alley performance.
Deprivation typically increased straight alley performance
primarily through its effect in increasing the probability of
completing the alley. This effect of deprivation appeared to
be related primarily to anticipation of feeding, but may have
also been related to per cent body weight.

Taken together these experiments suggest several roles for
exploration in alley running. First, at least some exploration of the maze environment appears highly correlated with a
state of caution or fear that interferes with rapid locomotion down the alley. Adapting to the environment reduces
the strength of this state of caution, and produces freer
movement with fewer specific exploratory responses in the
apparatus. Acquisition of straight alley running seems highly
related to this process; however, the dropping out of overt
exploratory responses that might compete with running appeared
less important than the apparent reduction of inhibition of
leaving the start box and moving rapidly (not to say gallop-
ing) down the alley. Support for the relation of reduced
fear or caution to acquisition is also provided by Barry,

Wagner and Miller (1962) who showed that tranquilized rats
acquired alley running more rapidly than controls, and by
Bower (1959) who showed that familiarity with an alley
increased acquisition speed.

Second, novel conditions that elicit approach and investigation
can facilitate acquisition of alley running, if the resultant
exploratory behavior does not compete with running. Thus,
placing a novel and highly interesting object at the goal end
of the alley produced faster acquisition of alley running.
This result might be used to argue for the reinforcement value
of exploration. However, this result can as readily be interpreted in terms of shifting the animal from a state of ex-
ploration/caution to one of directed exploration, thereby decreasing inhibition of running. Support for this interpretation is found in the increased initial activity of the animals
with the object at the end of the alley. Further, it is
doubtful that novelty associated with the end of the alley
could be the major determinant of alley running. A bare alley
end is little different than the alley, so as the alley decreased in novelty so should the end. Further, older rats
would be expected to run much more slowly than young rats,
and novel objects at the end of the alley should alter asymptotic run times, both outcomes contradicted by the present
results. Last, such an exploration/reinforcement account
cannot explain unrewarded acquisition of complex mazes because
there are dozens of competing alley ends, all equally novel.

Third, a general tendency to explore does not appear to be an
adequate explanation of either acquisition or asymptotic performance in the alley. According to Glanzer's (1961) realization of this view, rats should decrease their run times to
a point as novelty decreases, and then increase run times with
further decreases in novelty. But in this view using year-old
animals, running more than one massed trial per day, or increasing the total number of trials should alter adaptation to
alley novelty and markedly affect run times. This did not
occur. Animals run two trials per day reached asymptote more
rapidly, but neither they nor year old animals appeared to differ in performance from young animals run one trial per day.
Instead, asymptotic running appeared relatively invariant
across all manipulations. If unrewarded alley running is based
on exploration, it is a form of exploration relatively specific
to the alley.

In short, improvement in alley running appeared based on the
decrease in strength of an exploration/caution state which
inhibited free movement in the alley. Neither conditioned or
unconditioned reward based on handling or removal from the

alley, conditioned or unconditioned activity drive, exploration, nor play readily accounted for the stable levels of asymptotic performance achieved. The tendency to explore directly novel stimuli seemed the most important determinant of running manipulated in these experiments, but its effects seemed largely mediated either through its role in decreasing the strength of the exploration/caution state, or through direct competition. It appears that the understanding of straight alley running will rest on a further analysis of the variables that control it, rather than on the assumption that it is reinforced in some fashion. It is to this task that we turn in the next section.

Section II. The Environmental and Procedural Determinants of Alley Running

The experiments in the previous section showed that explanations of unrewarded alley running based on alternative sources of reinforcement such as association of goal-box removal with feeding and handling, or the intrinsic rewards of exploration or play left much to be desired. Such explanations did not appear sufficient to account for the data, did little to increase our understanding of alley running, and in many cases were difficult to test. The main purpose of the experiments in this section was to examine more carefully the potential stimuli that elicit and support unrewarded alley running in ad lib rats. In a sense these experiments treated alley running as a complex species-typical action pattern released, guided, and supported by the stimulus configuration presented by the alley and its physical and procedural "surround." Thus, we were concerned with the effects on running of modifying components of this "surround." We also further evaluated the role of the exploration/caution state in acquisition.

Experiment 6 examined the effect of the sensory inputs of illumination and odor trials in the runway. Experiment 7 tested the importance of the apparatus cues of length of runway, height of top, and presence of two walls. Experiment 8 examined the importance of the procedures involved in releasing the animal from the start box and removing it from the goal box. To increase the number of subjects typically completing the alley (thereby gaining a better notion of the variables directly affecting running), we changed the experimental procedures slightly beginning with Experiment 6. Any animal that had not left the start box by Trial 3 was placed in the middle of the alley for 2-4 minutes immediately after its trial. The same procedure was used again if it failed to start on Trial 4. This procedure reduced to zero the number of animals in the control groups not completing the alley at asymptote.

Experiment 6

This experiment examined the effect on alley performance of the
level of illumination and presence of odor in the apparatus.
Since rats in natural settings run trails in daylight and darkness, the illumination manipulation might be expected to have
little effect on asymptotic performance. However, to the
extent that acquisition of alley running is based on decreased
caution or fear it would be expected that brightness should
decrease acquisition speed and darkness should increase it
(Valle, 1970; Williams, 1971). Group Bright was run with full
room lights (two double tube fluorescent fixtures, 8 ft long)
plus two 100 watt flood lights focused on the alley about one-
third of the way from each end and located approximately six
feet above the alley. The flood lights were turned off between
trials to prevent temperature build up. Group Dark was run
with a single 25 w red bulb placed 6 feet from the alley and
approximately parallel to the start box.

Many investigators have argued for the importance of odor
trails in determining trailing behavior in rats (e.g., Barnett,
1975). However, Calhoun (1962), in observations of an outdoor
rat colony, reported that odor trails were unnecessary to obtain systematic running in a natural setting. Following
development of a snow cover, a sleet freeze covered existing
trails thereby probably eliminated all odor cues. Following
another snow, Calhoun observed that rats broke the new snow
to form the same trails they had previously used. Further,
work on the sensory control of maze running has indicated that
odor was not essential to maze performance (e.g., Olton and
Samuelson, 1976; Watson, 1907). In the present study Group
Odor received the odor of all previous animals run that day
in the runway, an average of five animals. The first animal
always followed an experienced animal from a previous study.
The Control Group received the standard treatment of a thorough
cleaning between subjects. The number of subjects in each
group varied from 9 to 10.

The results, shown in Figure 8, suggest that alley performance
is readily acquired under all conditions, though darkness
appeared to facilitate acquisition and bright light and odor
appeared to interfere with asymptotic performance. Run times
and activity significantly differed over groups, $Fs(3,33)$ =
6.00 and 5.67, $p<.01$, and trials, $Fs(8,264)$ = 12.77 and 20.64,
$p<.001$. Tests of the means of Groups Odor, Bright-Light, and
Dark versus the control mean revealed significant differences
for all groups on run times, Dunnett's $ts(4,33)$ = 5.58, 6.05,
and 5.16, $ps<.01$, and activity, Dunnett's $ts(4,33)$ = 3.82,
6.20, and 5.39, $ps<.01$. At asymptote (the last four trials)

the control group did not differ significantly in running time
from Group Dark, $t(16)<1$, $p>.10$, but ran significantly faster
than either Group Odor or Group Bright Light, ts(16 and 17)
= 2.94 and 2.33, $p<.01$ and .05.

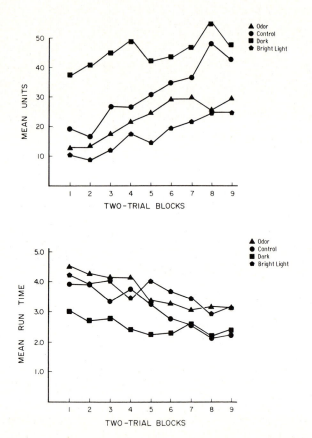

Figure 8. Average $\log_e$ run times (bottom) and units
    entered (top) over two-trial blocks for Groups
    Odor, Bright Light, Dark, and Control (Experiment
    6).

In short, rate of acquisition of alley running was facilitated
by darkness, an outcome that would be expected if acquisition
speed was inversely related to caution or fear. Further,
though odor cues may guide animals to particular locations,
presenting novel odors in the runway in the absence of food
slightly inhibited the development and asymptote of running

behavior, apparently through elicitation of directed sniffing
of the alley. Last, the bright light appeared to slow
acquisition slightly and inhibit asymptotic performance,
primarily through its effect on the slowest running animals.

Experiment 7

This experiment manipulated length of alley, height of roof,
and presence of two sides to the alley to determine to what
extent the physical organization of the alley environment controlled alley running. The length of the alley was expected to
be important on two grounds. First, to the extent that exploratory attractiveness is an important determiner of activity,
shortening the alley from approximately 8 ft to 4 ft should
reduce units entered and possibly running rate. Second, to
the extent that acceleration and braking make up different
proportions of the distance over which run times are measured,
we would expect slower running in the shorter alley.

Trail running in natural environments occurs both above ground
and underground. For Group Low Roof, the height of the roof
was lowered from 12- to 4-in to more closely duplicate some
of the features of underground passageways. It was hypothesized that this resemblance might reduce some of the initial
fear or caution and thus produce more rapid acquisition. However, no facilitation in asymptotic performance was expected.

The last manipulation consisted of removing the Plexiglas side
of the alley and moving the alley to the edge of the cabinet
top on which it rested. This was done to determine the importance of vertical visual and tactile stimuli in producing
maze running. Evidence from elevated mazes and observations
of rats in natural environments demonstrates that animals can
learn to run completely exposed trails. However, Montgomery
(1955) in a study of exploration found considerably less locomotion in an elevated maze than in an enclosed maze. Further,
Lester (1965) reported that, given a choice between enclosed
and open portions of maze, rats were reluctant to enter the
open portion. Since the present experiment provided the
animals with the possibility of remaining in an enclosed start
box, we expect slower acquisition and less activity for Group
Side and Top Off. The number of subjects in each group ranged
from 10 to 11.

The results, seen in Figure 9, supported most of these predictions. There was a significant effect of groups on both
run times and activity, $Fs(3,38)$ = 19.09 and 21.00, $p<.01$, and
a significant effect of trials, $Fs(8,304)$ = 8.43 and 17.50,
$p<.01$. Tests of the simple main effects of trials for each
group, prompted by significant trials x groups interactions,

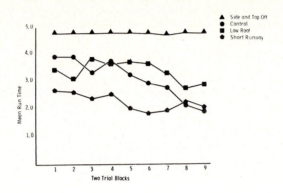

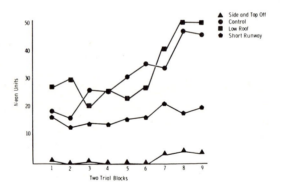

Figure 9. Average log$_e$ run times (top) and units entered
(bottom) over two trial blocks for Groups Side and Top
Off, Short Alley, Low Roof, and Control (Experiment 7).

revealed significant changes in performance (ps<.01) for all
groups except Group Side and Top Off. Tests of the mean of
Groups Side and Top Off, Short Alley, and Low Roof versus
the control mean showed significant differences in run times
for all groups, Dunnett's ts(4,38) = 15.7, 6,54, and 2.88,
ps<.01 and .05, and significant differences in activity for
all but Group Low Roof, Dunnett's ts(4,38) = 21.0, 11.3, and
1.34, first two ps<.01. Last, consideration of the first two
trial blocks showed run times for Group Low Roof that were
significantly faster than those for the control, t(15) = 1.82,
p<.05.

In short, the absence of a side and top on the alley essentially deterred rats from entering the alley. They would sniff
around the start box door and over the edge of the cabinet,
but they would rarely enter the open "alley." The short runway markedly reduced activity and increased run times relative
to the control (the run time scores are indistinguishable from
the control group at asymptote though Group Short Alley ran
approximately half the distance). Clearly a short runway does
not support or elicit running as well as a longer alley. Last,
as anticipated, the low roof had a small facilitating effect
on initial run times which disappeared at asymptote. Apparently the low roof decreased the initial exploration/caution
state through some change in tactile and visual cues.

Experiment 8

This experiment tested the effect on alley performance of varying two experimental procedures, delay in the start box prior
to opening the door, and rapid removal from the goal box. A
four minute delay in the start box was imposed to test the
possibility that alley performance was partly determined by the
boosted activation level produced by handling the animals
shortly before they ran. Black, Fowler and Kimbrell (1964)
showed that heart rate of rats is strongly affected by handling,
and this effect does not habituate over a 20 day period of
handling. The effect does, though, decrease sharply over time
since handling. After four minutes heart rate is about half
of its original increased rate. If heart rate reflects general
arousal or escape responses, then delaying the animal in the
start box prior to its release should decrease that arousal and
slow the animals performance.

Fowler (1963) inadvertantly provided evidence of this effect in
a series of studies attempting to show the importance of a boredom drive. If animals were confined to the start box of a
straight alley for different periods of time and then handled
just before they were released they showed a marginal direct
relation between alley performance and length of time confined,
as predicted from a boredom hypothesis. However, if the animals were not handled just prior to release there was a clear
negative relation between time in start box and alley performance, as would be predicted from an arousal-handling analysis. Thus, the present experiment can be seen as a further
test of the relative importance for alley running of boredom
produced by confinement versus activation produced by handling.

A frequent procedure in straight alley studies is to remove the
animal from the goal box within a very brief time of its entry
(less than 10 sec). The reasons for this are largely practical,

the less time spent with each animal the less time it takes to
complete the study each day. This rapid removal appears to
have little effect on animals munching the remains of their
reward as they are lifted from the goal box; however, it may be
more noxious to animals that have never received food. Such a
possibility might account for some studies that find no acquisition effect with unrewarded controls. In the present study
rats were removed (gently) approximately 8 sec after entering
the goal box (the time it took the experimenter to write down
the clock times, get up and reach the goal box). The number
of subjects varied from 9 to 10.

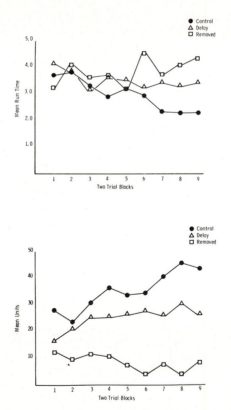

Figure 10. Average $\log_e$ run times (top) and units
    entered (bottom) over two trial blocks for Groups
    Delayed, Removed, and Control (Experiment 8).

The results, displayed in Figure 10, show that both delaying
the animal in the start box and removing the animal from the
alley had inhibitory effects on run times and activity. Both

run time and activity showed significant groups and trials effects, $Fs_{(2,26)}$ = 3.58 and 11.69, $ps < .05$ and .01 for groups, and $Fs_{(8,208)}$ = 4.64 and 8.39, $p < .01$ for trials. Comparisons of all means versus the control showed significant differences for Group Removed and Group Delay on both run times, Dunnett's $ts_{(3,26)}$ = 12.2 and 8.77, $p < .01$, and activity, Dunnett's $ts_{(3,26)}$ = 13.4 and 2.31, $p < .01$ and .05. A test of simple main effects of each group over trials (encouraged by significant interaction terms) showed that Group Delay significantly decreased run times while Group Removed significantly increased run times, $Fs_{(1,208)}$ = 17.7 and 38.6, $p < .01$.

In short, both rapidly removing the animal from the goal box and delaying the start time decreased unrewarded performance in the alley. Rapid removal actually decreased the tendency to run the alley over trials. The delay manipulation seemed to inhibit the tendency of animals to run the alley, but over trials they still ran and improved their performance. Thus, a common removal procedure used in the alley may inhibit the development of unrewarded running. probably because the disruption of movement and handling is noxious to at least some animals. Further, it appears that handling the animals just before they run produces an increased tendency to run, probably through an activation effect of some sort. However, the evidence is clear that handling does not account for all aspects of improved running. Also, it is conceivable that something operationally related to a "boredom" drive might become important as the duration of exposure to the start box increased further, thereby allowing the heart rate to decline more and become less important.

Discussion

Treating running as produced by a characteristic interaction of the organism with the stimuli and procedures of alley experiments proved a fruitful approach. A number of procedural and environmental variables affected acquisition of running, including side and top off, odor, darkness, bright light, low roof, immediate removal from the goal box, and delay in the start box. These variables can be classed into three sets, one affecting the caution or fear state of the animal, the second related to the amount of directed exploration elicited, and the third related to the activation of the animal. Darkness and a low roof appeared to reduce the amount of initial caution, while bright light and removal from the goal box appeared to increase the exploration/caution state, even at asymptote. The effects of increased caution in the case of removal showed up as a disruption of the animal's running, as they began to avoid the end of the alley where they were removed. These

inhibitory effects showed clearly around Trial 12. Logan
(1960) in a similar study of the effects of removal after 3
sec of confinement in a goal box did not show an inhibitory
effect on unrewarded alley running until around 30 trials.
Though the findings are in the same direction, the reason for
the differences in point of onset are not clear.

Side and top off also produced considerable caution and exploration around the end of the enclosed start box accompanied
by a refusal to enter the alley. It should be clear that
visual exposure to room cues was not critical to this latter
effect since the wall removed was made of Plexiglas. There
was a moderate change in illumination from the start box to
the alley because the start box had a roof; however, it was
the absence of the wall which elicited the most attention.
Given that rats in the wild run along open paths such as ropes
and wires, it seems likely that with more trials, an incentive
in the goal box, or reason to leave the start box, running
would have occurred in the present situation.

After its initial role in reducing exploration/caution, low
roof elicited its own set of competing responses, much like
the odor condition. Odor, perhaps because of its daily variation in amount and make-up, continued to elicit exploratory
behavior which competed with alley running even at asymptote.
The last variable to affect running was time since handling
(delay in the start box). These data, in conjunction with
those of Fowler (1963), seem to argue that an important
component of maze performance is the activation level of the
subject. (Charles Flaherty has shown me similar data for
rewarded animals.) However, there seems no reason to assume
that activation is the primary determinant of maze running, or
that it could explain unrewarded running of complex mazes.
Whether handling and deprivation represent a single drive or
activation dimension remains to be seen.

In sum, an examination of the sensory, apparatus, and procedural "surround" of unrewarded maze running revealed a number of complex determinants of alley running. Bright lights,
the transition from closed to partly open runways, immediate
removal from the goal box, delay in the start box, and the
presence of novel odors in the alley all inhibited to some
degree acquisition of alley running. The basis for this inhibition appeared to be either an increase in a caution or
fear state, the eliciting of specific exploration reactions
that competed with running, or a reduction in the activating
effects of handling. Alley running was facilitated, at least
initially, by darkness, recent handling, low roof, and a long
alley. This facilitation appeared to be based on instigating
exploration of the end box or alley, or on reducing, or

superceding, the initial exploration/caution state. Last, though there were several variables that affected asymptotic levels of responding, nonetheless, there was considerable similarity across most manipulations in the way alley running occurred and developed.

Section III. A Comparison of Rewarded and Unrewarded Running

To this point we have examined reinforcement explanations of unrewarded maze running and found them wanting. Further examination of the stimuli instigating, eliciting, and controlling maze running revealed a complex set of determinants of alley running related to exploration/caution, directed exploration, and some form of activation. The purpose of the experiments in the present section was to explore the relation between rewarded and unrewarded alley running.

If unrewarded alley running plays a large part in rewarded alley running, the variables that control and instigate it become important in understanding and modeling rewarded alley running. However, it may be that our procedures have produced a unique phenomenon that bears little relation to alley running as it is ordinarily studied. Or rewarded and unrewarded alley running may be determined by relatively independent sets of determinants. For example, the one is intrinsically motivated, whereas the other is extrinsically motivated. Experiment 9 determined how adding different amounts of reward to our present procedures affected alley performance. Experiment 10 compared rewarded and unrewarded animals under procedures more typical of runway studies to determine to what extent rewarded and unrewarded running resemble each other in this context.

Experiment 9

The present experiment determined the effect of 0, 1, or 10 pellets of reward presented in a small cup milled from 3/4 in. round aluminum stock and fastened to the floor of the alley 5 cm from the back wall. There were 10, 9, and 10 animals in the 0, 1, and 10 pellet groups, respectively. Except for the presence of reward, the groups were run under standard procedures. Figure 11 shows that all groups readily acquired the running response, decreasing start and run times, and increasing activity $Fs(8,245)$ = 35.7, 32.7, and 7.64, all $p<.01$. The groups also differed significantly on start and run times and units entered, $Fs(2,26)$ = 9.15, 4.62, 4.64, $ps<.01$ and .05, with the rewarded groups starting and running faster, and moving about less. The extinction results essentially mirrored those of acquisition.

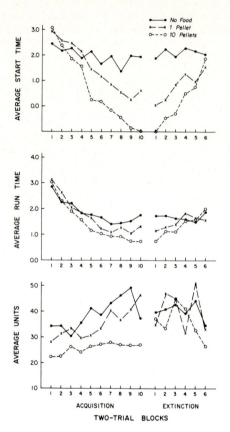

Figure 11. Average log$_e$ start times (top), run times (Middle), and units entered (bottom) over two trials blocks during acquisition and extinction for Group 1 pellet, 10 pellets, and 0 pellets (Experiment 9).

These results strongly suggest that rewarded running under the conditions of these studies has determinants very similar to those of unrewarded running. The run times, in particular, were only slightly different in absolute value and essential-ly parallel in their development between rewarded and unre-warded subjects. By far the greatest effect of reward was on the transition from the start box to the alley (the start time). In the absence of reward there was only a small and variable improvement in start times over trials. However, in the presence of reward there was a large effect on start times over trials that was highly related to the amount of reward.

The effects of reward on activity appeared to confirm previous speculations about the inhibitory effects of reward on subsequent responding (e.g., Catania, 1973) because the decrease in activity was greater than would be expected simply on the basis of the overt eating times.

Experiment 10

Despite the clear indication in Experiment 9 that unrewarded alley running forms the basis for rewarded running, it still might be argued that the results have little relation to the majority of straight alley studies because of differences in the procedure and the absence of an actual goal box. In fact, several studies in the straight alley literature that have used an unrewarded control group and more common procedures and apparatus have not shown acquisition of running in this group (e.g., Cox, 1976, though he reports only goal measures). The purpose of the present study was to compare rewarded and unrewarded running in an alley with a goal box, using the typical procedure of confining the animal for 30 sec in the goal box followed by removal to the home cage.

Two groups of 10 animals each were used, one rewarded with 10 pellets, the other unrewarded. Both groups were run for 20 acquisition and 12 extinction trials. The apparatus was modified to include a goal box by inserting a guillotine door 12 in. from the end of the runway. Figure 12 shows clearly that both rewarded and unrewarded animals decreased their start and run times over acquisition trials, $Fs(9,153) = 28.2$ and 18.8 $p<.01$. Rewarded animals also performed better, $Fs(1,17) = 11.9$ and 10.8, $ps<.01$. During extinction the rewarded animals increased their start and run times, $Fs(1,17) = 6.93$ and 7.96, $p<.01$. For both start and run times, the final level of performance did not differ between the rewarded and unrewarded animals. Tests of the simple main effects of reward on the last extinction trial (prompted by significant interactions) were insignificant, $Fs(1,17) = .54$ and 1.75, $ps>.10$. Activity was not measured in this study since the animals were confined and removed.

In brief, even using procedures typical of straight alley experiments, there is clear acquisition of running by the unrewarded animals. Further, there is no apparent reason to presume that reward markedly changes the set of variables determining alley running. Reward primarily seems to facilitate performance based on the primary appetitive structure shown in unrewarded running.

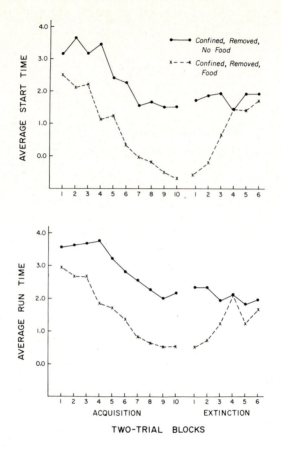

Figure 12. Average log$_e$ start times (top) and run times (bottom) over two trial blocks during acquisition and extinction for Groups No Food and Food (Experiment 10).

Discussion

In these experiments the presence of reward in the goal box had a large effect on start times, and a smaller effect on run times. Previous comparisons of rewarded and unrewarded animals showed an additional effect of reward on running rate in the very final segment of the alley, the goal speed (e.g., Logan, 1960. We did not take this measure, but the effect may have been due in part to a shorter alley, and to the fact that animals less often go directly to the end wall of the alley if no food is present near it. In Experiment 9 reward also appeared to inhibit activity to a degree greater than would be expected on the basis of eating time. In sum, food in the goal box influences running, but the primary basis of running appears already to be present and is expressed in the absence of food reward.

General Discussion

The Determinants of Unrewarded Straight Alley Running

The modern study of maze running has been far too restricted in its focus on overt and conditioned reward. The present experiments show that alley running is elicited, controlled, and maintained by variables other than overt or conditioned reward for completing the alley. The procedure of removing animals from the maze at a fixed time appeared successful in breaking any differential association between removal and a particular portion of the apparatus. The resultant random relation between location and removal makes it difficult to argue that improvement in speed of traversing the alley was due to any conditioned or unconditioned attribute of handling or removal.

Unrewarded alley running appeared related to four major factors, a caution or fear state (exploration/caution), directed exploration, the activation or drive level of the animal, and the expectation of the experimenter. The acquisition of alley running appeared most related to the decrease in strength of an exploration/caution state in the animal. In our studies this state was expressed primarily by cautious sniffing of the apparatus, and slow movements with body extension. The high initial levels of specific sniffing shown in Experiment 4 did not appear to compete directly with running once it began, but rather reflected the existence of a state in which anything but slow movement was unlikely to occur. In completely ungentled animals introduced to the alley we find a stronger fear state with very little exploration and a great deal of freezing or startled dashing. This state usually eventually passes into the exploration/caution state as the animals begin to acquire alley running.

Variables that increase caution or fear, such as bright lights, quick removal from the end of the alley, transition from enclosed to open environment, and loud sounds (unpublished research) all inhibit or prevent acquisition of alley running. Variables that decrease fear or caution, such as darkness, increased number of trials in a day, tranquilizers (Barry, Wagner and Miller, 1962), and extensive pretraining in the apparatus (Bower, 1959), increase speed of acquisition.

Directed exploration had a complex effect on straight alley
running. Stimuli worthy of exploration or other forms of
attention (food) located at the end of the alley facilitated
acquisition of alley running. Stimuli worthy of exploration
in the start box (Timberlake and Birch, 1967), or in the alley
itself (novel odors and maybe low roof), inhibited acquisition
and maintenance of alley running. Thus, directed exploratory
behavior facilitated or inhibited running depending on whether
its expression was compatible with moving rapidly down the
alley.

The third major factor related to unrewarded alley running was
the drive or activation level of the animal. The first two
experiments showed that drive in the form of conditioned
anticipation of feeding or loss of body-weight facilitated
alley performance primarily through an increase in the probability of completing the alley. Logan (1960) also reported
an effect of drive on unrewarded alley performance. These
findings suggest a role for drive in the acquired organization
of alley running that does not depend on mediation by reward.
Animals simply exposed to the alley under different drive
states acquired somewhat different behaviors, and modified
their behavior over trials differently (see Experiments 1, 2,
and 4). Further, this acquisition appeared to represent a
relatively permanent change in the animals behavior in the
maze environment independent of drive state. Animals that
acquired maze running under a typical deprivation schedule on
which they were fed after running showed greater levels of
performance than the ad lib. group even when tested later
under the same ad lib conditions (unpublished research).

Drive or activation also was implicated as a determinant of
unrewarded performance by the discovery in Experiment 8 that
animals confined to the start box for 4 minutes before the
beginning of the trial tended to run more slowly and were less
active than control animals. This finding suggested that
activation produced by handling the animal is an important
factor in straight alley performance. It was not a necessary
condition for improvement, but it definitely had an effect on
the amount of improvement and activity. Compatible results
were reported by Fowler (1963).

The fourth determinant of runway behavior is related to the
expectations of the experimenters that unrewarded running
will occur. Let me hasten to add that experimenter expectations were not critical in our obtaining the basic effect of
unrewarded running. As noted in the methods section, the
initial experiments were performed by several naive experimenters who had no clear idea of what it expect. However,

once unrewarded running was solidly established we modified
our procedures to encourage the animals to run in much the same
way other experimenters have encouraged food-rewarded animals
to run. In the second and third set of experiments we took
animals that refused to start and placed them in the alley
for several minutes following the third and fourth trials.
Other researchers have placed recalcitrant subjects in or outside the goal box and allowed them to eat, prodded them down
the alley, or removed them from the study entirely. Our more
modest procedures were successful in increasing the number of
subjects that completed the alley. It is clear that experimenters who do not expect unrewarded alley running could inhibit its development simply by using some combination of short
elevated alleys with enclosed start boxes, immediate removal
from the goal box, long delay in the start box, bright lights,
the presence of odor cues in the runway, and possibly more
subtle cues, (Rosenthal, 1966).

In sum, straight alley running is affected by reduction of an
initial exploration/caution state, by the nature and location
of the cues eliciting directed exploration and movement, by
the level of activation of the animal, and by the expectation
of the experimenter, expressed in arranging conditions appropriate to its emergence. However, none of these determinants
quite captures the essential basis of straight alley running.
What is missing is a clear picture of the appetitive structure
of response patterns and stimulus sensitivities that underlies
alley running and provides the organization which allows the
experimenter's manipulations to have their effect.

The evidence presented here suggests that the appetitive structure underlying alley running is elicited by vertical stimuli
and long clear paths, and is probably sensitized in a general
way by motivational variables related to foraging or escape.
The repeated elicitation of this locomotor structure in the
straight alley results in the development of the integrated
sequence of responses and stimuli that we call alley running.
Its expression in behavior is affected by cues controlling
caution, directed exploration and movement, activation, and
probably memories or hints of food in particular settings.
The results do not mean there is a basic alley running drive,
or that alley running is intrinsically reinforcing. They do
imply that there is a unique appetitive structure that shapes
the stimulus sensitivities, behavioral expression, and
learned relations of alley running.

The Role of Reward in Alley Running

What then is the role of overt and conditioned reward in alley
running? Comparison of experiments in the first two sections
of this paper with results obtained in previous studies using
food reward was confounded by procedural differences in removing subjects from the alley. It might have been argued
that these differences made our results largely irrelevant to
the study of rewarded maze running. However, the experiments
in the third section of the paper suggested strongly that rewarded running is not fundamentally different from unrewarded
running in its determinants or development. The major effect
of reward in these studies was to increase the consistency
with which an animal completed the alley and to influence
markedly the speed with which it began the trip down the alley.

Other studies using both rewarded and unrewarded animals run
in straight alleys have also found basic similarities and
some differences in performance. Logan (1960) and Spear and
Spitzner (1967) found that reward decreased run times. Wong
and Traupman (1973) showed that reward increased the effect
of deprivation on running. Mote and Finger (1942) reported
that reward increased the persistence of running over multiple
daily trials. Logan (1960) and Wong (1971) found that an
unrewarded group improved only start and run speeds, not goal
speeds. However, this was most likely due to the failure of
unrewarded animals to rapidly cover the last small distance to
the end wall of the alley, something they had no strong reason
to do. Other effects have been noted, such as Weinstock's
(1958) observation that unrewarded pretraining appeared to
reduce the subsequent effects of intermittent reward on runway behavior, and Logan's (1960) report that delay in the goal
box had no significant effect on unrewarded running. On the
basis of these studies the clearest effect of reward is to
make the animal more efficient in traveling from the start box
to the end of the goal box, and to sustain running under circumstances that might ordinarily inhibit it.

The present results strongly suggest that models used to predict acquisition of alley running need to consider the contribution of unrewarded running to that performance. The effect
of deprivation on unrewarded running and the effect of acquisition drive on performance under ad lib test conditions both
raise questions about the importance of incentives in mediating the effects of deprivation on acquisition (Bolles, 1967,
Ch. 12). The effect of deprivation in increasing the probability of completing the alley may be related to the report
by Cotton (1953) that deprivation reduced the tendency of rats
to interrupt running in the alley.

Learning and Performance

The present results when coupled with the notion of appetitive
structure raise some classic questions about the necessary
conditions for learning, and the distinction between learning and performance. Traditional theories have separated
learning and performance, considering the former as an abstract
change in cognitions or peripheral stimulus-response connections, and the latter as the result of energizing this structural change. Within this framework, considerable argument
has been addressed to whether the structural change can occur
in the absence of reward.

In the present view there is no question that learning can
occur in the absence of reward, so long as reward is defined
as overt or conditioned reward associated with the goal box.
There is some question, though, if we assume (as many appear
to) that lasting performance change cannot occur without reward. It could then be argued that since performance change
occurred in the present experiments, reward must have been
present. However, this approach is neither satisfying nor
productive. Assumptions of intrinsic or hidden rewards do not
lead to studies clarifying the complex elicitation and control
of behaviors such as alley running, nor do they conceptualize
the contribution of the organism in a way that encourages contact with its selection environment and genetic make-up. We
are left instead to investigate a poorly defined general
ability to profit from and seek whatever constitutes reward.
To me, it is much preferable to begin with a less encompass-
ing mechanism whose attributes can be analyzed by careful
manipulation of the complex set of cues controlling behavior.

What then should we make of the distinction between learning
and performance? First, it seems clear that reward is not
necessary to produce either learning or performance. Performance, like learning, is simply a product of the interaction of an appetitive structure of the animal with the appropriate stimulus conditions set by the evolutionary context.
An appetitive structure, though usually evolved to increase
the chances of obtaining a reward, does not always require that
reward to produce learning or performance.

Second, within the present view, the distinction between structure and motivation is often arbitrary. Appetitive structures
contain elements of both organization and motivation. Rarely
do animals learn an abstract copy of the environment, and then
use it as they are energized by a current motivational variable. More often the current expression of appetitive structure depends on the circumstances under which it was developed.

In the present studies the nature of performance differed as a
function of the variables that were present during learning.
Learning to run an alley for food reward, like learning to run
in the presence of drive states or exploratory incentives,
means learning something different than would be learned in
the absence of these conditions. It is the delineation of
the complexities of such learning that should be our goal,
not the development of an artifically clear distinction between
learning and performance.

The present results also raise important questions about the
nature of latent learning. Most theorists have treated latent
learning as a form of unexpressed perceptual or cognitive
learning (e.g., Thistlewaite, 1951). Though such perceptual
learning may be important, the present approach suggests,
first, that latent learning effects may not be so latent, and
second, that perceptual learning has limitations and performance tendencies built into it by the nature of the appetitive
structure within which it occurs. In fact, as indicated in
the introduction, several experimenters have found evidence
of acquisition in groups run unrewarded in complex mazes.

Blodgett's (1929) classic study of maze performance found a
large drop in errors in Day 7 for the group shifted to reward
on Day 7, hardly what one would expect unless the drop were
ready to occur in the absence of food reward. Tolman and
Honzik (1930) found that an unrewarded group fed three hours
after running (and thus fairly insulated from the effects of
conditioned reward) decreased their errors and run times over
trials in a 14 unit multiple T-maze (see also Anderson, 1941).

Brant and Kavanau (1965) found that field mice living in an
environment containing several complex mazes, each approximately 75 sq ft in area, showed dramatic drops in run times
over fewer than 10 self administered trials. Finally, Battig
and Schlatter (1979) found clear improvement in the efficiency
with which rats and mice patrolled complex mazes containing no
reward. This improvement over trials were not related to a
concurrent measure of locomotor activity in the maze. Though
it can always be argued that there are latent or secondary
sources of reinforcement in such studies, the weight of the
evidence suggests more than a reasonable doubt that maze learning and maze performance require any reward at all.

The Nature of Unrewarded Running and Appetitive Structure

The present experiments have been concerned with the verifi-
cation and analysis of a locomotor appetitive structure
evolved to get rats around safely, rapidly, and productively

in their selection environment. In a sense this structure
is a general purpose sub-routine that can be called by a number
of appetitive programs and functional systems. The structure
is engaged by vertical stimuli and long clear paths, and its
expression in behavior is related to a variety of determinants
including caution, directed exploration and movement, and
activation. The straight alley is a nearly ideal environment
for eliciting this structure. The only better environment
might be the circular mazes popular with behavioral pharma-
cologists that present unending vertical walls and clear
paths. As is commonly known, these mazes are very effective
in eliciting movement.

The locomotor appetitive structure underlying alley running
should be elicited by any environment that remotely resembles
the critical features of a maze. Over "trials" in this
environment running will emerge, released and directed by
alley-like cues, and integrated by experience with the circumstances (for a somewhat similar conception of how bill-pecking
emerges in gulls, see Hailman, 1967). To explain a rat's
ability to find its way rapidly about an environment after a
brief exposure, we have only to presume that a salient aspect
of locomotor structure is the remembrance of and tendency to
choose the longest uninterrupted path; this is, of course, the
path to food in most complex maze studies, and often the path
to safety in natural environments.

From an evolutionary perspective, the ability of the rat to
increase efficiency of movement in maze-like environments
without overt or conditioned reward is not surprising. A rat
capable of learning and performing rapid movements in particular environments is more likely to survive to reproduce than
a rat that performs only after receiving appropriate incentives
at particular locations. The former animal will forage more
efficiently for food in a given time of exposure outside the
burrow, return to the burrow more rapidly and by more paths
(if pursued by a predator), and escape the burrow more
efficiently if pursued by a predator underground. An animal
that learns the environment at a cognitive or "latent" level,
but makes all its performance adjustments only after it has
found food or avoided a predator a few times will find itself
at an evolutionary disadvantage.

The presence of an evolved appetitive structure suitable for
maze running does not mean that rats will be unable to learn
to move about in locations without walls or paths. However,
the speed of acquisition and performance typical of maze running should be reduced. Further, I would expect animals,
whenever possible, to choose a trail through a given environment based on the presence of vertical stimuli and clear paths.

Given this sort of flexibility, the appetitive structure
studied in maze learning should serve rats in many contexts.

The appetitive structure approach outlined here has potential
relevance for all learning. At the least it suggests that the
organization the subject brings to the situation is an important determinant of what is learned. At best, it argues
that all learning in artificial situations, no matter how
arbitrary it seems, is based on underlying appetitive structures. The manipulations of the experimenter merely set the
stage for the expression of these structures. Whatever the
correct conception, the present approach raises a number of
questions concerning the nature of learning that must be faced
by those desiring a general theory.

The widespread use of reinforcement concepts to account for
behavior developed as an alternative to careless explanations
couched in terms of instincts. The abandonment of instinct
as an explanatory device was easy because the concept of
instinct was largely an unanalyzed assumption. However, now
may be an appropriate time to look more carefully at the careless use of reward to explain behavior. In a sense we have
simply switched from explaining behavior in terms of inadequately analyzed causal constructs located in the organism to
inadequately analyzed causal constructs located in the experimenter's procedures. Alley running was for many years a key
instrumental response in the development of general, context-
free theories of learning. The present results suggest we did
not adequately understand it. Reward in the goal box of an
alley is not the only, or even the primary, determinant of
alley running in rats.

Footnotes

Supported in part by NSF Grant 79 11159. The first study is
based on a paper by William Timberlake and Pat House read at
the meetings of the Midwestern Psychological Association,
Detroit, 1971. Experiments 6, 7, and 8 were presented in a
paper by William Timberlake at the meetings of the Animal
Behavior Society, Champaign-Urbana, Illinois, 1974. Thanks
to Sue Baird, James Deich, Joe Head, Kirk Heimlich, Pat House,
Dale Johnson, Kevin King, Katy O'Shea, Laryn Peterson, Donna
Washburne and Mark Wozny for their assistance in data collection and analysis, and to Gary Lucas for his critical comments.
I also thank the reviewers and editors who indirectly provided the opportunity to put this work together in a single
package.

References

1 Anderson, E. E. The externalization of drive. III. Maze learning by non-rewarded and by satiated rats. Journal of Genetic Psychology, 1941, 59, 397-426.

2 Baerends, G. P. On drive, conflict, and instinct, and the functional organization of behavior. In M. A. Corner & D. F. Swaab (Eds.), Perspectives in Brain Research, 1976, 45, 427-447.

3 Barnett, S. A. The rat: A study in behavior. Chicago: University of Chicago Press, 1975.

4 Barry, H., III., Wagner, A. R. & Miller, N. E. Effects of alcohol and amobarbital on performance inhibited by experimental extinction. Journal of Comparative and Physiological Psychology, 1962, 55, 464-468.

5 Battig, K. & Schlatter, J. Effects of sex and strain on exploratory locomotion and development of nonreinforced maze patrolling. Animal Learning & Behavior, 1979, 7, 99-105.

6 Berlyne, D. E. Conflict, arousal, and curiosity. New York: McGraw-Hill, 1960.

7 Black, R. W., Fowler, R. L. & Kimbrell, G. Adaptation and habituation of heart rate to handling in the rat. Journal of Comparative and Physiological Psychology, 1964, 57, 422-425.

8 Blodgett, H. C. The effects of the introduction of reward upon the maze performance of rats. University of California Publications in Psychology, 1929, 4, 113-134.

9 Bolles, R. C. Theory of Motivation. New York: Harper & Row, 1957.

10 Bolles, R. C. Species-specific defense reactions and avoidance learning. Psychological Review, 1970, 71, 32-48.

11 Bower, G. H. Correlated delay of reinforcement. Unpublished doctoral dissertation, Yale University, 1959.

12 Brant, D. H. & Kavanau, J. L. "Unrewarded" exploration and learning of complex mazes by wild and domestic mice. Nature, 1965, 204, 267-269.

13 Calhoun, J. B. The ecology and sociology of the norway
    rat. Bethesda, Maryland: U. S. Department of Health,
    Education, and Welfare, 1962, Publication 1008.

14 Catania, A. C. Self-inhibiting effects of reinforcement.
    Journal of the Experimental Analysis of Behavior, 1973,
    19, 517-526.

15 Chapman, R. M. & Levy, N. Hunger drive and reinforcing
    effect of novel stimuli. Journal of Comparative and
    Physiological Psychology, 1957, 50, 233-238.

16 Clark. J. & Miller, S. B. The development of rapid running
    in T-Mazes in the absence of obvious rewards. Psychonomic
    Science, 1966, 4, 127-218.

17 Collier, G. Body weight loss as a measure of motivation
    in hunger and thirst. Annals of the New York Academcy
    of Sciences, 1969, 157, 597-609.

18 Cotton, J. W. Running time as a function of amount of food
    deprivation. Journal of Experimental Psychology, 1953,
    46, 188-198.

19 Cox, W. M. Eight drive-reward combinations: A test of
    incentive-motivational theory. Bulletin of Psychonomic
    Society, 1976, 7, 121-124.

20 Daly, J. B. Learning of a hurdle-jump response to escape
    cues paired with reduced reward or frustrative non-reward.
    Journal of Experimental Psychology, 1969, 79, 146-157.

21 Ewer, R. F. The biology and behavior of a free-living
    population of black rats (Rattus rattus). Animal Behavior
    Monographs, 1971, 4(3).

22 Finger, F. W., Reid, L. S. & Weasner, M. H. Activity
    changes as a function of reinforcement under low drive.
    Journal of Comparative and Physiological Psychology,
    1960, 53, 385-387.

23 Fowler, H. Exploration motivation and animal handling:
    The effect on runway performance of start-box exposure
    time. Journal of Comparative and Physiological Psychology,
    1963, 56, 866-871.

24 Fowler, H. Satiation and curiosity. In G. H. Bower (Ed.),
    The psychology of learning and motivation (Vol. 1). New
    York: Academic Press, 1967.

25 Furchgott, E., Wechkin, S. & Dees, J. W. Open field exploration as a function of age. Journal of Comparative and Physiological Psychology, 1961, 54, 386-388.

26 Galef, B. G., Jr. & Clark, M. M. Mother's milk and adult presence: Two factors determining initial dietary selection by weanling rats. Journal of Comparative and Physiological Psychology, 1972, 78, 220-228.

27 Glanzer, M. Changes and interrelations in exploratory behavior. Journal of Comparative and Physiological Psychology, 1961, 54, 433-438.

28 Goodrick, C. L. Activity and exploration as a function of age and deprivation. Journal of General Psychology, 1966, 108, 239-252.

29 Hailman, J. P. The ontogeny of an instinct. Behaviour Supplements, 1967, 15.

30 Hull, C. L. The rat's speed of locomotion gradient in the approach to food. Journal of Comparative Psychology, 1934, 17, 393-422.

31 Jennings, H. S. Behavior of the lower organisms. New York: Columbia University Press, 1906.

32 Johnston, T. D. Contrasting approaches to a theory of learning. The Behavioral and Brain Sciences, 1981, 4, 125-139.

33 Kimball, R. C., Kimball, L. T. & Weaver, H. E. Latent learning as a function of the number of differential cues. Journal of Comparative and Physiological Psychology, 1953, 24, 274-280.

34 King, D. L. & Applebaum, J. R. Effect of trials on emotionality behavior of the rat and mouse. Journal of Comparative and Physiological Psychology, 1973, 85, 186-194.

35 Lester, D. The relationship between fear and exploration in rats. Psychonomic Science, 1969, 14, 128-130.

36 Ligon, E. A comparative study of certain incentives in the learning of the white rat. Comparative Psychology Monographs, 1929, 6, No. 28.

37 Logan, F. A.   Incentive.   New Haven: Yale University
    Press, 1960.

38 Mackintosh, N. J.   The psychology of animal learning.
    London: Academic Press, 1974.

39 Montgomery, K. C.   The relation between fear produced by
    novel stimulation and exploratory behavior.   Journal of
    Comparative and Physiological Psychology, 1955, 48, 254-
    260.

40 Morgan, C. Lloyd.   Habit and Instinct.   London: Arnold,
    1896.

41 Mote, F. A. & Finger, F. W.   Exploratory drive and second-
    ary reinforcement in the acquisition and extinction of a
    single running response.   Journal of Experimental Psycho-
    logy, 1942, 31, 57-68.

42 Muller-Schwarze, D.   Ludic behavior in young mammals.   In
    M. B. Sterman, D. J. Mcginty & A. M. Adinolfi (Eds.),
    Brain Development and Behavior.   New York: Academic Press,
    1971.

43 Munn, N. L.   Handbook of psychological research on the rat.
    Boston: Houghton-Mifflin, 1950.

44 Olton, D. S.   Characteristics of spatial memory.   In S. H.
    Hulse, H. Fowler & W. K. Honig (Eds.), Cognitive processes
    in animal behavior.   Hillsdale, New Jersey: Lawrence
    Erlbaum, 1978.

45 Olton, D. S. & Samuelson, R. J.   Remembrance of places
    passed: Spatial memory in rats.   Journal of Experimental
    Psychology: Animal Behavior Processes, 1976, 2, 97-116.

46 Paul, L.   High performance obtained with no reward.
    Psychonomic Science, 1969, 14, 224-225.

47 Paul, L. & Calabrese, F.   The development of self-rein-
    forcing running in satiated rats.   Animal Learning &
    Behavior, 1973, 1, 268-272.

48 Rosenthal, R.   Experimenter effects in behavioral research.
    New York: Appleton-Century-Crofts, 1966.

49 Rozin P. & Kalat, J.   Specific hungers and poison avoidance
    as adaptive specializations of learning.   Psychological
    Review, 1971, 78, 459-486.

50 Schneider, G. E. & Gross, C. G. Curiosity in the hamster. Journal of Comparative and Physiological Psychology, 1965, 59, 150-152.

51 Seligman, M. E. P. On the generality of the laws of learning. Psychological Review, 1970, 77, 406-418.

52 Shettleworth, S. J. Constraints on learning. In D. S. Lehrman, R. A. Hinde & E. Shaw (Eds.), Advances in the study of behavior (Vol.4). New York: Academic Press, 1972.

53 Simmons, R. The relative effectiveness of certain incentives in animal learning. Comparative Psychology Monographs, 1924, 7.

54 Skinner, B. F. Science and human behavior. New York: Macmillan, 1953.

55 Small, W. S. An experimental study of the mental processes of the rat. American Journal of Psychology, 1900, 11, 133-165.

56 Spear, N. E. & Spitzner, J. H. Effect of initial non-rewarded trials: Factors responsible for increased resistance to extinction. Journal of Experimental Psychology, 1967, 74, 525-537.

57 Sperling, S. E. & Valle, F. P. Handling gentling as a positive secondary reinforcer. Journal of Experimental Psychology, 1964, 67, 573-576.

58 Steiniger, F. von. Beitrage zur soziopolgie und sonstigen biologie der wanderratte. Zeitschrift fur Tierpsychologie, 1950, 7, 356-379.

59 Telle, H. J. Beitrage zur kenntnis der verhaltensweise von ratten, vergleichend dargestellt bei, Rattus-norvegicus und Rattus rattus. Zeitschrift fur Angewandte Zoologie, 1966, 48, 97-129.

60 Thistlewaite, D. A critical review of latent learning and related experiments. Psychological Bulletin, 1951, 48, 97-129.

61 Timberlake, W. A molar equilibrium theory of learned performance. In G. H. Bower (Ed.), The Psychology of Learning and Motivation (Vol. 14). New York: Academic Press, 1980.

62 Timberlake. W. The functional organization of appetitive
    behavior: Behavior systems and learning. In M. D. Zeiler
    & P. Harzem (Eds.), Advances in analysis of behavior
    (Vol. 3): Biological factors in learning. Chichester:
    Wiley, In press.

63 Timberlake, W. & Birch, D. Complexity, novelty, and food
    deprivation as determinants of speed of shift of behavior.
    Journal of Comparative and Physiological Psychology, 1967,
    63, 545-548.

64 Timberlake, W., Wahl, G. & King, D. Stimulus and response
    contingencies in the misbehavior of rats. Journal of
    Experimental Psychology, 1982, 8, 62-85.

65 Tinbergen, N. The study of instinct. New York: Oxford
    University Press, 1951.

66 Tolman, E. C. Purposive behavior in animals and men.
    New York: Appleton-Century-Crofts, 1932.

67 Tolman, E. C. & Honzik, C. H. Introduction and removal
    of reward and maze performance in rats. University of
    California Publications in Psychology, 1930, 4, 257-275.

68 Valle, F. P. Effects of strain, sex, and illumination on
    open field behavior of rats. American Journal of Psycho-
    logy, 1970, 83, 103-111.

69 Valle, F. P. Rats' performance on repeated tests in the
    open field as a function of age. Psychonomic Science,
    1971, 23, 333-335.

70 Watson, J. B. Kinaesthetic and organic sensations: Their
    role in the reactions of the white rat. Psychological
    Review Monographs, 1907, 8(2).

71 Williams, D. I. Maze exploration in the rat under dif-
    ferent levels of illumination. Animal Behavior, 1971,
    19, 361-363.

72 Weinstock. S. Acquisition and extinction of a partially
    reinforced running response at a 24-hour intertrial inter-
    val. Journal of Experimental Psychology, 1958, 56, 151-
    158.

73 Welker, W. I. An analysis of exploratory and play behavior
    in animals. In D. W. Fiske & S. R. Maddi (Eds.), Functions
    of varied experience. Homewood, Illinois: Dorsey Press,
    1961.

74 Wong, P. T. P. Coerced approach to shock, punishment of competing responses, and resistance to extinction in the rat. Journal of Comparative and Physiological Psychology, 1971, 76, 275-281.

75 Wong, P. T. P. & Traupmann, K. L. Extra-acquisitional factors in extinction: The effect of feeding activities of neighboring rats. Psychonomic Science, 1971, 23, 359-360.

76 Wong, P. T. P. & Traupmann, K. L. Residual effects of food deprivation on food consumption and runway performance in the satiated rat. Journal of Comparative and Physiological Psychology, 1973, 84, 345-352.

ANIMAL COGNITION AND BEHAVIOR
Roger L. Mellgren, editor
© *North-Holland Publishing Company, 1983*

MAZES, SKINNER BOXES AND FEEDING BEHAVIOR

Roger L. Mellgren and Mark W. Olson

University of Oklahoma and Coe College

Historically, two major methods and pieces of apparatus have dominated the study of appetitive behavior in animal psychology. The "Skinner Box", or operant conditioning apparatus is one method-apparatus and the maze, in the simpliest case the straight alley and in more complex (at least in terms of constructing the apparatus) cases the T-maze and the radial-arm maze, is the other. In this chapter we will argue that the different responses required by each piece of apparatus may be affected in different ways by rewarding events due to each responses different biological function.

Some researchers have used only operant conditioning techniques, others have used only the maze procedure throughout their published career. Adherence to one or the other procedure is governed by the same phenomenon the researchers study in their experiments; the law of effect. Thus studies evaluating the persistence of behavior in the form of resistance to extinction predominantly utilize the maze procedure, but studies evaluating stimulus generalization predominantly utilize the operant procedure. In each case the particular procedure results in orderly data, encouraging the further use of the particular procedure to evaluate the particular behavioral phenomenon. Occasionally a behavioral phenomenon is studied using both procedures. Positive and negative contrast effects seem to be one example, although reviewers claim it is difficult to know whether such effects represent the same psychological process across the two procedures, or not (e.g., Dunham, 1968).

Not only is there a segregation of behavioral phenomena into different methodological procedures, there is also an apparent segregation of the importance of the explanation of these phenomena. Investigators utilizing the operant methodology often act as if maze data and theory did not exist and vice versa. This division of thought reflects the fact that it is more than a piece of apparatus which defines these approaches and makes them different. One purpose of this chapter is to explore the nature of the differences and to provide the beginnings of a synthesis of the approaches. Some of these differences are listed in Table 1, with the most important points of comparison being those numbered 3, 4 and 5 for the present discussion.

Table 1

Points of Comparison Between Operant and Maze Procedures

Operant	Maze
1. Single subjects, long observation time per session	1. Groups of Subjects, short observation time per session
2. Averaged individual data with criteria for stability of data	2. Averaged group data with statistical analysis
3. Continuous trials in one location	3. Discrete trials involving spatial locomotion
4. Subject initiated trials	4. Experimenter enabled trials
5. Physical contact with manipulandum as the response	5. Movement through space as the response
6. Steady-state behavior	6. Transitional behavior
7. Descriptive, atheoretical approach non-associative accounts of responding	7. Theoretical, hypothesis testing approach mechanistic, associative theory
8. Pigeon as predominant subject, rats secondarily	8. Rats as predominant subject
9. Complex schedules of rewards	9. Simple schedules of rewards
10. Automated procedures, little or no human contact with subject during experiment	10. Semi-automated, human contact to start and finish a trial

The attempt to bring together these two approaches is not novel. For example, Logan and Ferraro (1970), Platt (1971) and Shimp (1975) have discussed this issue. The most straightforward way of attempting a synthesis is to make the analogy of one step in a runway to one press of the lever in the Skinner box. This brings about a kind of response equivalence and it then becomes an empirical question to evaluate whether responses using the two procedures are analogous to each other as a function of independent variable manipulations. There are numerous problems with such an approach, not the least of which is that it is awkward to use time based schedules in a maze. For example, on a fixed interval 30 sec. schedule the rat might be placed in the maze and the first step after 30 sec. had elapsed would result in a food pellet being delivered. The practical difficulties of actually doing this are obvious.

Perhaps differences between operant and maze approaches are insignificant and with suitable cleverness empirical phenomena and theoretical propositions can be translated from one to the other. Many of the current textbooks on animal learning and motivation would suggest this to be the case by the lack of discussion of the topic. Tarpy (1982) discusses the fact that the traditional operant approach is a "black box" approach and is anti-theoretical as opposed to a "mechanistic strategy" where the black box is opened and its contents evaluated (pp. 11-13). This distinction is one aspect of the operant-maze differences, but not a very important aspect for the current discussion. Hulse, Egeth and Deese (1980) are silent on the topic. Bitterman, LoLordo, Overmier and Rashotte (1979) mention the fact that discrete trials are different from a free operant procedure (p. 13), and latter suggest that discrete trials encourage an associative analysis of learning and the operant approach does not. They note that, "there is a growing interest in examining the relation between the determinants of performance in the trials (maze) and free-responding (operant) procedures" (p. 130). Domjan and Burkhard (1982) have a fairly extensive consideration of the differences and similarities of the two approaches and conclude that some questions are more easily answered using one procedure and others are more easily answered using the other and they will discuss both types (pp. 127-131). Hill (1981) in commenting on operant and maze procedures says "the fact that the reinforced response does not end the trial makes it (Skinner box) somewhat different from the case of the rat in the runway" (p. 22). But subsequently Hill states that, "for the most part, what we have already said about instrumental learning in which a single reinforcer is applied after each trial applies also to instrumental learning without separate trials" (p. 23). Fantino and Logan (1979) discuss the philosophical (theory or no theory) and practical differences (e.g., individual subjects vs. groups of subjects) of the two approaches, but are not concerned about

generalizing between them since, "Having almost exhausted the
other views, experimental psychology turned to the more purely
empirical approach of Skinner, and operant conditioning (also
termed the experimental analysis of behavior) became the pre-
dominant approach to the study of behavior change" (p. 37).
Maze data may be ignored. Mackintosh (1974) shows greater concern for the differences between the two approaches and states
that, "the same procedures may produce different outcomes in
different types of apparatus and ... we are no less ignorant
about the causes of these differences" (p. 144).

From this nonrandom survey of textbooks it can be seen that all
manner of opinion exists concerning the appropriateness of even
asking the question of whether operant and maze approaches provide a picture of the organism which is generalizable or not
across the approaches, and, if that question is appropriate
to pose, there is no agreed-upon answer to it.

What form would an experiment take which would be relevant to
answering questions about the interprocedural generalizability
of operant and maze data? First, it is necessary to have behavior measured on a commensurate scale. There has been much
chest-pounding and flag-waving over the supposed superiority
of rate of response over other possible measures as a dependent
variable. In fact, operant and maze researchers both use a
rate measure as a matter of common practice. In the operant
situation the dependent variable, rate of response, R, is
calculated by the formula

$$R = \frac{n}{t} \qquad (1)$$

where n = number of responses and t = time allowed for responding. In the maze situation, rate of response, R, is calculated
by the formula

$$R = \frac{d}{t} \qquad (2)$$

where d = distance traveled and t = time taken to travel d.
Formal identity between equations 1 and 2 could be achieved by
determining the distance of one rat-step and using this length
as our standard measure of distance. Under this condition,
number of steps is equal to distance, or n=d. The purist would
certainly be unhappy with this arrangement and would be quick
to point out that a rat-step will vary in distance as function
of the speed of the rat at the time of measurement. Of course
formal identity is not really needed for purposes of the type

comparison involved in the interprocedural generalizability
question. The simple fact that both procedures measure rate of
response is sufficient to allow comparisons between them.

Figure 1 illustrates the kind of experimental design and results
which are relevant to assessing operant-maze generalizability.
Panel A of Figure 1 shows a situation where the independent

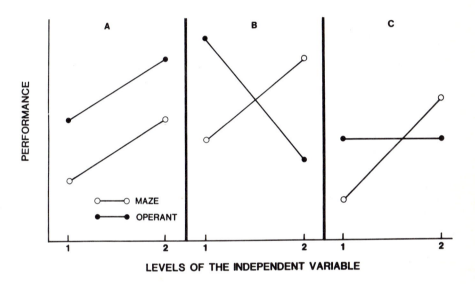

Figure 1. Possible outcomes of an experiment utilizing
    both maze and operant responses (see text for
    discussion).

variable has similar effects for both operant and maze performance and supports a position that the two are generalizable.
Panels B and C show situations (not the only ones) where generalization between one procedure and the other is not appropriate. In Panel B operant performance decreases by moving from
level 1 to level 2 of the independent variable, but maze performance increases in rate. In Panel C the independent variable had no effect on operant performance, but did have an effect on maze performance.

Manipulation of an independent variable in an analogous fashion
in the operant and maze situation should provide a strategy for
evaluation of the similarities of differences between each
procedure. One of the problems with a simple application of
the strategy summarized in Figure 1 is that it is not always

clear that the same independent variable is operative in operant
and maze studies. For example, suppose the independent variable
of interest is delay of reinforcement. The rat presses the
lever 10 times (FR-10) and 5 sec. latter is given a food pellet.
The analogy in a maze is to allow the rat to run to the goal
box, but require him to wait 5 sec. before the food pellet is
presented. The rat in the operant box might press the lever a
number of times after the requirement for food delivery had been
met. Thus the separation between response and food pellet may
be less than 5 sec. and a comparison of behavioral effects between the operant and maze situation is contaminated by the
"extra" lever presses occurring during the delay interval. Perhaps removing the lever for the 5 sec. delay when the response
requirement is met could prevent such confounding. The removal
of the lever, being a discrete change in stimulus conditions,
might come to function as a conditioned reinforcer--something
which the maze-rat does not experience in an analogous manner.
The point is not to decide how this experiment should be done,
but to emphasize the difficulty of translating a fairly straightforward experimental strategy into an unambiguous experimental
design.

One way of removing such ambiguity is to embed both maze and
operant characteristics into a single situation. If both maze
and operant behaviors (or components of them) are required to
produce a given outcome for the subjects, then differences in
the patterns of behavior as a function of different outcomes
must be due to the difference in the nature of maze and operant
behaviors. With regard to Figure 1, if the same animal experiences level 1 of the independent variable (a particular outcome)
but is required to engage in both operant and maze behavior to
get the outcome and another subject must also engage itself in
those same behaviors to get a different outcome (level 2 of the
independent variable), then results such as those shown in
Panels B and C of Figure 1 must be caused by different processes
operating on operant and maze behaviors. Of course if the subject is required to engage in both types of responding they will
form some sort of chain and therefore the properties of the
chain itself will need to be considered.

Foraging Analogy

Psychologists are becoming increasingly aware of an area of
inquiry which closely parallels and often intersects traditional
animal psychology, called Behavioral Ecology (e.g., Krebs and
Davies, 1978). Of particular interest is the part of behavioral
ecology known as foraging behavior (e.g., Kamil and Sargent,
1981) and theories of feeding strategies and optimal foraging

theory (e.g., Schoener, 1971). Several of the chapters in this volume reinforce the increased awareness of ecological theories and variables by psycologists.

There are several different types of predators discussed by ecologists such as the "sit and wait" type whose foraging consists of watching for food, monitoring water, territorial invaders, and potential predators on itself all at the same time; filter feeders who actively forage for uniformly distributed prey; and active foragers who encounter prey in discrete clumps or patches, to name a few. Perhaps the most commonly discussed predator type is the active forager in a patchy environment. This is due to the fact that such a pattern of predation covers a wide variety of species. Examples abound. Many birds, for example, fly to a beach or wet area and then walk on the ground pecking for worms or insects. Ants march out in well-defined columns to a food source where they harvest a bit of food and transport it back to their home. A foraging bear plunges through woods and thickets, encountering a berry bush where it then gorges itself on the ripe fruit.

Because animals are adapted to feed in certain ways it seems likely that experimental procedures must, either through design or differential reinforcement of the experimenter, be consistent with such adaptation. The evolutionary function of a behavior has been proposed as an important determiner of its appropriateness as a learned behavior to produce food or other important outcomes (e.g., Shettleworth, 1975).

The analogy between foraging behavior as it occurs in nature and operant and maze behavior as it occurs in laboratories is quite obvious. The active forager in a patchy environment must search for food in space, and upon encountering it, harvest it in some fashion. Movement through space from a nonfood source to a food source (or potential one) is the major component of maze procedures. Harvesting food from the food source is the major component of operant procedures.

The argument is, that the foraging behavior of many mammals is characterized by search and harvesting (procurement) and these processes are represented in the laboratory of experimental psychologists by mazes and operant chambers, respectively. In the maze procedure the rat runs to a goal box where food is available in a food cup. Here the relevant process is "search" with virtually no emphasis on the procurement process. Indeed, the standard practice in maze experiments is to use specially produced food pellets (Noyes pellets) which are soft to the degree that a mature male rat can "inhale" 8 or 10 in a very short time as any rat-psychologist can attest. This practice minimizes the harvesting cost often encountered in naturalistic conditions called "handling time" (e.g., Krebs, 1978). In the operant procedure quite the opposite is true. The subject is placed directly in the food source and is required to procure the available food through the repetitive act of pressing the lever. The search component is virtually nonexistent as the subjects simply positions itself at the lever and presses-away.

The question naturally comes up as to whether search and procurement are governed by the same psychological processes or not. Behavioral ecologists seem to have no clearer answer to this question than experimental psychologists had concerning operant-maze differences. One concept which may suggest that search and procurement involve different processes is the phenomenon of "area-restricted search" (Tinbergen, Impekoven and Frank, 1967). When food is encountered and eaten, a stickleback (gasterosteus aculeatus) will increase the "tortuosity" of its swim-path over the next seconds relative to not finding food (Thomas, 1974). This dramatic change in behavior may reflect a switch from a general food search to a procurement expectancy, although such an interpretation is clearly speculative on our part.

Experimental Manipulations

The experiments we report all utilize schedule of food availability as the independent variable. In one condition the animal discovers food on every entry to the food location and in the other condition food is available only half the time. Following repeated experiences of one of these two types (continuous reinforcement, CR, and partial reinforcement, PR, respectively) the food source abruptly dries-up (extinction). This procedure was chosen because of the extensive documentation of the effects of schedule on the persistence animals show in responding during extinction. In maze studies extensive theoretical development (e.g. Amsel, 1967; Capaldi, 1967) has shown that both motivational factors and memory of past events are important in accounting for the greater tendency to continue responding following PR than CR.

One reason for choosing extinction of responding as the dependent variable of interest is the fact that operant procedures designed to be analogous to maze procedures (e.g., maze procedures of Haggbboon and Williams, 1971; Mellgren and Dyck, 1972) have sometimes yielded results somewhat similar to the maze data (Bitgood and Platt, 1971; Haddad, Walkenbach and Goeddel, 1980) and sometimes not (Collier, Steil, Pavlik, 1978; Pavlik and Carlton, 1965; Pavlik and Collier, 1977). The lack of consistency suggests the possibility that maze and operant

procedures may produce different results under extinction conditions. Many operant procedures which are not explicitly designed to be directly analogous to maze procedures, but may be
interpreted in such a fashion, yield results which are opposite
to maze results (Nevin, 1979). Anomolies beg for explanation.

A second reason for using extinction is the fact that transfer-
of-training methodology is often a very effective strategy for
revealing the operation of different processes which are not
observable at the time they occur. For example, at the time
they are watching, children who view a violent television program may exhibit no observable difference from children who
view a nonviolent program but latter, when playing, the
children who watched the violent program may show greater
aggression than those that watched the nonviolent show. Rats
in mazes often are very similar in performance during PR and
CR schedules, but in extinction show very marked differences in
the persistance of responding.

Experiment 1

Application of the strategy represented by Figure 1 requires
that a comparison between operant and maze characteristics be
made with respect to an independent variable. In this experiment the characteristic of operant procedures which was included is the object-contact involved in lever pressing. The
object in the present experiment was a rubber ball with a small
metal screw implanted in it. The characteristic of maze procedures was movement through space. The actual apparatus is
shown in Figure 2. The rats were required to enter the tunnel
leading to the food box. The rubber ball was positioned in the
tunnel and a metal rod protruding through the ceiling of the
tunnel prevented the ball from being pushed forward. Thus the
rat was required to remove the ball by pulling it back to the
start box which then allowed her to run through the unblocked
tunnel and obtain the food which was available in the goal box.

Female hooded rats, 80 days old, served as subjects. After
their weight was reduced to between 85 to 90% of ad lib they
were given 20 pretraining trials spaced across four days (5
trials per day). The reward consisted of four sucrose pellets
weighing 48 mg. each. During pretraining the ball was not in
the tunnel, but was placed in the start box. Nine rats, randomly chosen, experienced food in the goal box on 50% of the
trials (2 or 3 per day) and no food on the other 50% (PR).
Eight rats found food on every trial (CR).

On the first acquisition day all rats received two reinforced
trials, the first trial with the ball in the middle of the
tunnel and the screw implanted in the ball facing the rat.

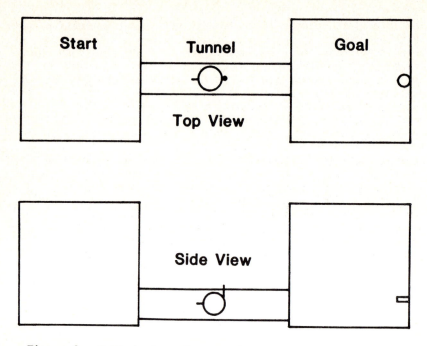

Figure 2.   Ball-in-tunnel apparatus.

Although all the rats actively contacted the ball, no rat
actually removed the ball from the tunnel within 5 min. At
this time the experimenter moved the ball to the edge of the
tunnel so that a slight contact with it would ensure that it
would roll away from the entrance and allow the rat access to
the tunnel and goal chamber. On the second trial of the day
the ball was placed on the threshold of the tunnel to ensure
that the subjects would be able to remove it relatively easily
(a few rats pushed the ball into the tunnel, in which case the
experimenter replaced it at the threshold for the rat). On
Day 2, 3 trials were given with the ball moved approximately
3 cm into the tunnel from the threshold on each successive
trial. Some of the rats were again given help (i.e., the ball
was moved closer to the threshold) when it seemed necessary.
On the third day 5 trials were given with the ball approximately 7 cm into the tunnel on trial 1 and half way in (11 cm) on
all subsequent trials. On Days 4-12, 5 trials were given with
the ball in the half-way position on all trials. Extinction
followed immediately and lasted for 5 days, 5 trials per day.
Across acquisition training the PR group received a 50% schedule
of reinforcement with nonreward consisting of a 12-15 sec. confinement in the goal chamber.

The three time measures were tunnel entry (time from the rat being placed in the start chamber facing the back corner until a photocell beam just inside the tunnel was broken), ball removal (time from the rat being past the first photocell until the ball was removed from the tunnel), and run (time from the ball being removed until a photocell beam just in front of the food cup was broken). The times were reciprocated to yield a rate (speed) of response measure, and both the median speeds and mean speeds were subjected to analyses of variance. Since the results were basically the same for the two types of analyses, the median data will be reported, except as noted.

At the end of acquisition training the PR group was faster than the CR in the tunnel entry measure, but slightly inferior in the ball removal measure, but none of these differences approached an acceptable level of significance in separate analyses of variance including the last 3 days of acquisition as a repeated factor. About half the rats in each group learned to remove the ball by grasping the screw in their teeth and pulling backwards until the ball was out, while the other half learned to "paw" the ball, from the top, but in a skilled manner so that the ball did not become wedged in the tunnel because of the screw which was implanted in the ball.

In extinction a partial reinforcement effect was clearly evident in the run measure (see Figure 3) as indicated by the significant Groups main effect and the Groups X Days interaction ($p < .025$ in both cases). It is mildly surprising to see the CR group show no apparent decrease in speeds on the first day of extinction. A Day 1 separate analysis including the 5 trials as a factor revealed that although there was not a significant main effect for Groups, The Group X Trials interaction was significant ($p = .05$), reflecting the fact that both groups were performing at the same level for the first 3 trials, but on the last 2 trials the CR group fell off badly (mean reciprocals = .67 and .49, see Figure 3 for comparison). Thus it can be safely said that the rats in this experiment demonstrated the usual robust partial reinforcement effect as measured by their tendency to run from the start chamber to the goal chamber.

The ball removal and tunnel entry speeds show a very different picture. The CR group was numerically more resistant to extinction than the PR group on the ball removal measure, but the difference was not significant ($F = 0.93$). The ball removal component did undergo extinction as shown by the significant Days effect ($p < .001$), and there was a marginally significant Group X Days interaction for the analysis of mean speeds ($F = 2.07$; $p = .09$), but not for the analysis of medians ($F = 1.20$). In contrast, for the tunnel entry measure there was no decrease

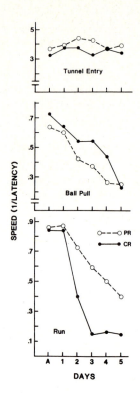

Figure 3.   Speed of responding on the last day of acquisition (A) and the five days of extinction.

in speeds, as is evident in Figure 3, and the analysis (F =
1.70 for Days). The difference between the groups was also not
significant for the tunnel entry measure (F = 2.40).

During extinction a record was kept of the occurrence of certain
types of behaviors (e.g., retracing, or stereotyped responses).
The most salient characteristics of the rats behavior was the
appearance of "play" with the ball in the start chamber for the
CR rats. This play consisted of climbing over the ball, chasing
it, biting the screw and tossing the head back and forth, and on
a few occasions, dragging the ball into the tunnel. Summed
across all extinction days, the CR group showed this behavior
on 41% of the trials, but the PR group showed it on only 15% of
their trials. Only 4 of the 8 PR rats had one or more trials
where they showed this behavior, while every CR rat had more
than one trial of such behavior. Furthermore, the CR group

showed a clear "acquisition" curve across extinction, going from 9% play trials to 58% play trials from Day 1 to 3, while the PR group showed a less dramatic increase from 2% to 17% over the same period. Of course, the use of the word "play" to describe this behavior is merely for convenience, and refers only to the behaviors listed above.

Experiment 2

The suggestion from Experiment 1 is that manipulative, object contact responding is not sensitive to schedule of reward in the same way as locomotion, approach-to-a-goal responding. Of course there are many possible explanations of why the ball pull and running measures differed in resistance to extinction following partial and continuous reward training. One possibility is that the two measures reflect different underlying processes analogous to the processes involved in maze and operant studies. Another possibility is that the differences are due to a blocking effect (e.g., Kamin, 1969).

Blocking is the phenomenon defined by the lack of association between a stimulus (or behavior) and an outcome as a function of another stimulus (or behavior) already having been associated with the outcome. So if a light is paired with a shock for some number of trials and then a new stimulus such as a tone is presented in compound with the light, test presentation of the tone indicates it was not associated with the shock, and is said to be "blocked from association" by the light.

Theories of the partial reinforcement effect generally postulate an association between memories of nonreward and behavior (Capaldi, 1967) or between motivational arousal (frustration) and behavior (Amsel, 1967). In the first experiment the subjects were pretrained to run through the tunnel with the ball absent from the tunnel. When the ball was later placed in the tunnel and the ball-pull response was required for successful performance, the ball pull response may have been blocked from entering into association with the memory of nonreward or frustration because of the prior association of running through the tunnel with the memory of nonreward or frustration.

A test of the blocking hypothesis is straightforward; train subjects from the start of the experiment with the ball in the tunnel and compare their performance to subjects given pretraining with no ball present prior to their ball pull training.

Male albino rats, 80 days old, served as subjects. They were reduced to 85% of their ad lib weight over a two week period. They were randomly divided into 4 equal groups of 12 each (one rat became ill during the experiment and he was excused). All

subjects were given two placements in the goal box on the two
days immediately preceeding the start of the experiment. Food
rewards was present and consisted of 6 pellets (45 mg. each)
produced by the P. J. Noyes Company (Formula A).

On succeeding days pretrained subjects received running trials
where the ball was in the start box, but did not block the
tunnel, while nonpretrained subjects experienced the ball in
the tunnel blocking access to the goal box. Three days of two
trials each were used, with the nonpretrained subjects receiving
"shaping" by placing the ball closer to the tunnel entrance if
they were not successful on the previous trial. On the first
of these three days both trials were rewarded, but on the last
two days only one of the two trials was rewarded for the partial
reinforcement groups. Another four days at 4 trials per day
were given with the nonpretrained groups having the ball placed
fully in the tunnel. Half of each pretraining condition
experienced partial reinforcement (50%) and half experienced,
continuous reinforcement. Thus the four groups are designated
(first letter = reinforcement schedule, second letter = pretraining condition); CN (continuous reinforcement, non-pre-
trained, PN (partial reinforcement, nonpretrained), CP (continuous reinforcement, pretrained), and PP (partial reinforcement, pretrained).

At the end of this phase the subjects in the pretrained conditions experienced the ball in the tunnel for the first time.
Two trials were given to all subjects during first day of
this phase under the assumption that there would be an extended
period of unsuccessful responding on the part of the pretrained
subjects (those experiencing the ball blocking their progress
to the goal for the first time). Over the next 12 days (4
trials per day) all subjects were required to remove the ball
from the tunnel and half received reward on every trial and half
received the partial reinforcement schedule.

The final phase consisted of 5 days of extinction, 4 trials per
day.

The apparatus for this experiment was similar to that used in
Experiment 1, but differed in detail (Exp. 1 was performed at
Cambridge University, Cambridge, England; Exp. 2 was performed
at the University of Oklahoma, Norman, Oklahoma, USA). The
most salient difference of the apparatus was that the ball used
in Exp. 1 was a rubber-type, but the ball in Exp. 2 was covered
with paper material and taped with black vinyl tape, completely
covering the paper stuffing. The taping was necessary to prevent the rat from slipping between the available sized ball and
the tunnel wall.

Compared to the subjects in the first experiment these subjects
learned to remove the ball from the tunnel with little difficulty. Among the pretrained subjects (those most closely replicating Exp. 1) there were only three subjects who failed to remove the ball within three minutes on the very first trial it
was encountered in the tunnel and those three were successful
on the second trial. There are several possible variables which
might account for the surprisingly effective performance of subjects in Experiment 2 including strain of rat and sex of rat.
Several theorists have suggested that sex may determine different foraging patterns in nature with females being characterized as energy maximizers and males being characterized as
time minimizers (Hixon, 1982; Schoener, 1971). The general idea
is that males, because they must spend time defending a territory and a mate, will be more selective in their feeding patterns, searching for rich sources of food which will allow them
to minimize the time spent feeding as compared to females who
will be more willing to utilize a relatively lean food source
and be less concerned with time. One implication of this view
is that males must have the ability to search rapidly and in
novel ways in order to effectively minimize time, while females
would not have as much pressure to do so. That such differences
in foraging patterns exist in rats and that they would account
for the differences in the ball removal capacity of males and
females in this experimental situation remains only an interesting speculation.

At the end of acquisition training the subjects given no pretraining (starting ball-pull training almost immediately) were
generally numerically superior to those given initial pretraining (see Figure 4) although the differences were not statistically significant ($F(1,43)$ = 1.81, 2.99 and .02 for tunnel entry,
ball pull and run, respectively). Schedule also had an effect
with CR resulting in more rapid running and ball pulling than
PR. The differences were marginally significant in the ball
pull measure, $F(1,43)$ = 3.92, p = .051, and not significant
for tunnel entry, F = 1.81, and run, F = 1.72.

The tunnel entry and run measures showed the results usually
seen in straight alley studies of partial reinforcement. The
main effect of schedule and the schedule by trials interaction
were significant in all cases. For the ball pull measure schedule by itself was not a significant source of variance, $F(1,43)$
= .03, but pretraining condition was, $F(1,43)$ = 9.71, p < .01.
There was an obvious interaction between schedule and trials,
$F(4,172)$ = 17.92, p < .001, reflecting the initial superiority
and ultimate inferiority of CR groups as compared to PR groups.

The pattern of persistence seen in this experiment fails to
replicate the pattern seen in the first experiment. Certainly

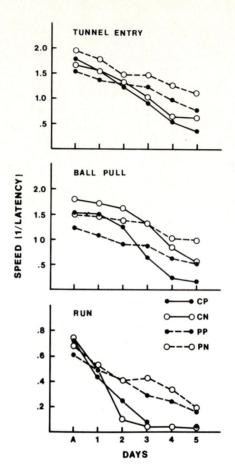

Figure 4.   Speed measures for acquisition (A) and
    extinction in the second ball-pull experiment.

the run measure replicated well, but the ball pull did not since
the PR subjects were more persistent than CR subjects as measured by the slope of the extinction curve (i.e., the interaction of schedule by trials).   Once again the source of the
differences in results is speculative.

Interactions with the ball once it had been removed were also
measured in this experiment.   Whereas, in the first experiment
such interactions were characterized as "play" because the
response topography seemed most closely described by that word,
in this experiment virtually all the interactions with the ball
involved chewing or gnawing on it.   The ball chew measure consisted of a score ranging from 0 to 4 each day depending on the

number of trials on which it was observed to occur. The CR subjects showed a much higher incidence of ball chewing than the PR subjects, $F(1,43) = 30.49$, $p < .001$, as is obvious from Figure 5. According to some views of the effects of schedule

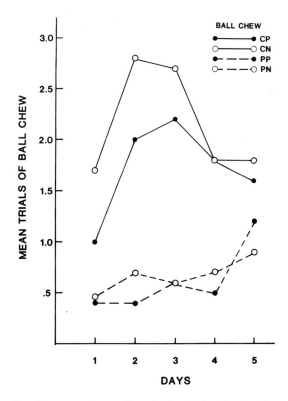

Figure 5. Mean number of trials of ball chewing during the five days of extinction.

on extinction performance, frustration will be an integral part of the subject's responding. The PR subjects should have associated instrumental behavior with frustration, but CR subjects should not. Thus the frustrated CR subjects should show unconditioned responses which appear as biting and ball chewing while PR subjects should not.

The results of these two experiments are tantilizing suggestions that different behaviors measured in the same experimental situation may produce different functional relationships with the independent variable. It will be recalled that the hypothesized underlying differences between ball pull and run were the manipulation - object contact vs. locomotion - approach-to-goal dimensions. Although the ball pull requires object manipulation and contact it also requires a degree of locomotion. Perhaps a more definitive procedural distinction between these two types of behaviors which are hypothesized to reflect the difference between operant and maze behavior would be useful in producing differential effects than the procedures used in the first two experiments.

Experiment 3

The distinguishing feature of operant and maze procedures that we are attempting to evaluate is the fact that operant procedures involve contact and manipulation of an object(s) and maze procedures involve locomotion and spatial location. It is clear, however, that operant and maze procedures are not "pure" cases of these different types of responses. In the operant situation the subject must contact and manipulate the lever, but that lever is also located in a discrete place in space. In the previous experiments the subjects were required to contact and manipulate the ball in order to remove it from the tunnel. Clearly this task also has a very important spatial component to it. In order to get a purer distinction along the spatial - manipulative dimension a new procedure was developed.

In nature and in laboratory situations rats are known to be burrowing animals and well adapted for digging (e.g., Boice, 1977). Taking advantage of their ability to dig, we constructed an alley which led to a box containing sand, with food being buried in the sand. The spatial component of this situation is the run from the start box to the box containing sand. The manipulative component in the digging in the sand for food. The advantage of using sand as the manipulandum is that it completely surrounds the subject and once the subject is in the "patch" there is no spatial differentiation of the manipulandum --its everywhere.

The experimental strategy was again to use schedule of reward as the independent variable. In this case the PR subjects found food buried in the sand on half the trials and no food on the other half. The CR subjects found food in the sand on every trial. The food consisted of hard pellets of Purina Startina of irregular sizes, but a total of 18 pieces which weighed 3 g. in total. The pellets were mixed with the sand and the surface was smoothed. If any pellets were exposed they were pushed firmly into the sand. The sand was 3.5 cm deep and the goal box (patch) was 30 x 37 cm and 18 cm high. The straight alley was 150 cm long and 10 cm wide.

A trial consisted of placing the subject in the startbox with a door separating it from the rest of the apparatus. After a 5 sec. pause the door was opened and the subject was allowed a total of 5 min. for a trial. We recorded three measures of approach-to-patch via photocells located 6 cm from the start-box door (start), 105 cm (run) and 156 cm (goal). Once in the patch a hand held switch was used to record the cumulative time the subject spent digging and the number of bouts of digging. The number of food pellets obtained was also recorded. It was decided to have 18 pellets in the patch on each reward trial to insure that the subject could not deplete the patch to a zero level in the 5 min. trial.

The subjects were given pretraining for digging in sand for food in a different room using a 10 gal. aquarium as the "patch". In this experiment pretraining lasted for 12 days with the food pellets gradually being placed deeper into the sand across days. Such elaborate pretraining is not necessary.

The 10 male albino rats were randomly divided into two groups, PR and CR, and given 4 trials per day for 10 days of acquisition. For PR subjects food was buried in the sand on 2 trials and no food was present on 2 trials for each day. For the CR sub-jects food was present on every trial during this phase. For the next 4 days all subjects experienced no food in the patch (extinction). Once again a total of 5 min. was allowed for a trial. As soon as the subject entered the patch he was re-quired to stay in it by the experimenter lowering a door block-ing access to the alley. The intertrial interval was approxi-mately 6 min. during all phases of the experiment.

In common with many studies using a straight alley, schedule of reward had little effect on running speeds during acquisi-tion, with analyses of variance failing to reject the null hypothesis for start run and goal speed differences.

The digging behavior analysis likewise revealed no significant differences between Pa and CR conditions. In order to evaluate time spent digging we subtracted "handling time" (time required to eat a pellet) and the time spent getting to the patch from the total time available. Thus digging time was converted to proportion of time spent digging relative to the total amount available for digging. Analysis of variance on the last day of acquisition showed no significant difference due to schedule ($F(1,8)<1.0$), with the mean proportion of time spent digging being .58 for the CR group and .47 for the PR group (if raw time spent digging is considered, these nonsignificant dif-ferences are even smaller). As another evaluation of the effect of nonrewarded-trials a correlated t-test was performed on the proportion of time spent digging on the first nonreward and first reward trial for the PR subjects. The difference was not significant $t(9)=1.78$ although the proportion was higher on the rewarded trial than the nonrewarded (.59 and .42). The mean number of bouts of digging (bouts ended with a pellet being found, or if the subject paused for more than 2 sec. or changed location while not continuing digging) was also not different between PR and CR conditions at the end of acquisition, mean bouts per trial being 13.45 and 14.10 for PR and CR, respectively.

Figure 6 shows the effect of schedule of reward on the approach measures during extinction. Analyses of variance confirmed the pattern of results shown in the figure: the PR subjects were

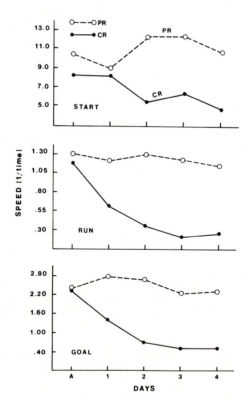

Figure 6. Approach to the goal measures on the last acquisition day and four days of extinction.

more resistant to extinction than the CR subjects in all measures.   In fact, the PR subjects show no decline in speeds and
even an increase in the start speed measure across the four days
of extinction.   This is a surprising result, particularly when
placed in the context of all subjects having had 5 min. per
trial for 4 trials per day with no food present.   By day 4 of
extinction the subjects had experienced 60 min. of cumulative
time with no food present, but continued to run as fast as they
had when food was present.

Figure 7 shows the decline in proportion of time spent digging
across extinction as a function of days and trials within days.
The PR subjects showed an abrupt decline in digging on the first
day and a gradual decline across days, while the CR subjects
showed a decline within trials of a day and across days.   The
pattern of results shown in Figure 7 should be reflected in an
analysis of variance as a significant interaction of schedule
with trials and with days, both of which were significant at the
.01 level or better.

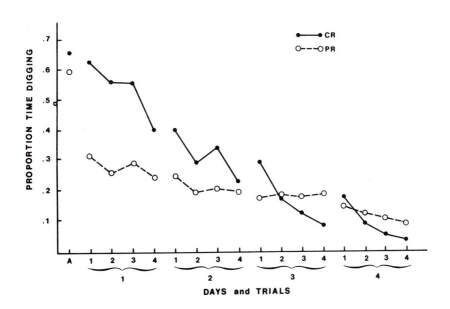

Figure 7.   Proportion of time spent digging across the
    four trials of each day of extinction.

Although the proportion of time digging was greater for the CR
condition than the PR, the number of digging bouts during
extinction did not differ as a function of schedule, with the
analysis of variance confirming significant effects only for
the general decline in the number of bouts across days.

The results of this experiment show three different effects
in extinction due to the schedule manipulation. The PR subjects
run faster to the patch than CR subjects, the PR subjects spend
less time digging in the patch than CR subjects, and the PR and
CR subjects show equivalent number of digging bouts. It seems
difficult, if not impossible, for a single-factor explanation
to fit these data. For example, the faster running speeds of
PR subjects than CR might be attributed to greater "hope" or a
more persistence expectancy that food will eventually appear
because of its inconsistent appearance in the past. If the PR
subjects are more "hopeful" why don't they dig more than CR
subjects? If the two groups of subjects are not equal in the
persistence of their expectancies why do they initiate equivalent numbers of bouts of digging?

Discussion and Implications

Mazes and Skinner boxes create analogies to different aspects
of natural foraging situations. The maze is analogous to the
search for a food source and the operant chamber is analogous
to the harvesting or procurement of available resources. Just
as quickly as we have distinguished these two analogies, we
should note that search and procurement will not necessarily be
differently affected by an independent variable. If an animal
finds a very rich patch it might be more vigorous in approaching
that patch and in procuring the food in the patch on subsequent
foraging expeditions. Similarly, a rat that finds many food
pellets at the end of a maze may run more rapidly than a rat
that finds only one food pellet, and the bar pressing rate of a
rat that receives many pellets for a given number of bar presses
may be faster than a rat that receives only one food pellet.
When will the maze procedure produce differences as compared
to the operant procedure and when will they be the same? At
this point we have no definitive answer to this question.

Extinction is one procedure which will produce different, or at
least inconsistent, results in the maze and operant situation.
The reason is clear; the maze data depends solely on approach or
"search" processes, but the operant data depends on both the
search and the procurement processes. In the operant situation
the subject breaks contact with the manipulandum during the
course of the extinction procedure. Once this occurs some degree of approach to the lever (search process) is required

before the subject again makes contact with the manipulandum
(procurement process). Thus we would predict that a variable,
such as schedule of reinforcement, in the operant situation
would produce fairly variable responding during extinction.
The CR subjects would show a declining tendency to approach and
press the lever, but a greater tendency to press for a long
time once in contact with the lever than subjects given PR
training. Extinction may look particularly "variable" in the
operant situation because of the fact that it encourages the
subject to leave the manipulandum, while steady-state reinforced responding does not induce leaving and is often remarkable in the consistency of responding that is seen.

Nevin (1979) in summarizing evidence concerning "response
strength", defined as "the resistance of responding to change",
found a striking discrepancy between the results of free operant
studies and maze studies. He notes that, "generalizations advanced above may apply only to single-subject research in the
free-operant situation with extended training. Alley-way
studies, which typically involve independent groups, spaced
discrete trials, and relatively brief training, give results
exactly opposite to those summarized here." (p. 124). Variables such as magnitude, frequency and delay of reward all had
effects on resistance to extinction of operant responding which
were exactly opposite to those found in straight-alley experiments. Such effects are not inconsistent with the current findings and the suggestion by Nevin that, "A considerable program
of research is required to...bridge the gap between the single-
subject, free-operant procedures described here and the in-
dependent-groups, discrete-trial methods that have dominated
the study of extinction." (p. 124) should clearly involve the
nature of the response--locomotion vs. manipulation--as one
dimension of study.

In our sand digging experiment and ball removal experiments the
subject is in contact with the "manipulandum" (sand or ball)
and schedule of reinforcement has either no effect or the reverse effect that it has when the subject is locomoting in the
same situation. We argue that rats in operant chambers may show
analogous effects. Bar presses involve contact with the manipulandum, but during the course of extinction the rat will
break contact with the manipulandum and this will necessitate
a reapproach to the manipulandum for further bar pressing to
occur. An implication of this view is that pauses between
bursts of response should account for differences due to
schedule manipulations. Mowrer and Jones (1945), for example,
show cumulative records (Figure 1, p. 298) of rats in extinction
following CR or FR-4 training and the rate of response seems
high in both groups, but the CR group shows longer pauses

accounting for the fewer total responses in extinction.
Overmann and Denny (1974) demonstrated that approaches to the
food cup elicited by a discriminative click presented on a
FR-6, -10 or -14 schedule during extinction resulted in an
increasing number of total bar presses but an equal number of
food cup approaches and, by implication, an equal number of
re-approaches to the lever.

Of course existing data are not definitive since pauses in a
cumulative record may represent the subject breaking contact
with the manipulandum only to rear above the lever with its
forepaws while still maintaining contact with the wall; it may
represent the rat holding the lever in a depressed mode for a
period of time, it may represent grooming at the lever, or it
may represent moving-off to a location removed from the lever.
Presumably it is only when the rat moves away from the lever
that he has broken contact and a re-approach (search process)
is necessary for subsequent responding. Measurement in this
type of detail as a function of schedule of reinforcement does
not currently exist to our knowledge.

It is clear that an important component of any study of extinction in the operant situation will be influenced by the size
of the chamber and the ability of the subject to move around in
it. Larger boxes should produce more variability than smaller
ones because the leave - re-approach component is more likely to
influence responding in the larger than the smaller. Perhaps
this was discovered by early workers on extinction such as
Rohrer (1947; 1949) whose "Skinner Box", as he called his
apparatus, required the rats head to be in a stock and its body
restrained into a sitting position by covering it with a mesh.
Thus restrained, the rat was on a wood platform which could be
moved forward to the manipulandum or retracted from it by the
experimenter.

Another aspect of our data which bears some discussion is the
fact that the procedures we used all involve chains of behavior. It is uniformly reported that extinction of a chain
schedule in free-operant studies results in more rapid decreases
in responding in the initial link than in the terminal link of
the chain. Catlin and Gleitman (1973) found, for example, that
following a chain schedule with FR components extinction was
more rapid in the initial than the terminal link. When only a
single operant response is used such as key pecking or lever
pressing, resistance to extinction or response strength may be
greatest the closer the subject is to the terminal link of the
chain. Our results with the ball-removal experiments suggest
that the chain schedule results may not be generalizable beyond
the homogeneous response situation of the operant conditioning
procedure. Our results also seem incompatible with a

conditioned reinforcement explanation of chain effects. Such
an explanation would hold that stimuli correlated with the
primary reinforcer would acquire conditioned reinforcing
qualities which would maintain earlier links of the chain. In
the ball-removal experiments the initial link of the behavioral
chain is the tunnel entry measure. Not only is its resistance
to extinction insensitive to the strength of the next component
of the chain (ball removal) it shows greater response strength
(in Nevin's terms) than any of the other components of the
chain. Since the conditioned reinforcing value of a stimulus is
usually indexed by its capacity to control or evoke responding
in its presence (e.g., Hull, 1943; Wycoff, 1959), the fact that
responding in the ball removal measure had decreased quite
dramatically while the tunnel entry measure was unaffected
seems incompatible with any straightforward application of a
conditioned reinforcer explanation. Similarly, the fact that
the PR subjects in Experiment 3 show a very large decline in
the terminal link of their behavior chain (digging), the relative lack of decline in the initial link (running) seems quite
incompatible with a conditioned reinforcer explanation.

It seems very likely that the conditioned reinforcement explanation of chain schedules is appropriate for the operant
situation, but is lacking in generality for explaining chains
of nonhomogeneous responses. Presumably the temporal or
spatial proximity of a response to the biologically important
event will be important, but it is clear from the present results that the topography and function of the response is also
very important in determining the persistence of the response.

Mechanism and Function

As is appropriate for experimental psychology we have been concerned with mechanisms operating to produce a pattern of
foraging behavior. These mechanisms or proximal causes
operate on the individual subject's biology, and the interaction
of environmental constraints and biological predispositions
will result in the final pattern of foraging behavior. Obviously, mazes and Skinner Boxes fall in the category of
environmental constraints. Biologists typically are not so
focused on how environmental constraints shape behavior,
but are more concerned with how a behavior is related to the
inclusive fitness of the organism. That is, the biologist is
concerned with the function a behavior has for promoting survival of the individual and the genetic input it has to subsequent generations.

Krebs and Davies (1981) warn against functional explanations
of behavior which are circular and little more than a relabeling

of a known phenomenon. These "explanations" go something like,
"Observe organism A doing X. Behavior X is observed because
it increases fitness in ways M, N and P." To be specific,
suppose we observe young animals engaged in rough and tumble
play. The function of play is to prepare the individual for
later aggressive encounters and without such play an individual
would be unprepared for male-male competition at mating time
and therefore loose his genetic input to the next generation.
Krebs and Davies call these type of explanations the "just so"
story, and warn against their use, but their influential
text, An Introduction to Behavioral Ecology, is liberally
sprinkled with just so stories. Rather than regarding a just
so story as a final explanation, it may be regarded as a
tentative hypothesis, or a logical possibility. Taken at this
level, perhaps it would reduce the negative reactions experimental psychologists have traditionally had toward functional
explanations of behavior.

Most environments are variable with respect to food availability.
Variability of available resources is brought about by such
factors as seasonal variations, predator effects, short-term
weather fluctuations, etc. Learning where food sources are
currently located must be an important component for most
foraging animals. Once a food source has been located, harvesting the food and assessing the value of the source relative to
other possible sources becomes an important consideration. The
two processes are usually correlated to a degree, in the sense
that if the animal must search for long periods to find a food
patch then it will utilize more of that patch than it would if
there were many patches and short search periods required to
find them. The two processes are also independent, to a degree.
If a predator finds a patch of preferred food, its utilization
of that food involves different behaviors and sensory events
than those which got it to the patch. One of us avoids going
to MacDonald's Hamburger Stores. When forced by circumstances
to approach this "patch" he knows exactly what to do--order the
fish sandwich. In this example the approach tendency to the
patch is not strong but patch utilization is immediate and
predictable.

An important implication of the idea that patch approach and
patch utilization involve different processes is that different
neural circuits are involved. As the neurosciences come to
depend on behavioral theory for knowledge of processes, a major
distinction between spatial-maze processes and harvesting-
operant processes may be of prime importance. Research and
thinking along these lines is long overdue.

References

1   Amsel, A. Partial reinforcement effects on vigor and persistence. In K. W. Spence & J. T. Spence (Eds.), The Psychology of Learning and Motivation, Vol. 1. New York: Academic Press, 1967, 1-65.

2   Bitgood, S. C. & Platt, J. R. A discrete-trial PREE in an operant situation. Psychonomic Science, 1971, 23, 17-19.

3   Bitterman, M. E., LoLordo, V. M., Overmier, J. B. & Rashotte, M. E. Animal Learning. New York: Plenum Press, 1979.

4   Boice, R. Burrows of wild and albino rats: effects of domestication, outdoor raising, age, experience and maternal state. Journal of Comparative and Physiological Psychology, 1977, 91, 649-661.

5   Capaldi, E. J. A sequential hypothesis of instrumental learning. In K. W. Spence & J. T. Spence (Eds.), The Psychology of Learning and Motivation, Vol. 1. New York: Academic Press, 1967, 67-156.

6   Catlin, J. & Gleitman, H. Pattern of disruption in the extinction of chain FR7 FR7 in pigeons. Animal Learning and Behavior, 1973, 1, 154-156.

7   Domjan, M. & Burkhard, B. The Principles of Learning and Behavior. Monterey, CA.: Brooks/Cole, 1982.

8   Dunham, P. J. Contrasted conditions of reinforcement: A selective critique. Psychological Bulletin, 1968, 69, 295-315.

9   Fantino, E. & Logan, C. A. The Experimental Analysis of Behavior. San Francisco: W. H. Freeman, 1979.

10 Haddad, N. F., Walkenbach, J. & Goeddel, P. S. Sequential effects on rats' lever-pressing and pigeons' key-pecking. American Journal of Psychology, 1980, 93, 41-51.

11 Haggbloom, S. J. & Williams, D. T. Increased resistance to extinction following partial reinforcement: A function of N-length or percentage of reinforcement? Psychonomic Science, 1971, 24, 16-18.

12 Hill, W. F. Principles of Learning. Palo Alto, CA: Mayfield, 1981.

13 Hixon, M. A. Energy maximizers and time minimizers: Theory and reality. The American Naturalist, 1982, 119, 596-599.

14 Hull, C. L. Principles of Behavior, New York: Appleton-Century-Crofts, 1943.

15 Hulse, S. H., Egeth, H. & Deese, J. The Psychology of Learning. New York: McGraw-Hill, 1980.

16 Kamil, A. C. & Sargent, T. D. (Eds.) Foraging Behavior. New York: Garland, 1981.

17 Kamin, L. J. Predictability, surprise, attention, and conditioning. In R. Church & B. Campbell (Eds.) Punishment and Aversive Behavior. New York: Appleton-Century, Crofts, 1969.

18 Krebs, J. R. Optimal foraging: Decision rules for predators. In J. R. Krebs & N. B. Davies (Eds.) Behavioral Ecology. Oxford, England: Blackwell Scientific Publications, 1978.

19 Krebs, J. R. & Davies, N. B. An Introduction to Behavioural Ecology. Sunderland, MA: Sinauer Associates, 1981.

20 Logan, F. A. & Ferraro, D. P. From free responding to discrete trials. In W. N. Schoenfeld (Ed.) The Theory of Reinforcement Schedules. New York: Appleton-Century-Crofts, 1970.

21 Mackintosh, N. J. The Psychology of Animal Learning. New York: Academic Press, 1974.

22 Mellgren, R. L. & Dyck, D. G. Partial reinforcement effect, reverse partial reinforcement effect, and generalized partial reinforcement effect within subjects. Journal of Experimental Psychology, 1972, 92, 339-346.

23 Mowrer, O. H. & Jones, H. M. Habit strength as a function of the pattern of reinforcement. Journal of Experimental Psychology, 1945, 35, 293-311.

24 Nevin, J. A. Reinforcement schedules and response strength. In M. D. Zeiler & P. Harzen (Eds.) Reinforcement and the Organization of Behavior. New York: John Wiley & Sons, 1979.

25 Overmann, S. R. & Denny, M. R. The free-operant partial reinforcement effect: A discrimination analysis. Learning and Motivation, 1974, 5, 248-257.

26 Pavlik, W. B. & Carlton, P. L. A reversed partial reinforcement effect. Journal of Experimental Psychology, 1965, 70, 417-423.

27 Pavlik, W. B. & Collier, A. C. Reinforcer magnitude effects on a within-subjects reversed PRE. Bulletin of the Psychonomic Society, 1973, 233-234.

28 Platt, J. R. Discrete trials and their relation to free-behavior situations. In H. H. Kendler & J. T. Spence (Eds.), Essays in Neobehaviorism. New York: Appleton, Century-Crofts, 1971.

29 Rohrer, J. H. A motivational state resulting from non-reward. Journal of Comparative and Physiological Psychology, 1949, 42, 478-485.

30 Rohrer, J. H. Experimental extinction as a function of the distribution of extinction trials and response strength. Journal of Experimental Psychology, 1947, 37, 473-493.

31 Schoener, T. W. Thoery of feeding strategies. Annual Review of Ecology and Systematics, 1971, 2, 369-404.

32 Shettleworth, S. J. Reinforcement and the organization of behavior in golden hamsters: Hunger, environment, and food reinforcement. Journal of Experimental Psychology: Animal Behavior Processes, 1975, 104, 56-87.

33 Shimp, C. P. Perspectives on the behavioral unit: choice behavior in animals. In W. K. Estes (Ed.) Handbook of Learning and Cognitive Processes, Vol. 2. Hillsdale, N. J.: Lawrence Erlbaum Associates, 1975.

34 Tarpy, R. M. Principles of Animal Learning and Motivation. Glenview, Ill.: Scott, Foresman, 1982.

35 Thomas, G. The influences of encountering a food object on subsequent searching behavior in gasterosteus aculeatus 1. Animal Behavior, 1974, 22, 941-952.

36 Tinbergen, N., Impekoven, M. & Franck, D. An experiment on spacing-out as a defense against predation. Behaviour, 1967, 28, 307-321.

37 Wyckoff, L. B. Toward a quantitative theory of secondary
    reinforcement. Psychological Review, 1959, 66, 68-78.

ANIMAL COGNITION AND BEHAVIOR
Roger L. Mellgren, editor
© North-Holland Publishing Company, 1983

STUDYING FORAGING
IN THE PSYCHOLOGICAL LABORATORY

William M. Baum

University of New Hampshire

E. O. Wilson, in his book Sociobiology, predicts that both
ethology and comparative psychology "are destined to be cannabalized by neurophysiology and sensory physiology from one end
and sociobiology and behavioral ecology from the other" (1975,
p. 6). "Amalgamated with" might have been a less provocative
phrase to use than "cannabalized by", but the thrust of Wilson's
statement is undeniable; research on behavior within biology
has converged with research on behavior within psychology.
Researchers from both disciplines have for years jointly
scrutinized nervous systems and sensory systems in search of
mechanisms underlying behavior. That will not concern us
here. The newer convergence, between sociobiology and behavioral ecology, on the one hand, and the psychological study
of behavior, on the other, must concern us, because it is
bound to affect the future course of research on behavior.

Continuity of Species

When Charles Darwin first proposed his theory of evolution, he,
like Wilson, saw that it would affect the study of behavior
and, through that, psychology. Darwin recognized that evolutionary theory implied a biological continuity between animal
species and the human species; man might be unique, but no more
than any other species. If man evolved from animal ancestors,
Darwin reasoned, then one should find not only correspondences
between animal and human anatomy, but also between animal and
human behavior and mental life. His book, "The Expression of
Emotions in Man and Animals" (Darwin, 1965) pursues exactly
this theme.

Romanes (1882) took Darwin's reasoning a step farther, arguing
that one could generally infer rudimentary human mental processes in animals that behave in ways comparable to the ways
of humans. He interpreted the continuity of species to mean,
"As we study human behavior, so we should study animal behavior."

In the United States, the functionalists, such as James, Dewey, and Angell, followed much the same line of reasoning. If mind and behavior evolved through the mechanism of natural selection, they must therefore be useful to the organism that possesses them in promoting survival. (Today we would say "enhancing fitness.") Behavior and mental processes will best be understood in light of the function they serve in the organism's environment (i.e., ecological niche).

Watson, trained as a functionalist, embraced the evolutionary framework. He rebelled, however, against the inference of mental processes in animals that Darwin, Romanes, and his teachers proposed. He complained at being forced into "the absurd position of attempting to construct the conscious content of the animal whose behavior we have been studying" (1913, p. 159). He rejected the idea that, having studied an animal's behavior, "we should still feel that the task is unfinished and that the results are worthless, until we can interpret them by analogy in the light of consciousness" (p. 160). Such reasoning by analogy was uncertain at best. Whereas behavior could be observed, consciousness could only be inferred. The former had a place in scientific inquiry; the latter did not.

Watson identified the origin of the error: "the position that the standing of an observation upon behavior is determined by its fruitfulness in yielding results which are interpretable only in the narrow realm of (really human) consciousness" (p. 160). The mistake was "to make consciousness, as the human being knows it, the center of reference of all behavior" (p. 161). He compared this position to the attitude of biologists when first confronted with Darwin's theory, when "the wealth of material collected...was considered valuable in so far as it tended to develop the concept of evolution in man" (p. 162). The situation in biology eventually changed, however:

> The moment zoology undertook the experimental study of evolution and descent, the situation immediately changed. Man ceased to be the center of reference. I doubt if any experimental biologist today, unless actually engaged in the problem of race differentiation in man, tries to interpret his findings in terms of human evolution, or ever refers to it in his thinking. (p. 162)

The lesson was clear. Just as biology had given up anthropocentrism in studying evolution, so psychology must give up anthropocentrism in studying behavior. Just as biologists

study general processes of evolution that allow one to understand human evolution as one special case, so psychologists
should study general processes of behavior that allow one to
understand human behavior as one special case. Indeed, only
by studying behavior in general could we hope to gain a better
understanding of human behavior in particular. Anthropocentrism is an impediment to the very goal it advocates.
Watson therefore interpreted the continuity of species in the
opposite way to Darwin and Romanes, as if to say, "As we study
animals, so we should study man."

Learning and Anthropocentrism

Psychologists, in particular, but behavioral scientists in
general, have been slow to take Watson's point. One still
finds it said in introductory psychology textbooks that the
reason for studying animal behavior is that one may find there
simple models of human behavior. Ask any social or clinical
psychologist why we study animals, and chances are this will
be the answer. Why so much resistance? Why so much clinging
to anthropocentrism? Because, despite Darwin, scholars still
widely hold the seventeenth-century view that human behavior
is fundamentally different from the behavior of animals. At
stake are the idea of free will and all the social institutions
on which it is founded. Darwin's and Watson's views imply
that human behavior, just like that of animals, is constrained
by genes, physiology, and the structure of the body. When we
add to this list the influence of environmental factors, no
room remains for free will.

The logic of the Darwinian Revolution remains unassailable,
however. If the human species evolved in the same way as all
other species, then it must be subject to the same sorts of
laws. Unless one denies that human beings are the product
of natural selection, then one must accept that human behavior is determined by the same sorts of factors and to the
same extent as animal behavior. In predicting the "canna-
balizing" of comparative psychology, Wilson predicted only
the further progress of the Darwinian Revolution. He might
as well have included social psychology, sociology, and
anthropology.

The mixing of behavioral ecology with the psychological study
of behavior has begun, and it could not have come at a better
time. For fifty years following Watson, the study of behavior in the United States emphasized learning. One may
argue that this is because Americans tend to be practically
oriented, that studying how to modify behavior would probably
lead to practical applications. Or one may argue that studying

learning would be most compatible with the spirit of democracy.
Whichever way one views it, this emphasis on learning was a
form of anthropocentrism. No doubt human beings are unique
among species in the amount they learn. To study animal behavior only with a view to understanding learning is to study
animals as if they were merely simpler human beings. As
Watson predicted, it now appears that this anthropocentrism has
impeded progress toward understanding human behavior--this at
least is an implication of Wilson's book and of sociobiology
in general. The emphasis on understanding (really human)
learning tended to blind psychologists to the importance of
other factors: genetic, physiological, and environmental.

General Laws of Learning

The study of learning sought for general laws. The word
"general" took on a special meaning, however. From Pavlov and
Skinner, we got the idea of so-called "arbitrary" stimuli,
responses, and reinforcers. This meant, presumably, that one
could pick freely, without constraint ("arbitrarily") among
stimuli, responses, and reinforcers when designing experiments.

Then came a series of embarrassments for such a view: autoshaping, polydipsia, taste avoidance. Evidence accumulated
rapidly that there were no "arbitrary" stimuli, responses,
and reinforcers. Old evidence that had been ignored, such as
"instinctive drift" (Breland & Breland, 1961), received new
attention. The idea of "general" laws of learning was called
into question (Seligman, 1970).

There is much room for misunderstanding here. The special
generality that is invalidated along with the notion of
"arbitrariness" refers to generality across stimuli, responses,
and reinforcers. Other types of generality are possible. In
particular, the evidence in no way precludes the idea of general
laws or paradigms of learning that govern learning whenever
and to whatever extent it occurs. No matter how constrained
the range of possible stimuli, responses, or reinforcers,
learning, when it occurs, could still proceed by association
(classical conditioning) and selection by consequences (operant
conditioning). This is generality across situations and
species (cf. Staddon and Simmelhag, 1971).

Some researchers have suggested, mistakenly, discarding the
law of effect (e.g., Collier and Rovee-Collier, in press).
The suggestion arises from a confusion of the two types of
generality. If one assumes the law of effect to be wedded to
the notion of arbitrary stimuli, responses, and reinforcers,
then it has been invalidated. But there is another view.

The law of effect has defied attempts at theoretical restate-
ment. No attempt to reduce stimuli responses, or reinforcers
to abstract qualities, such as energy or movement or need
reduction, has ever succeeded. The law of effect was and remains circular, just as the laws of thermodynamics are circular
(see Pratt, 1939, and Rachlin, 1971, for further discussion).
This is probably what Skinner (1950) meant when he wrote,
"...the Law of Effect is no theory. It simply specifies a
procedure for altering the probability of a chosen response."

Learning and Evolution

Although the concepts we bring to bear on learning--reward,
punishment, and association--remain valid, they may be a great
deal less important than we used to think. The flood of
evidence that learning almost always occurs within a variety
of major constraints (e.g., Hinde and Stevenson-Hinde, 1973;
Seligman and Hager, 1972) must be integrated into our thinking. It reminds us that the study of behavior must always take
account of phylogeny, of selection and niche. Learning, like
any adaptation, has both benefits (e.g., flexibility) and costs
(e.g., development of underlying mechanisms and unreliability;
Alcock, 1979). Often speed and reliability matter more than
flexibility--as, for example, when potential prey responds
to a predator. If natural selection is an optimizing process
(Wilson, 1975), then the flexibility represented by learning
ought to occur only to the extent that and in the context in
which it contributes to fitness (Alcock, 1979). In other words,
we should expect that learning will generally be constrained; a
behavioral outcome will be as reliable as is compatible with
the advantage of flexibility.

The interests of experimental psychologists and behavioral
ecologists overlap substantially--more perhaps than either
group yet sees. If we are to understand behavior in general,
we will have to consider how it contributes to fitness. This
will be the way also to understanding human behavior as a
special case. No doubt, for example, the acquisition of
language involves learning. But if verbal behavior increases
fitness, what is the likelihood that its ontogeny is unconstrained? The degree of reliability in the acquisition--
virtually every child acquires the language to which it is
exposed within the first two years of life, without special
instruction--can only reflect major genetic and environmental
constraints. To this extent, Chomsky's (1959) attack on
Skinner's (1957) Verbal Behavior was justified, but only to
this extent. Its inaccuracy and wrong-headedness put linguists
and behavioral psychologists unnecessarily at odds

(MacCorquodale, 1969; 1970). Verbal behavior can well be
treated as operant behavior, but like any operant behavior, it
develops and occurs within constraints. If we hope to understand language, we must face both of these propositions.

When I was a graduate student, my fellow students and I often
speculated about the "real" world of nature and attempted to
relate our experiments to it. This tendency, which has probably persisted from Darwin's day to the present, has kept
psychological research on behavior more or less on the right
track. We can build on what we have done, but we need to re-
mind ourselves that the behavior we study was made possible
only by a long history of evolution.

Natural selection favors organisms that are more fit because
they can better gain resources (e.g., food, nest sites, mates,
territories), better avoid danger (e.g., predators, heights,
freezing), and better respond to signs of potential resources
and dangers (e.g., sights, sounds, smells). When fitness is
enhanced by some degree of flexibility in the behavior needed
to gain a resource, the means for positive reinforcement are
selected for. The range of responses possible and the nature
of the reinforcer, of course, will be selected as well. When
fitness is enhanced by some degree of flexibility in the behavior needed to avoid danger, the means for negative reinforcement are selected for. As with positive reinforcement, the
range of responses possible and the nature of the aversive
stimulus will also be selected and, therefore, constrained.
For the sake of completeness, similar statements could be made
about positive punishment, in which the responses that produce
an aversive stimulus (punisher) decrease in frequency, and
negative punishment, or omission training, in which the responses that interfere with gaining a resource (reinforcer)
decrease in frequency. Like positive and negative reinforcement, either of these would be selected for, if it enhanced
fitness in the organism's natural habitat. Finally, responding to "signs of potential resources and dangers" refers to
association--that is, to classical conditioning and stimulus
control. If fitness is enhanced by flexibility as to which
signs of a resource or a danger produce changes in behavior
that prepare for the resource or danger or that make resource
more attainable or the danger less likely, then the means for
such sign-flexibility will be selected for. The range of possible signs (conditional or discriminative stimuli), as well as
the reinforcer or aversive stimulus, also will be selected and,
therefore, constrained. In every situation, the advantages of
flexibility (learning) are balanced (constrained) by the
advantages of reliability. Psychologists need to balance the
study of flexibility with the study of reliability.

To understand how selection has favored reliability, as well as flexibility, the so-called "arbitrary" stimuli, responses, and reinforcers will probably prove inadequate. Reliability in behavior depends on features of the organism's habitat being predictable, and predictable over many generations. To study the interaction between the laws of flexibility and the laws of reliability, therefore, we need more closely to approximate the predictable features of organisms' natural habitats. For "arbitrary", one might as well read "artificial". Indeed, the artificiality of our experimental situations is the issue here. If the experiments are to be more "natural", we need to know how to make them more "natural".

Three Artificialities

The typical experiment on operant behavior contains at least three artificialities: (1) it occurs inside a small box, rather than outdoors, (2) it occupies only a small portion of the organism's active hours, and (3) it presents food on a schedule that bears little resemblance to the occurrence of food in nature. No doubt this list could be expanded; for now, we consider just these most obvious departures from nature.

The Little Box

Many invertebrates and most vertebrates hunt for their food. They cover spaces that are large relative to the size of their bodies. The typical experimental chamber provides no comparable space; the possibilities for locomotion are limited.

The steps required to overcome this limitation seem obvious: build bigger experimental spaces, or better yet, work outdoors. Some experimenters have made moves in this direction. Graft, Lea, and Whitworth (1977) and Goldstein (1981a and b) studied rats in larger than usual spaces. Graft et al. studied choice between concurrent variable-interval schedules in which the response alternatives were feet apart. The results seemed to compare to those from more typical studies (Baum, 1979). Goldstein (1981b) varied the work requirements (fixed-ratio schedules) at eight different feeding stations and found high sensitivity to small changes.

Similar published studies with pigeons are remarkably absent. One outdoor experiment (Baum, 1974c) studied a flock of wild pigeons, free to come and go as they pleased. The choice relations describing pecks at two adjacent response keys, each producing food according to various variable-interval schedules, compared favorably with results from more typical experiments. In fact, the data were more orderly and less variable than usual. More needs to be done, particularly to investigate the

effects of spacing the response alternatives far enough apart
that the pigeons must walk or fly (which? it may make a difference) from one to the other. The general paradigm of making
apparatus available to freely roving wild or farm animals has
many advantages for all sorts of experiments. In particular,
since the animals can obtain sustenance from other sources and
have their own living quarters, one need not provide nearly as
much caretaking - no colony, no cages, etc.

Even if an experimenter lacks room in the laboratory for larger
experimental spaces and has no access to outdoor arrangements,
by a simple modification he can introduce locomotion even into
the typical little box. It requires only adding a partition
down the middle of the space. One can, for example, require
a pigeon to walk substantial distances by reinforcing pecks at
two keys, one on either side of the partition. Or one could
have a platform on either side of the partition to sense the
bird's presence, in analogy to the two-platform box used by
Baum and Rachlin (1969). Using a partition has the advantage
over the Baum-Rachlin type of arrangement that it requires only
one feeder; a small opening in the partition can allow a pigeon
to eat from either side.

Though simple technically and conceptually, adding a partition
opens a wide variety of possibilities. It allows the experimenter to define and reinforce operant behavior to include
locomotion. In particular, it allows analogies to the kind of
hunting animals do for their food in nature. With locomotion
required, an animal can be made to hunt for its food even in a
little box.

Using a partition has implications and, as it turns out, profound effects on experiments on choice between concurrent
schedules, because the partition requires that the animal
travel from one alternative to the other. In one experiment
(Baum, 1982), two groups of three pigeons each had to move
around a partition to change response keys in choosing between
otherwise standard concurrent variable-interval schedules.
Group 1 had to go around an eight-inch partition and, in
addition, jump over a hurdle. Group 2 had to go around just
a four-inch partition. Figure 1 illustrates the basic results.
The data were averaged within groups. For each group, the
data were fitted by the equation:

$$\log \frac{B_1}{B_2} = a \log \frac{r_1}{r_2} + \log b \qquad\qquad (1)$$

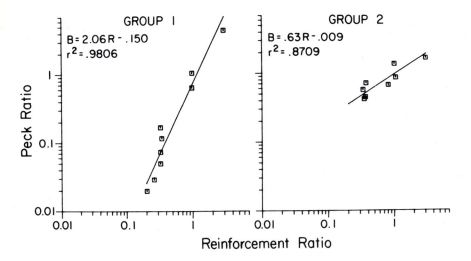

Figure 1. Choice relations for two groups of pigeons
with different travel requirements. Group 1, with
the larger requirement produced overmatching,
whereas Group 2, with the smaller requirement
produced undermatching.

the logarithm of the power law of choice:

$$\frac{B_1}{B_2} = b \left(\frac{r_1}{r_2}\right)^a \tag{2}$$

where B_1 and B_2 are measured here as pecks at key 1 and key 2,
r_1 and r_2 are reinforcement obtained from keys 1 and 2, b represents any inherent bias toward one key or the other, and a
represents the sensitivity of the distribution of behavior to
the distribution of reinforcement between the two alternatives
(Baum, 1974b). Conformance to the matching law means a equal
to 1.0. Experimental results frequently reveal undermatching
($a < 1.0$), rarely overmatching ($a > 1.0$). Figure 1 shows that
Equation 1 fitted both groups' data well. Although neither
group exhibited much bias (b approximately 1.0), they differed
substantially in the slope a. Group 2, with the four-inch
partition, produced results comparable to those of more typical
experiments ($a \leq 1.0$), whereas Group 1 produced strong overmatching, an atypical result.

Why does partition length change the choice relation this way?
The analogy to foraging suggests that increased travel should
affect behavioral allocation by affecting duration of stays

at the alternatives ("patches" or keys). Increasing the travel
requirement lengthened the stays at both alternatives, whether
measured by time of pecks (Baum, 1982). Perhaps to understand
the choice relations we must seek for order in the stays at
the alternatives.

My student, Doug Lea, has been studying closely the duration
of stays on two platforms on either side of a 21-cm partition.
Figures 2A and 2B give a sample of his findings. Each inset

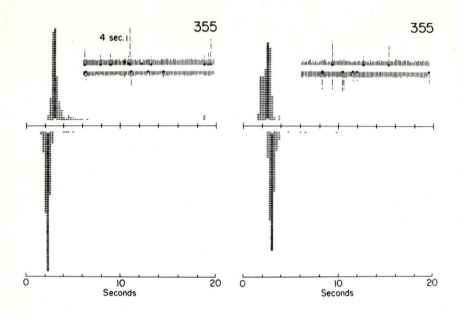

Figure 2A. Stay times of Pigeon 355 in two sessions
    from two different conditions. Insets show portions
    of the sessions in sequence from left to right.
    Distance upward indicates duration of stay on one
    side; distance downward, the other side. Dots
    indicate reinforcement. The main panels give
    frequency distributions, over the whole session,
    of stay times at the two sides. Frequency in-
    creases in an upward direction for one side and
    increases in a downward direction for the other.

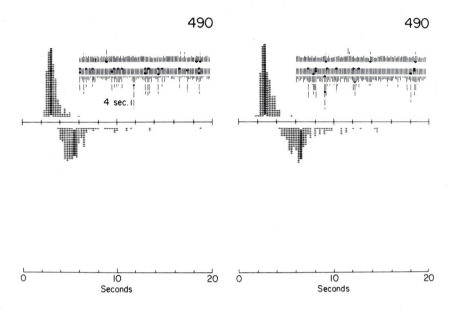

Figure 2B. Stay times of Pigeon 490 in two sessions, from
two different conditions, as in Figure 2A.

shows a portion of the record of a single session. Time goes
from left to right. The lengths of lines going up from the
time line represent the durations of stays at one side, the
lengths of lines going down represent durations of stays at
the other side. Dots indicate occurrences of reinforcement.
Except for those stays that include reinforcement, the durations of stay vary remarkably little. Reinforcement almost
always lengthens the stay, but in the absence of reinforcement, the vast majority of stays vary only over a range of
about a second. The main panels in Figure 2A and 2B make
this clearer. Each pair shows, for a single session, the two
frequency distributions of durations for the two sides. Duration is represented from left to right. Frequency is represented going upward from the axis for one side, downward for
the other. The distributions have strong modes and are remarkably narrow.

Figures 2A and 2B suggest that, when the choice alternatives
are separated by a partition, pigeons spend relatively constant
periods of time at the two sides, failing to switch only if
reinforcement occurs--a sort of "win-stay, lose-shift" strategy.

Questions remain. What determines that even unreinforced stays
are longer on the richer side? How much does a reinforcer
lengthen a stay? And so on.

Putting a partition into the chamber produces some remarkable
results. Perhaps requiring still longer travel for changeover
will produce further surprises. How can we reconcile the results with previous experiments, which included no obvious
travel requirement?

The key to answering this question lies in the contingencies
surrounding changeover in typical concurrent schedules. Figure
3 shows the usual sequence of events, with time going from left
to right. At first, the pigeon is pecking at key 1. After a

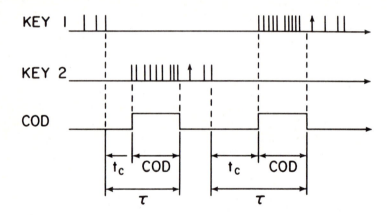

Figure 3. Sequence of events in the typical experiment on concurrent schedules of reinforcement.
Lines for Key 1 and Key 2 indicate responses
at the keys. COD stands for changeover delay.
The interval t_c is the time necessary to change
from responding at one alternative to responding at the other. See text for further explanation.

while, it changes over to pecking key 2, taking a small amount
of time (labelled t_c) to make the switch. At the first peck
on key 2, the changeover delay (COD) begins. During the COD,
no reinforcement can be obtained. The first peck that may produce reinforcement, after the COD, is symbolized with an arrow.
The pigeon responds at a higher rate during the COD, by a
factor of two or more, than after (Baum, 1974a, Silberberg and
Fantino, 1970). The same kind of sequence repeats when the
pigeon switches back to key 1.

Figure 3 makes it clear that, for every changeover, a period of time τ goes by, equal to t_C + COD, during which no reinforcement can be obtained. Since the change of keys must be obvious and the pigeon responds at a higher rate during the COD, we can consider this period of nonreinforcement τ to be discriminated.

The time period τ has at least three characteristics in common with travel: (1) it is obligatory for changeover, (2) it is time during which no feeding occurs, and (3) it is distinct from time during which feeding does occur. It differs from travel, of course, in that it requires neither locomotion nor exertion. Although only further research can ascertain how important these differences are, some evidence suggests that they may not be crucial. First, excluding responses made during the COD from the response ratios, either by recording them separately or by preventing them with black-out during the COD, changes the choice relation from undermatching or matching to general overmatching (Baum, 1974a; Silberberg and Fantino, 1970; Todorov, 1971). Requiring a moderate number of responses on a separate key for changeover, and excluding those responses from the response ratios, likewise leads to overmatching (Baum, 1982; Pliskoff, Cicerone, and Nelson, 1978; Pliskoff and Fetterman, 1981). These parallels with the results in Figure 1 suggest that the changeover time τ in Figure 3 should be treated as analogous to travel and, therefore, excluded from the measures of the stays at the alternatives. One would logically conclude, then, that experiments on choice should typically produce, not matching, but overmatching.

Two kinds of evidence appear to go against such a conclusion. First, some experiments produced matching of time ratios to reinforcement ratios, even though the COD, which was signalled, was subtracted from the times (Baum and Rachlin, 1969; Baum, 1973). Since these data appear to be corrected for τ, it might seem they could not be described as overmatching. Since these experiments were conducted in a two-panel box, however, one cannot ignore t_C, the time to move from one alternative to the other. Even if t_C is only a fraction of a second, its exclusion can change apparent matching to strong overmatching, particularly when strong preference leads to brief stays at the nonpreferred alternative. Second, some experiments have produced matching without a COD (Baum, 1974c; Heyman, 1979). Since, t_C went unmeasured in these experiments, they would be subject to the same criticism as the two-panel experiments, were it not that changeover between two adjacent keys requires only a fraction of a second. In such a situation τ should be negligible and the results could not be described as overmatching. In nature, however, although travel (τ) is generally

significant, it does vary. The matching obtained in these
situations with negligible τ may represent a boundary condition;
as τ grows, the choice relation moves to more and more overmatching.

Overmatching, as a description of choice is less simple, hence
less attractive, than matching. If we pursue the analogy
to nature, the way back to simplicity doubtless lies, not
through such molar analysis, but through a more molecular
examination of durations of stay, as illustrated in Figures 2A
and 2B.

Brief Sessions

An organism may behave a certain way during a one-hour experimental session, but what is it doing the other twenty-three
hours? Can one assume that the one-hour sample is representative? How might its behavior change if it had access to
the experimental situation 24 hours a day? Organisms in the
wild typically forage for food for substantial portions of the
day at certain times of day. I find, for example, that pigeons
with 24-hour access to a variable-interval schedule peck at the
response key mostly in the early morning and late afternoon.
Usual experimental procedures take no account of such diurnal
patterns.

A number of experiments with 24-hour access have been done.
Collier and his associates have studied a variety of species
responding on variable-ratio schedules for food or water
(Collier, 1980; Collier, in press). Hursh (1980), Graft et al.
(1977), and I (Baum, 1972, 1974c) have studied continuously
available concurrent variable-interval schedules.

We have learned from these experiments that behavior in the
"closed economy" (the 24-hour experiment) differs relatively
little from behavior in the "open economy" (the one-hour
session). Animals still respond at high rates on variable-
ratio schedules and still match response ratios to ratios of
food reinforcement.

New questions arise, however, in the study of the closed
economy, questions that would be unlikely or impossible in
the study of the open economy. In particular, questions about
the organism's need become crucial. If a variable-ratio
schedule is increased in a closed economy, the animal must
work harder to meet its need. Collier and his associates find
that behavior rises to the challenge, even leading to responding on schedules that would fail to maintain behavior in an
open economy. Conventional experimental procedure guards

against the organism's meeting its need, against satiation,
whereas 24-hour experiments make contact with foraging in the
wild by bringing consideration of the predator's need to the
fore (Baum, in press).

To understand the role of need, 24-hour experiments are
essential. But other, perhaps more difficult, questions arise
about the closed economy and foraging in the wild. By what
rules do organisms exploit and manage the resources in their
environment? Here psychologists' interests coincide with behavioral ecologists' (e.g., Baum, in press; Krebs, 1974, 1978).
If 24-hour experiments have thus far shed relatively little
light on these questions, it is probably because they have
ventured too little distance from conventional procedures.
Most studies of closed economies have concerned individual
animals living isolated in little boxes. Prudence dictated
small departures from established paradigms, but now further
departures will be necessary. What happens, for example, when
locomotion is required in a 24-hour experiment? We need to
get the closed economy out of the little box.

The Artificiality of Schedules

To what extent do schedules of reinforcement model the obtaining of food in nature? There are many points of difference.
Foraging typically depletes the food source in nature; schedules of reinforcement typically remain invariant through time.
Viewed on a molar level, foraging resembles operant behavior,
responding interspersed with eating; but at a molecular level
foraging consists of a repeated cycle of searching, capturing,
handling, and eating. No obvious scheme exists for mapping all
of these components onto typical schedule-controlled behavior.

Several investigators have attempted to model the foraging cycle
with various chain schedules (Abarca and Fantino, 1982; Collier
and Rovee-Collier, 1981; Kamil, in press; Lea, in press;
Snyderman, Note 1). All include arbitrary and questionable
features in the procedure. Collier, for example, obliges the
animal to wait thirty seconds when passing up a more costly
reinforcer (i.e., one that requires more effort to obtain).
Does a rat pause for such a duration when it passes up a relatively inaccessible bit of food in nature? Such artificialities
are only more obvious, however, than the artificiality of
schedules with which we need to be concerned.

Some procedures model handling with an interval schedule, some
with a ratio schedule, and others with a delay. Of these,
the ratio seems most appropriate, because handling, whether
it be subduing a mouse, opening a nut, or chewing up some
foliage, represents a job to be done, with a beginning and end.

But even the ratio schedule may be inappropriate, because handling consists in manipulating a food item, not pressing a lever or pecking a key in the absence of food, as in the search phase. For a species like the pigeon, moreover, for which capture and handling naturally consist in nothing more than picking up a small seed, one may question the sense of trying to introduce substantial handling requirements at all.

Following a similar line of reasoning, one may question the sense of the typical method of presenting food to a pigeon: raising a hopper full of grain long enough for the eating of several pieces. In nature, pigeons find scattered grains or bits of food. To mimic this, one would do best to present single seeds as reinforcers.

Like handling, search has been modelled both with interval schedules and ratio schedules. Fixed-interval or fixed-ratio schedules would both be inappropriate here, because search always implies an uncertain outcome. Is search like a variable-ratio schedule? Animals that lie in wait for their prey (web-spinning spiders, ant lions, etc.) might receive food on a variable-interval schedule--more properly a variable-time schedule, since no response is needed to find the food. Search, in contrast, requires effort. As with a ratio schedule, the more one searches the more one finds. In nature, however, search generally leads to depletion; few sources of food replenish fast enough to keep up with a forager. Variable-interval schedules share to some extent the properties of depletion and replenishment; food is more likely to be obtained at the beginning of a period of responding, after an absence from the schedule, than subsequently. The rise in likelihood of reinforcement with absence resembles replenishment; the decrease in likelihood once the first reinforcer is obtained resembles depletion.

Search provides food neither on a variable-interval nor a variable-ratio schedule, but some combination of the two. Search is like a variable-ratio schedule that increases as a function of amount eaten (i.e., depletes) and decreases as a function of time (i.e., replenishes). (For a more detailed analysis see Baum, in press). To my knowledge, no one has studied such adjusting variable-ratio schedules.

Watching Animals Eat

For understanding the similarities and differences between laboratory situations and foraging in nature, there may be no substitute for spending some time watching how animals actually eat. Behavioral ecologists have been doing this for years (e.g., Burger, 1982; Gibb, 1958; Krebs, Ryan, and Charnov,

1974; Smith and Sweatman, 1974; Zach and Falls, 1976a, b, c).
Psychologists can do it too, and when they do the questions
and methods are apt to differ from those of ecologists.

To begin with, the two groups appear to differ over what
constitutes an adequate quantity of data. Psychologists, because their experiments are run and recorded by automatic
control equipment, consider a relatively large amount of data
to be adequate for publication. Psychological experiments
typically consist of several conditions, each in effect for ten
to as many as sixty days or more. With four or six animals
generating a thousand or more responses per day, such experiments commonly involve 100,000 to a million responses. Ecologists seem to publish with far fewer data. Doubtless the reason is that field work requires far more time and labor to
gather an appreciable quantity of data. Even in the laboratory,
the more natural the experimental setting the less easily it
lends itself to automation.

The psychologist, by training, will want to distinguish the
transient phase of performance in a condition from the stable
phase -- that is, adjustment to the condition from maintained
performance in the condition. Approaching the same problem,
the psychologist is likely to run more trials than the ecolo-
gist, and to chart the progress of performance across trials.
Since he will probably be concerned also about reliability, he
will feel compelled to replicate some conditions. And, if he
is concerned with quantitative relations, the psychologist will
look for a way to gather data from enough different conditions
to plot a functional relation.

A Sample Experiment

By way of example, I will describe the first experiment I did
by watching pigeons eat.

Ecologists use the term "patch" to refer to a local aggregate
of prey. The term "prey" refers to a range of items from
animate and inanimate food to a parasite's hosts. By "foraging"
or "predation" is meant the activity involved in seeking and
handling (i.e., subduing or otherwise interacting with) prey.
Broadly speaking, "prey" corresponds to "reinforcement,"
"foraging" to "instrumental behavior," and "patch" to "programmed source of reinforcement." Defining the correspondence
between patches and schedules constitutes the central problem
in bridging the gap between psychological and ecological research.

One practical approach to this problem is to model patches physically under laboratory conditions that are as natural as possible. This should allow gathering of detailed quantitative data on both groups and individual animals. The studies of Smith and Dawkins (1971), Smith and Sweatman (1974), Krebs et al. (1974), and Zach and Falls (1976a, b, c) succeeded with artificial patches in enclosed open areas. Although their data suggest that orderly relations can be found, the experiments were too brief to be more than suggestive. Longer-term parametric studies need to be done.

The subjects in this experiment were members of a flock of 20-25 mixed White Carneaux-Silver King pigeons that live in a coop on the roof of the psychology building at the University of New Hampshire. Attached to the coop is a flyway made of wire netting, approximately 8 feet wide, 8 feet high, and 60 feet long. A small door controls the pigeons' access to the flyway. Doors in the flyway allow portions of it to be shut off.

The basic experimental paradigm was adapted from previous studies (Krebs et al., 1974; Smith and Dawkins, 1971; Smith and Sweatman, 1974). An individual animal or a small group is released into an open area containing one or more discrete patches of known food density. To insure that it assesses density only by foraging in the patches, and not by cues at a distance, the prey are made cryptic. After a period of time foraging, the animal is removed from the area. In previous experiments, this time was kept short, presumably to prevent too much depletion. With larger patches and no extra cues of depletion (e.g., covers removed from cups), it was possible to conduct longer sessions.

Pigeons are ground feeders. Although they fly between patches, within a patch they walk, searching for relatively cryptic bits of food (e.g., seeds) on the ground. Accordingly, the experimental patch consisted of a plywood sheet (4 feet by 4 feet) placed on the floor of the fly pen. It contained 52 1/2 inch holes 1/2-inch deep and was painted gray. A known quantity of hemp seed was placed randomly in the holes, which are too deep to allow sight from a distance and shallow enough for easy capture. Pigeons were released from the coop in groups of five. (Difficulty in getting the pigeons to feed individually suggested the use of groups. Subsequent efforts show that with patience it is possible to get pigeons to feed individually.) An observer recorded the time of each change in the number of pigeons foraging on the patch, along with the number. A trial ended when all pigeons ceased foraging. The amount of grain left was then recorded. The amount of grain distributed varied unpredictably from day to day. Trials were run in June, July, and August of 1980, weather permitting.

The data from each trial provide a record of the number of pigeons foraging as a function of time. The lower panels of Figures 4, 5, and 6 show visual representations of such records. As the trial proceeded, the number of birds dropped, with several reversals, from five to zero.

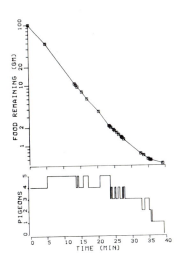

Figure 4. Record of a trial in which 100 g. of grain was distributed. Below: number of pigeons foraging as a function of time. Above: estimated grain re- maining as a function of time. See text for further explanation.

The pigeons did not simply eat up all the grain, they general- ly quit while some grain still remained, even when the amount distributed was small. The lower panel of Figure 7 shows the relation between amount eaten and amount distributed. Each point represents one trial. The diagonal line indicates the locus of equality (i.e., all grain eaten). When large quanti- ties of grain were distributed, the pigeons' intake reached an upper limit of about 125 grams; increasing the amount dis- tributed produced no systematic increase in amount eaten be- yond this amount. These results suggest that the pigeons ceased foraging under two conditions: (1) when they had eaten their fill, and (2) when some criterion of scarcity was met.

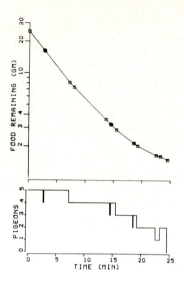

Figure 5.   Record of a trial in which 25 g of grain was distributed, as in Figure 4.

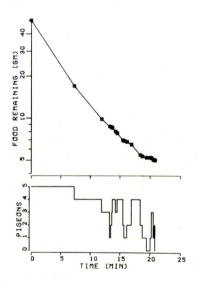

Figure 6.   Record of a trial in which 50 g of grain was distributed, as in Figure 4.

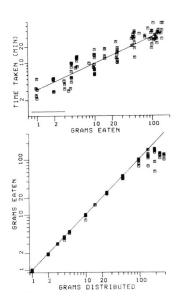

Figure 7. Foraging relations across trials. Each point
  represents one trial. Below: amount of food eaten
  as a function of amount distributed. Line indicates
  equality (all food eaten). Above: time taken to
  eat as a function of amount eaten. Horizontal line
  indicates time spent foraging with no food present.
  Diagonal line was fitted by inspection.

The upper panel of Figure 7 shows the relation between time
taken (trial duration) and amount eaten. In these logarithmic
coordinates, the relation appears roughly linear. Its slope,
however, is less than 1.0. The line drawn through the points
has a slope of .493. The time taken to eat an amount of grain,
therefore was not directly proportional to the amount; instead,
it was approximately proportional to the square root of the
amount. The horizontal line indicates the lower limit of time
spent foraging: the time spent when no grain was distributed
(1.21 min; median of 6 trials). Subtracting this time from
every trial duration fails to eliminate the lack of proportionality between time and amount.

To be able to test theories of foraging, we must be able to
estimate the amount of grain remaining in a patch at various
times (e.g., times of switching or giving up). Since, with
the present method, one cannot determine the amount of food
remaining without terminating the trial and driving off the
birds, potentially disrupting the experiment, it would be

desirable to try to estimate the amount of grain remaining with a model. The model could be verified by interruption at a later time.

If we assume that a patch resembles an adjusting ratio schedule, we need to know how the ratio adjusts. This will depend both on the animal's behavior and on the physical characteristics of the patch. One can reasonably suppose (ecologists generally do) that the probability of capturing prey is directly proportional to their density. For a patch of area A with a population size F, and probability of capture on encounter p, the probability of eating a prey (reinforcement) is pF/A. Accordingly, the rate of predation (reinforcement) r is given by:

$$r = \frac{pFs}{A} \qquad\qquad (3)$$

where s is the predator's rate of searching. Supposing A, p, and s to be constant, we can simplify Equation 3 to:

$$r = aF \qquad\qquad (4)$$

This states that the rate of reinforcement (predation) is directly proportional to the food remaining on the patch. Since the rate of reinforcement equals the rate of depletion, we can use Equation 4 to write:

$$\frac{dF}{dt} = -anF \qquad\qquad (5)$$

where a is the same constant of proportionality as in Equation 4 (ps/$\overline{A}$) and n is the number of pigeons on the patch. This differential equation leads directly to the exponential relation:

$$F = F_o e^{-ant} \qquad\qquad (6)$$

where t is time since the beginning of an interval during which n remains constant, a is the time constant of the depletion, and F_o is the amount of food at the beginning of that interval. Every time n changes, t is reset to zero and F_o is set to the amount at the time of the change. By this reasoning, and with the knowledge of the amount of food distributed, the amount of food eaten, and the durations of the various intervals (Figures 4, 5, and 6), it is possible to derive the value of a that best predicts the amount of food eaten in a trial. A computer program was written to do this. The upper panels of Figures 4,

5, and 6 show the amount of food remaining, estimated by this procedure, through the course of the three trials depected.

Figure 8 shows the values of a for a series of trials in which the amount of grain distributed varied from one gram to 150 grams. The first four trials produced unusually low values of

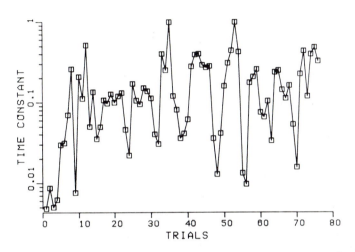

Figure 8.   Time constant a from Equation 6 across trials.

a; they were omitted from subsequent analysis. Figure 8 indicates that a varied considerably from trial to trial; it was by no means constant.

Figure 9 shows, across trials, the value of a as a function of the amount of grain eaten in the trial. The two are clearly related. In these logarithmic coordinates, a straight line indicates a power function. The line drawn has a slope of -.71, indicating that the points approximate a power function with that exponent.

Figure 9 shows only how a sort of average value of a over a trial relates to the total amount of grain eaten in the trial. It does not show how the momentary value of a changed within a trial, but it suggests that, far from remaining constant, a decreased gradually within a trial.

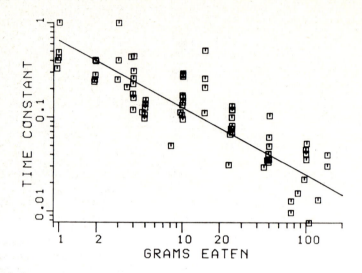

Figure 9. Time constant a from Equation 6 as a function
of amount of food eaten. Each point represents one
trial.

When a pigeon ceases foraging, we can describe that by saying
that a decreases to zero. The upper limit on food intake
in Figure 7 suggests that a must decrease to zero after
approximately 125 grams have been eaten. Below 125 grams, when
a decreases to zero, it must be because food has become scarce.
The value of a gradually decreases within a trial presumably
because the amount of food gradually decreases. This is to
say that a in Equations 4, 5, and 6 is not constant, but rather
is a direct function of F.

Since a equals ps/A, we may speculate as to which of these
three parameters might change as F decreases. Watching the
birds indicates that, as food becomes scarce, they tend to move
about more rapidly -- that is, s increases. This coincides
with observations of many vertebrate and invertebrate species
showing that search becomes more intensive when a prey is found
and more diffuse (covering more ground) when no prey is found
(e.g., Jones, 1977; Smith, 1974a, b). At the same time that
search rate (s) increases, it appears that p decreases; the
pigeons appear more likely to pass by a seed in the path of
movement. Increasing s and decreasing p tend to cancel one
another. A decrease in a would require that p decrease by a
greater factor than s increases. The only other way that a

could decrease would be for $\underline{A}$ to increase. Although the area of the patch remains fixed, the effective area, over which foraging actually takes place, might change. When five pigeons are foraging, perhaps each pigeon forages over only one fifth of the area of the patch. Then four might each forage over one fourth, and so on. In other words, $\underline{A}$ would equal the area of the patch divided by $\underline{n}$. If only one bird were present, then, from beginning to end, $\underline{a}$ would remain constant. These possibilities remain to be investigated.

The experiment raises many further questions. The amount of food varied unpredictably from day to day. How would behavior change if the same amount were put out several days in a row? What will be the effects of varying group size? What happens when a second patch, a choice, is present? Though a simple paradigm methodologically, this type of situation is behaviorally complex and potentially rewarding.

Conclusion

In the history of psychological research on behavior, never has the possibility of making contact with behavior in nature been more possible or more needful than now. An old conception is passing, the idea that we study animal behavior only to understand human behavior. We should be glad to see it go. Now we can complete the Darwinian revolution within psychology. By studying behavior, learned or not. within the framework of evolutionary theory, in the light of the ecological factors that make behavior necessary and beneficial, we can hope to understand better the behavior of all organisms, including humans.

Acknowledgement

The previously unreported research in this chapter was supported by National Science Foundation Grant BNS-06852 to the University of New Hampshire.

Reference Note

1. Snyderman, M. Prey selection and self control. Talk given at the Fifth Harvard Symposium on Quantitative Analyses of Behavior: the Effect of Delay and of Intervening Events on Reinforcement Value, Cambridge, MA., June 1981.

References

1 Abarca, N. & Fantino, E. Choice and foraging. Journal of the Experimental Analysis of Behavior, 1982, 38, 117-123.

2 Alcock, J. Animal Behavior, 2nd ed. Sunderland, MA: Sinauer Associates, 1979.

3 Baum, W. M. Choice in a continuous procedure. Psychonomic Science, 1972, 28, 263-265.

4 Baum, W. M. Time allocation and negative reinforcement. Journal of the Experimental Analysis of Behavior, 1973, 20, 313-322.

5 Baum, W. M. Chained concurrent schedules: reinforcement as situation transition. Journal of the Experimental Analysis of Behavior, 1974, 22, 91-101 (a).

6 Baum, W. M. On two types of deviation from the matching law: bias and undermatching. Journal of the Experimental Analysis of Behavior, 1974, 22, 231-242 (b).

7 Baum, W. M. Choice in free-ranging wild pigeons. Science, 1974, 185, 78-79 (c).

8 Baum, W. M. Matching; undermatching, and overmatching in studies of choice. Journal of the Experimental Analysis of Behavior, 1979, 32, 269-281.

9 Baum, W. M. Choice, changeover, and travel. Journal of Experimental Analysis of Behavior, 1982, 38, 35-49.

10 Baum, W. M. Instrumental behavior and foraging in the wild. In M. L. Commons, R. J. Herrnstein & H. Rachlin (Eds.) Quantitative Analyses of Behavior, Vol. 2: Matching and Maximizing Accounts. Cambridge, MA: Ballinger, in press.

11 Baum, W. M. & Rachlin, H. C. Choice as time allocation. Journal of the Experimental Analysis of Behavior, 1969, 12, 861-874.

12 Breland, K. & Breland, M. The misbehavior of organisms. American Psychologist, 1961, 16, 681-684.

13 Burger, A. E. Foraging behaviour of lesser sheathbills Chionis minor exploiting invertebrates on a sub-antarctic island. Oecologia, 1982, 52, 236-245.

14   Chomsky, N.   Verbal Behavior, by B. F. Skinner.   Language,
     1959, 35, 26-58.

15   Collier, G. H.   An ecological analysis of motivation.   In
     F. M. Toates & T. R. Halliday (Eds.), Analysis of Motiva-
     tional Processes.   London: Academic Press, 1980.

16   Collier, G. H.   Life in a closed economy: the ecology of
     learning and motivation.   In M. D. Zeiler & P. Harzem
     (Eds.), Advances in Analysis of Behavior, Vol. 3: Biolo-
     gical Factors in Learning.   New York: Wiley, in press.

17   Collier, G. H. & Rovee-Collier, C. K.   An ecological per-
     spective of reinforcement and motivation.   In E. Satinoff
     and P. Teitelbaum (Eds.), Handbook of Neurobiology:
     Motivation.   New York: Plenum, in press.

18   Darwin, C.   The Expression of the Emotions in Man and
     Animals.   Chicago: University of Chicago Press, 1965.

19   Gibb, J. A.   Predation by tits and squirrels on the
     eucosmid Enarmonia conicolana (Heyl).   Journal of Animal
     Ecology, 1958, 27, 375-396.

20   Graft, D. A., Lea, S. E. G., & Whitworth, T. L.   The
     matching law in and within groups of rats.   Journal of the
     Experimental Analysis of Behavior, 1977, 27, 183-194.

21   Goldstein, S. R.   An operant arena for rats.   Behavior
     Research Methods & Instrumentation, 1981, 13, 37-39 (a).

22   Goldstein, S. F.   Schedule control of dispersion and density
     patterns in rats.   Behaviour Analysis Letters, 1981, 1,
     283, 295 (b).

23   Heyman, G. M.   A Markov model description of changeover
     probabilities on concurrent variable-interval schedules.
     Journal of the Experimental Analysis of Behavior, 1979,
     31, 41-51.

24   Hinde, R. A. & Stevenson-Hinde, J.   Constraints on Learning.
     New York: Academic Press, 1973.

25   Hursh, S. R.   Economic concepts for the analysis of be-
     havior.   Journal of the Experimental Analysis of Behavior,
     1980, 34, 219-238.

26   Jones, R. E.   Search behavior: a study of three caterpillar
     species.   Behaviour, 1977, 60, 237-259.

27 Kamil, A. To maximize or not to maximize: an ecological
    question. In M. L. Commons, R. J. Herrnstein, &
    H. Rachlin (Eds.), Quantitative Analysis of Behavior, Vol.
    2: Matching and Maximizing Accounts. Cambridge, MA:
    Ballinger, in press.

28 Krebs, J. R. Behavioral aspects of predation: In P. P.
    Bateson & P. H. Klopfer (Eds.), Perspectives in Ethology,
    New York: Plenum Press, 1974, 73-111.

29 Krebs, J. R. Optimal foraging: Decision rules for pre-
    dators. In J. R. Krebs & N. B. Davies (Eds.), Behavioural
    Ecology. Oxford: Blackwell Scientific, 1978.

30 Krebs, J. R., Ryan, J., & Charnov, E. L. Hunting by
    expectation or optimal foraging? A study of patch use by
    chickadees. Animal Behavior, 1974, 22, 953-964.

31 Lea, S. E. G. The mechanism of optimality in foraging. In
    M. L. Commons, R. J. Herrnstein & H. Rachlin (Eds.),
    Quantitative Analysis of Behavior, Vol. 2: Matching and
    Maximizing Accounts. Cambridge, MA: Ballinger, in press.

32 MacCorquodale, K., B. F. Skinner's Verbal Behavior: a
    retrospective appreciation. Journal of the Experimental
    Analysis of Behavior, 1969, 12, 831-841.

33 MacCorquodale, K. On Chomsky's review of Skinner's Verbal
    Behavior. Journal of the Experimental Analysis of Be-
    havior. 1970, 13, 83-99.

34 Pliskoff, S. S., Cicerone, R. & Nelson, T. D. Local
    response-rate constancy on concurrent variable-interval
    schedules of reinforcement. Journal of the Experimental
    Analysis of Behavior, 1978, 29, 431-446.

35 Pliskoff, S. S. & Fetterman, J. G. Undermatching and
    overmatching: the fixed-ratio changeover requirement.
    Journal of the Experimental Analysis of Behavior, 1981,
    36, 21-27.

36 Pratt, C. C. The Logic of Modern Psychology. New York:
    MacMillan, 1939.

37 Rachlin, H. On the tautology of the matching law. Journal
    of the Experimental Analysis of Behavior, 1971, 15, 249-251.

38 Romanes, G. J. Animal Intelligence. London: Kegan Paul,
    Trench, & Co., 1882.

39 Seligman, M. E. P. On the generality of the laws of learn-
    ing. Psychological Review, 1970, 77, 406-418.

40 Seligman, M. E. P. & Hager, J. L. Biological Boundaries
    of Learning. New York: Appleton-Century-Crofts, 1972.

41 Silberberg, A. & Fantino, E. Choice, rate of reinforcement,
    and changeover delay. Journal of the Experimental Analysis
    of Behavior, 1970, 13, 187-197.

42 Skinner, B. F. Are theories of learning necessary?
    Psychological Review, 1950, 57, 193-216.

43 Skinner, B. F. Verbal Behavior. New York: Appleton-
    Century-Crofts, 1957.

44 Smith, J. N. M. The food searching behavior of two
    European threshes: I. Description and analysis of search
    paths. Behaviour, 1974, 48, 276-302 (a).

45 Smith, J. N. M. The food searching behavior of two
    European threshes: II. The adaptiveness of the search
    patterns. Behaviour, 1974, 1-61 (b).

46 Smith, J. N. M. & Sweatman, H. P. A. Food searching be-
    havior of titmice in patchy environments. Ecology,
    1974, 55, 1216-1232.

47 Staddon, J. E. R. & Simmelhag, V. The "superstition"
    experiment: a re-examination of its implications for the
    principles of adaptive behavior. Psychological Review,
    1971, 78, 3-43.

48 Todorov, J. C. Concurrent performances: effect of punish-
    ment contingent on the switching response. Journal of the
    Experimental Analysis of Behavior, 1971, 16, 51-62.

49 Watson, J. B. Psychology as the behaviorist views it.
    Psychological Review, 1913, 20, 158-177.

50 Wilson, E. O. Sociobiology. Cambridge: Harvard University
    Press, 1975.

51 Zach, R. & Falls, J. B. Ovenbird (Aves: Parulidae) hunt-
    ing behavior in a patchy environment: an experimental
    study. Canadian Journal of Zoology, 1976, 54, 1863-
    1879 (a).

52   Zach, R. & Falls, J. B.   Foraging behavior, learning, and
     exploration by captive ovenbirds (Aves: Parulidae).
     Canadian Journal of Zoology, 1976, 54, 1880-1893 (b).

53   Zach, R. & Falls, J. B.   Do ovenbirds (Aves: Parulidae)
     hunt by expectation?   Canadian Journal of Zoology, 1976,
     54, 1894-1903 (c).

ANIMAL COGNITION AND BEHAVIOR
Roger L. Mellgren, editor
© North-Holland Publishing Company, 1983

CONDITIONED DEFENSIVE BURYING:
A BIOLOGICAL AND COGNITIVE APPROACH
TO AVOIDANCE LEARNING

John P.J. Pinel and Donald M. Wilkie

The University of British Columbia

Most rats will use bedding or other material to bury an object that has been the source of a single electric shock, a behavior that Terlecki, Pinel, and Treit (1979) have called conditioned defensive burying. This chapter is a discussion of recent research on the conditioned defensive burying phenomenon, a substantial proportion of which has been conducted at the University of British Columbia by ourselves and our collaborators -- most notably: MacLennan, Spetch, Terlecki, and Treit. In addition to summarizing the conditioned defensive burying experiments, the purpose of the present chapter was to illustrate the usefulness of studying animal avoidance learning from cognitive and biological perspectives. Although Hudson's (1950) original studies of conditioned defensive burying predate the current popularity of cognitive and biological analyses of animal learning, from the outset the study of conditioned defensive burying has been influenced by two assumptions: 1) a cognitive assumption that the conditioned defensive burying phenomenon reflects the acquisition of information about environmental events rather than the simple strengthening of a stimulus-response relationship, and 2) a biological assumption that it is useful to study animal learning in relation to the survival requirements of the species' natural environment.

The four major sections of the chapter include: 1) Hudson's (1950) original observations of conditioned defensive burying, 2) the more recent development of the conditioned defensive burying paradigm, 3) a summary of some of the major environmental and organismic variables that have been found to influence conditioned defensive burying, and 4) a description of some of the features of conditioned defensive burying that illustrate its cognitive basis. The use of the conditioned defensive burying paradigm to assess the behavioral effects of neurosurgical and neuropharmacological manipulations has been recently reviewed (Pinel and Treit, in press) and will not be discussed here.

The 1950 Monograph of Bradford B. Hudson

During the height of the stimulus-response (S-R) tradition, Bradford B. Hudson, A student of Tolman's, published as a monograph a series of studies of animal avoidance learning. One notable feature of these studies was that they did not reflect the tendency for psychologists of that era to study animal learning in situations that had little obvious relation to the natural environments of their subjects. Rather than trying to increase the generality of his results by minimizing the effects on learning of specific biological predispositions, Hudson purposely selected for study a form of learning that reflected the specific survival requirements of his subjects' natural environment.

The object of Hudson's study was one-trial avoidance learning in the rat. Although psychologists interested in animal learning had focused almost exclusively on forms of learning involving multiple-trial acquisition, Hudson concluded that one-trial avoidance learning warranted particularly careful experimental analysis.

> From the point of view of survival, an animal in its native environment cannot afford repeatedly to experience attacks from its natural enemies and to depend upon their cumulative effect to finally condition him to respond appropriately to the sight or sound of them; rather learning must necessarily be, and is, rapid under such circumstances. (Hudson, 1950, p. 103)

In Hudson's experiments, rats received a single shock to the mouth as they were feeding in their home cages from a mash-baited metal food holder. The foodholder and the black and white striped bakelite stimulus panel on which it was mounted were removed from the cage 4 min after the shock. Learning was assessed by later returning the stimulus panel with baited food holder to the wall of the home cage and recording each rat's reactions to it during the ensuing 5-min test (extinction) period.

During the test, most of the subjects incorporated their bedding material into an interesting defensive maneuver. They sent showers of wood shavings at the panel-mounted food holder with forward pushing motions of their forepaws. Although there was considerable variability in the amount of time that each rat spent spraying bedding at the food holder and in the degree to which the food holder was covered by the bedding, a full 95% of the shocked animals engaged in this behavior at least once during the brief test.

> Pushing of shavings proved to be a useful measure....
> It is usually accomplished by a movement of the
> forepaws which sends a shower of shavings in the
> direction of the object of avoidance. During the
> first extinction trials, although the range of
> individual differences is very large, many ani-
> mals proceeded to cover the pattern by piling the
> shavings from the floor of the cage against it,
> and having partially or fully accomplished this,
> would occasionally approach close enough to snatch
> food from the holder. Many animals would spent a
> large part of the 5-minute test period gathering
> material together and then flicking it with their
> forepaws or carrying it in their mouth to the
> pattern, then often stopping to pack it more firm-
> ly against it, after which returning to the end
> of the cage to gather more. Anything in the cage
> that was loose...would be carried to the pile.
> (Hudson, 1950, pp. 106-107).

The observation that rats can learn in a single trial to approach and bury a source of noxious shock was, and still is,
totally incompatible with any of the S-R theories of avoidance.
The essence of most S-R approaches to avoidance learning is the
attribution of the development of the avoidance response to
instrumental conditioning. Negative reinforcement is presumed
to act retroactively on the avoidance response during the learning trials. For example, according to one version of two-factor
theory (e.g., Mowrer, 1947), animals learn to flee from stimuli
associated with noxious stimulation, because fleeing is negatively reinforced by the decline in "anxiety" associated with
increased separation from these stimuli. In the case of conditioned defensive burying, however, such explanations are
clearly untenable. Because a substantial period of burying behavior is required to accumulate a pile high enough even to
reach the shock source mounted on the wall, it is difficult to
account for the initiation of the response by appealing to the
negative reinforcement emanating from any fear reduction that
might be a product of covering of the shock source (cf. Peacock
and Wong, 1982). Moreover, it is difficult to image how a
behavior as complex as burying could be shaped by reinforcement
contingencies in a single trial.

And S-R accounts of Hudson's observations that are based on
Pavlovian principles are no more plausible. Although one might
view the burying observed by Hudson as a conditioned response
to a conditioned stimulus, i.e. the food holder, such a hypothesis suffers because burying is not an unconditioned response to shock.

In spite of the fact that Hudson's findings were clearly pro-
blematic for the prevailing S-R theories, his research has had
relatively little impact on the study of avoidance learning.
Although the reasons for this are unclear, there are two problems with Hudson's work that may have contributed to its relative obscurity. The first is that Hudson did not report
any direct measures of burying behavior. The following statement provides the only basis for understanding how burying was
scored.

> The frequency of this behavior was scored by re-
> cording a symbol for it at an approximately
> constant rate while it was occurring. The number
> of these symbols was then used as the frequency
> measure. (Hudson, 1950, p. 107)

Furthermore, even this measure was not reported. Instead,
Hudson reported only a composite avoidance score that was, "the
sum of the recorded frequencies of pushing shavings and ap-
proach-withdrawals" (Hudson, 1950, p. 107). Thus, anecdotal
report provided the major support for many of Hudson's statements about burying behavior.

The second problem is that many of Hudson's experiments lacked
essential controls. For example, critical unshocked control
subjects were not included in any of the experiments in which
the burying behavior of shocked subjects was studied. However, the fact that most of the rats in Hudson's experiments
routinely sprayed a few shavings at the food holder when it
was first mounted on the cage wall, prior to shock, underlines
the necessity of including unshocked control subjects in these
experiments. In effect, Hudson concluded that the burying behavior that he observed following shock was a product of associative learning without including in the same experiments the
control measurements necessary to rule out more simple, nonassociative interpretations.

One can only guess the degree to which these violations of
experimental protocol have contributed to the relative obscurity of Hudson's monograph. Even the most well documented
evidence of a form of avoidance learning that so clearly
violated instrumental and Pavlovian principles may have been
ignored during the height of the S-R tradition. Although the
research in Hudson's monograph is not as methodologically
rigorous as it might have been, it is nonetheless a rich source
of interesting data, ideas, and anecdotal accounts, which are
surprisingly attuned to today's more cognitive and biological
approaches to the topic of animal learning.

Evidence That Conditioned Defensive Burying
Is a Product of Associative Learning

In 1978, Pinel and Treit published the first of an ongoing
series of studies of conditioned defensive burying conducted
at the University of British Columbia. The general objectives
of the initial studies of this series were: 1) to develop a
simple learning paradigm for reliably producing burying behavior, and 2) to conduct the controlled experiments necessary
to provide unequivocal evidence that the burying behavior observed in this paradigm is in fact a product of associative
learning. The discussion in this section will focus on these
two objectives.

The main stimulus for beginning this series of experiments was
a chance observation in 1970 made by one of us (Pinel) while
conducting a study of experimental epilepsy in the laboratory
of Stephan L. Chorover at the Massachusetts Institute of
Technology. The subjects in this study were eight rats housed
in individual clear plastic cages aligned on a table top
against one wall of the testing room. Both the arrangement of
the cages and the fact that the only access to the interior of
each was through a small hinged door on the front wall proved
to be of special significance.

To begin the study, each rat was captured by reaching into
the cage through the lone aperture. Then Chlorambucil, a drug
that can, after repeated administration, produce a condition comparable in some ways to petit mal epilepsy (Pinel and Chorover,
1972), was injected intraperitoneally. A routine check of the
animals an hour or so later revealed that each of the eight
subjects had systematically rearranged its bedding. Prior to
the injection, each subject had accumulated its bedding material
at the back of its cage to form a nest well away from the traf-
fic of the room, but after the injection each rat moved the
bedding from the back of its cage and heaped it against the
cage door at the front, ostensibly blocking further access to
the interior of the cage by the experimenter. Because of the
alignment of the cages at table height, this change provided
a particularly compelling visual illustration of the consequences
of burying behavior.

Developing a simple methodology for reliability producing conditioned defensive burying was not difficult. A few pilot observations led to the development of the methods first employed
by Pinel and Treit (1978), and these proved so effective that
most subsequent investigations of conditioned defensive burying
have employed the same general paradigm. First, all of the
subjects in each experiment were handled and pre-exposed in

groups of four or five to the 44x30x44 cm Plexiglas test chamber
for 30-min periods on each of 4 consecutive days. The chamber
floor was covered evenly during all phases of the experiments
with bedding (regular grade San-i-cel, a commercial bedding
material of ground corn cob) leveled to a height of 5 cm. On
Day 5, the rats were assigned to conditions. Those subjects
assigned to experimental (shock) conditions were placed individually in the center of the chamber facing away from the
6.5x5x.5 cm wire-wrapped wooden prod that had been attached following the habituation phase to the center of the wall at one
end, 2 cm above the level of the bedding material. When the
experimental rats first touched the prod with a forepaw, a brief
shock, initiated by the experimenter and terminated by the withdrawal of the subject, was delivered between the two uninsulated
wires. The shocks averaged 7.9 mA (SD = 1.47 mA) in intensity
and 42.9 msec (SD = 9.8 msec) in duration. Control rats were
not shocked when they touched the prod, but were otherwise
treated in the same manner.

In some experiments, contact with the prod has signalled the
beginning of the 15-min test period for both experimental and
control animals. In others, each animal has been removed from
the chamber as soon as possible after contacting the prod and
returned to the chamber for a 15-min test at some later time.

The test performance of may of the rats in the Pinel and Treit
(1978) experiments was quite stereotyped. During the first min
or so, most of the rats remained at the back of the cage (i.e.,
the end of the cage away from the shocked prod) always facing
the prod. During this initial period, the activity level of
the rats was low, but the animals did not display the catatonic
immobility characteristic of freezing. After this brief period
of inactivity, almost every shocked subject began to approach
the prod, investigating it with numerous approach-withdrawal
responses. This exploratory activity was punctuated in most
cases by periods of burying behavior. Each burying sequence
typically began with the rat facing the prod, from the back of
the apparatus. Then the rat moved directly toward the prod
spraying and pushing the bedding material toward it with rapid
alternating pushing movements of its forepaws, occasionally in
conjunction with shovelling movements of its snout. The fact
that burying is frequently seen in conjunction with approach-
withdrawal sequences is one piece of evidence that suggests
that burying serves as a defensive function in some situations
(Pinel, Hoyer and Terlecki, 1980).

Pinel and Treit (1978) measured the burying behavior of their
subjects in two different ways. First, they recorded the total
amount of time that each subject engaged in <u>directed</u> forelimb

sparying during the test period. The occasional movement of material with responses other than forelimb spraying and/or in directions other than at the prod was noted separately but did not contribute to the duration-of-burying score. This duration measure although somewhat subjective has proven to be very reliable. For example, Pinel, Treit and Wilkie (1980 and Davis and Rossheim (1980) have reported high correlations, .988 and .93 respectively, between the scores of independent observers. The second measure of burying, taken when the test session was complete, was the height of the bedding material accumulated at the point where the prod left the wall. Not surprisingly, the height and duration measures tend to be positively correlated (e.g., r = .89, Pinel and Treit, 1979).

It is important to note that the term "burying" as used by Pinel and Treit is more descriptive of the behavior than of its consequences. Whereas, directed forelimb spraying (i.e., burying) is a reasonably reliable response to prod shock in the presence of bedding material, not every subject engages in this behavior frequently enough during the usual 15-min test period to accumulate a pile of bedding sufficiently large to completely cover a prod mounted 2 cm above the original level of the bedding (e.g., Modaresi, 1982).

It is tempting to assume as Hudson did that rats bury shock sources because they associated the source and the shock. However, there is a simpler possibility -- one that does not involve an associative mechanism. Perhaps shocked rats bury the prod simply as an unconditioned response, elicited by shock and directed at the prod simply because it is the most salient feature of the test environment. Pinel and Treit ruled out the possibility of such nonassociative interpretations by showing that conditioned defensive burying is directed at the shock source even in the presence of comparable alternatives. In this experiment, 10 rats were shocked by one of two identical prods mounted at opposite ends of the chamber and were tested immediately. The burying scores of each of the 10 subjects is presented in Figure 1. Clearly each of the subjects directed their burying behavior towards the prod that had been the source of shock. All 10 rats spent time burying the shock prod; the only subject to spend any time at all directing bedding at the control prod did so only at the end of the session, after the shock prod had been completely covered. All 10 subjects also accumulated higher piles of bedding material at the shock prod than at the control prod. Because the only difference between the two prods was that one had been a shock source and the other had not, the only basis the subject could have used for their selective behavior was the association between the shock and the prod.

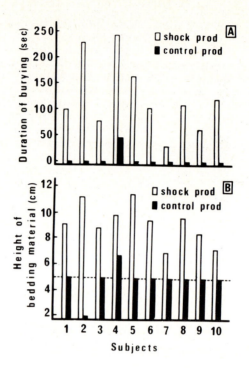

Figure 1.   Duration of burying directed by each subject
    at the shock prod and at a comparable control prod
    mounted on the opposite wall (Panel A) and the height
    of the bedding material accumulated at the two prods
    by the end of the 15-min test (Panel B).   Prior to
    the test the bedding was leveled at a height of 5 cm.
    (Pinel and Treit, 1978)

subsequent investigation of the stimulus control of conditioned
defensive burying by Pinel, Treit and Wilkie (1980) provided
further evidence of its associative nature.   In one experiment,
the amount of conditioned burying was significantly reduced if
either the position (front or back of the chamber) or brightness (black or white) of the prod was changed while the rat was
removed from the chamber for 1-min prior to the test.   And in
another, rats, shocked by one of two prods, white or black,
mounted at opposite ends of the chamber did not selectively
bury the shock prod if the positions of the prods had been
switched during the 1-min shock-test interval; they buried both
prods to the same degree.   The results of both of these experiments show that the prod burying of shocked rats is controlled
by an association between the shock and specific features of
the prod such as its brightness or position.

Clearly then, these initial studies accomplished their objectives. First, they demonstrated a method for producing burying
behavior that was both simple and effective; in some conditions
every rat displayed at least some burying during the 15-min
test. Secondly, they provided compelling evidence that the
burial of shock sources by rats is a product of associative
learning. Although Hudson did not provide strong evidence
for such a conclusion in his original studies, it now appears
that his associative view of the phenomenon was essentially
correct.

What theoretical framework provides the most satisfactory basis
for understanding conditioned defensive burying? Pinel, Treit
and Wilkie (1980) argued that conditioned defensive burying is
appropriately considered as a form of avoidance learning because the burial of a source of aversive stimulation potentially reduces the probability of subsequent contact with it
and any aversive stimulation that this contact might entail.
Although in the past it has been common to equate avoidance
with flight, studies of conditioned defensive burying demonstrate that rats can avoid sources of aversive stimulation
by approaching and coping with them. Because this form of
avoidance learning is so different from those that have served
as the basis for most theory development, assessment of the
degree to which theories of avoidance can account for conditioned defensive burying constitutes an interesting test of
their generality.

As argued in the previous section, S-R theories of avoidance
learning, be they instrumental or Pavlovian, have difficulty
accounting for conditioned defensive burying; burying is not an
unconditioned response to shock, and it occurs full blown before there has been a chance for it to be negatively reinforced
or for competing responses to be punished. In view of this
fact, it is important to remember that the descriptor "conditioned" is not intended to imply that the conditioned defensive burying phenomenon is a form of instrumental or
Pavlovian conditioning. The term was applied by Terlecki,
Pinel and Treit (1979) to emphasize that the phenomenon is a
simple form of associative learning and to distinguish the
burying that occurs after a learning trial from the unconditioned burying that sometimes occurs before it.

In contrast, the conditioned defensive burying phenomenon can
be accommodated readily by modern cognitive approaches to
learning, which view learning from an information processing
perspective (cf. Estes, 1975). Whereas S-R theories equate
changes in behavior that occur as the result of reinforcement
or stimulus pairing with the learning process itself, the

cognitive approach views animal learning as an <u>acquisition of
information</u> that has the potential to influence subsequent behavior. From this latter point of view, changes in behavior
merely indicate that new information has been assimilated (cf.
Bolles, 1975). Lack of behavior change, as latent learning
experiments (e.g., Tolman and Honzik, 1930) have shown, does
not mean that learning has not occurred.

Approached within a cognitive framework, the learning within
the burying paradigm is the assimilation of information about
the relation between the prod and the shock. The burying response is presumably akin to what Bolles (1970) has termed
species-specific defense reactions (SSDRs), innate reactions of
animals to aversive or novel stimuli.

General Features of Conditioned Defensive Burying

In the few years since Pinel and Treit's report, a number of
the environmental and organismic variables that influence conditioned defensive burying have been identified and studied.
In most cases, these studies resulted from a general interest
in the adaptive significance of this form of learning.

Conditioned Defensive Burying in Nonshock Situations

Wilkie, MacLennan and Pinel (1979) were the first to show that
conditioned burying is not limited to situations involving
electric shock. In one of their experiments, rats were allowed
to drink sweetened condensed milk for 30 min from a spout in
their home cages, after which they were injected intra-
peritioneally with the toxin lithium chloride and placed in a
holding cage for 5 min. During the 5-min interval, the bedding material covering the bottom of the home cage was levelled
by the experimenter. Then the subjects were returned to their
home cages. At intervals during the ensuing 24 hr, the height
of the bedding at the milk spout was compared with that at a
water spout mounted at the same height on the opposite wall.

A conditioned taste aversion to the condensed milk was established in all five subjects; whereas each rat drank milk before poisoning, they consumed no milk during the entire 24-hr
test. Moreover, every rat deposited substantially more bedding at the milk spout than at the water spout (see Figure 2).
In fact, only one rat failed to cover the milk spout completely
with the bedding. Rat 5 deposited the bedding material more
in front of the spout than did the other rats, and as a result
the milk spout remained exposued where it entered the cage,
although the tip itself was covered (see unconnected points,

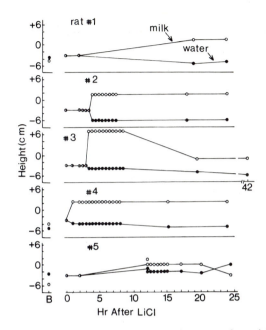

Figure 2. Height of bedding material at the milk and
   water spouts at various times after injection of
   lithium chloride. Height scores are cm above (+)
   or below (-) the point at which the spout entered
   the box. Points above B reflect the averages of
   the height of bedding at the two spouts during
   the baseline period. (Wilkie, MacLennan and
   Pinel, 1979)

Figure 2). In another experiment, Wilkie et al. found that
rats buried a spout containing Tobasco pepper sauce, but not a
concurrently available water spout.

It is interesting to note that Garcia, Rusinisk and Brett
(1977) in their studies of food-illness aversions in wild animals have made some similar observations. Coyotes, after eating mutton containing lithium chloride, sometimes buried it.

Wilkie et al. (1979) questioned whether the burial of noxious
food sources benefits the poisoned rat in its natural environment. Because rats so readily learn to avoid consuming aversive substances, it is difficult to see how an individual rat
could be protected by burying a noxious food that it would not
consume in any case. Wilkie et al. suggested that burying
noxious food might serve an altruistic function, i.e. that its

primary benefit might be to other members of the species, a notion receiving some support from Whillans and Shettleworth (1981).

Terlecki, Pinel and Treit (1979) also reported a series of conditioned defensive burying experiments in which stimuli other than footshock served as aversive stimuli. But paradoxically, it was during the conduct of these experiments designed to establish the generality of conditioned defensive burying that Terlecki, Pinel and Treit (1979) confirmed Hudson's observation that burying can also appear as an unconditioned component of a rat's neophobic reaction to some objects. In their first experiment, one of four different sources of aversive stimulation, a prod, a length of polyethylene tubing, a flashbulb in a protective collar, or a mouse trap with the springs loosened to lessen the impact, was mounted on an end wall following the usual 4 days of habituation to the test chamber. When subjects in each of the four experimental groups touched one of the test objects with a forepaw, they received either a shock, airblast, flash, or physical blow followed immediately by a 15-min test. Four control groups, one for each experimental condition, were treated in exactly the same fashion except that the aversive stimulation was not administered following contact with the test object.

The behavior of the control and experimental rats in the airblast condition was comparable to that of the rats in the shock groups. Control rats engaged in little or no burying, whereas almost every experimental rat buried the test object, i.e. the prod or air tube. However, in the trap and flashbulb conditions, control subjects spent an appreciable amount of time burying the test objects, and many of the experimental subjects began to bury the test object before the aversive stimulus could be administered. This unconditioned burying was such an effective defensive maneuver that it was impossible to spring the trap on four of the experimental subjects in the trap condition (n=10). One of these subjects triggered the trap from a distance by spraying it with the bedding material, and three others completely buried unsprung traps.

In their next experiment, Terlecki et al. eliminated the confounding effects of unconditioned burying by habituating the rats to the test objects as well as the test chamber during the first 4 days of the experiment. In each of the four conditions, two comparable test objects (prods, tubes, flashbulbs or traps), one white and one black, were mounted on opposite walls of the test chamber during all phases of the experiment, including the 4 days of habituation. In this experiment, there was no unconditioned burying, but following the conditioning

trial the rats in all four conditions directed their burying
almost exclusively at the source of aversive stimulation rather
than at the comparable control object (see Figure 3).

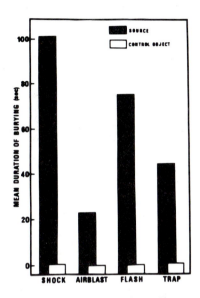

Figure 3. Mean duration of burying directed at the
    source of aversive stimulations and at a comparable
    control object mounted on the opposite wall of the
    test chamber in each of the four conditions.
    (Terlecki, Pinel and Treit, 1979)

Thus, although it is convenient to study conditioned burying in
situations where neutral objects serve as a source of electric
shock, conditioned defensive burying is not limited to such
artificial aversive stimuli. Moreover, regardless of the type
of aversive stimulation used in its induction, the burying behavior always seems to be selectively directed at its source,
even when comparable control objects are in the chamber.

Conditioned Defensive Burying in Rats Free to Escape

The standard 44x30 cm test chambers used in most burying experiments are not large enough to afford a rat an opportunity to
leave the immediate vicinity of the shock prod. This raised
the possibility that burying might occur as a defensive response only in those situations in which flight is not possible, or at least that the amount of burying might be inversely related to the viability of flight as a defensive maneuver.

To assess this possibility Pinel, Treit, Ladsk and MacLennan
(1980) compared the conditioned defensive burying of rats in
chambers of different size: 25x20 cm, 50x40 cm, 100x60 cm, or
200x80 cm. They found that the shocked rats in each of the four
conditions spent significantly more time burying the prod than
did the unshocked rats in the corresponding control groups; however, the shocked rats tested in the smaller chambers spent considerably more time burying than did those in the larger ones
(see Figure 4). Even when rats were shocked in one compartment
of a two-compartment box, they frequently entered the shock
compartment to bury the prod rather than staying in the safe
compartment. Thus, conditioned defensive burying is not
limited to situations in which rats are restricted to the
immediate vicinity of the test object, but less time is spent
burying when flight is a viable option.

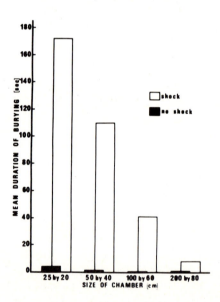

Figure 4. Mean duration of burying by separate groups
    of shock and no-shock subjects tested in chambers of
    different size. (Pinel, Treit, Ladak and MacLennan,
    1980)

Shock Parameters

Pinel and Treit (1978) argued that it is particularly difficult
for rats to respond effectively to shock administered through
a grid floor because the source of the shock is not readily

distinguishable from the environment, and attributed the rapid learning in the conditioned defensive burying paradigm to the fact that shock was administered from a well-defined, localized source, i.e. the prod. Although acquisition may be more rapid when the shock and prod are spatially contiguous, it is not necessary for the shock to emanate from the prod. Spetch, Terlecki, Pinel, Wilkie and Treit (1982) found that rats buried a prod whose insertion into the chamber was paired six times with shock to the back administered through wound clips.

Treit, Pinel and Terlecki (1980) studied the effect of four different intensities of prod shock on conditioned defensive burying, and Treit, Pinel and Fibiger (1981) included two different shock intensities in one of their experiments. In these experiments, conditioned defensive burying occurred at all shock intensities from .5 to 10 mA, however, the amount of burying was directly related to the intensity of the shock. Rats receiving more intense shock returned to the prod sooner, spent more time burying it, and accumulated higher piles of bedding around it.

Arnaut and Shettleworth (1981) showed that temporal contiguity between touching the prod and receiving a shock is important. Rats shocked once through a grid floor while touching the prod buried the prod, whereas rats that were shocked more than 1 min after touching the prod did not.

Organismic Variables

Most studies of conditioned defensive burying have been studies of adult, male, Long-Evans hooded rats; however, there is no considerable evidence that the phenomenon is not restricted to such subjects. The most comprehensive investigation of the effects of organismic variables on conditioned defensive burying was published by Treit, Terlecki and Pinel (1980).

Age and Gender. Treit et al. reported that conditioned defensive burying occurred in both male and female Long-Evans hooded rats aged 30, 60, or 90 days. Although almost every shocked rat in the study displayed some burying behavior, those rats that were only 30 days old buried significantly less than did the older subjects. Gender had no significant effect on burying.

Strain of Rat. Treit et al. compared the discriminated conditioned defensive burying behavior of three strains of rat: Long-Evans, Wistar and Fischer. Almost all of the rats in each of the three groups displayed some burying behavior, and it was directed for the most part at the shock prod (white or black) rather than at the control prod mounted at the opposite end to

the chamber. The Fischer rats buried the shock prod significantly longer than did the Long-Evans rats, but neither of these
strains differed significantly from the Wistar rats in this regard. Conditioned defensive burying has also been observed in
Sprague-Dawley (McKim and Lett, 1979) and Holtzman rats (Davis
and Rossheim, 1980). McKim and Lett reported the Long-Evans
rats engaged in more unconditioned burying than did Sprague-
Dawleys, but that there was no significant difference between
the two strains in terms of conditioned burying.

Strain of Mouse. Treit et al. also compared the discriminated
burying of three strains of mice: CD-1, CF-1, and BALB. Most
of the CD-1 and CF-1 mice spent some time burying the shock
prod, and in both strains significantly more time was spent
burying the shock prod (white or black) than the control prod
mounted on the opposite wall. On the other hand, only half of
the BALB mice displayed any burying at all and this was not
selectively directed at the experimental prod.

Although mice, of some strains at least, engage in conditioned
defensive burying, we have generally found mice to be poorer
subjects than rats in conditioned defensive burying experiments.
They generally spend less time burying than rats, and unlike
rats, they frequently engage in apparently random digging, thus
making the burying behavior itself more difficult to score.
Although it is tempting to conclude that the erratic performance
of mice in burying experiments may reflect some important difference between the abilities of the two species, it may simply
reflect the fact that the mice have been tested under conditions devised for rats.

Hamsters and Gerbils. A similar point can be made regarding
hamsters and gerbils. Although hamsters and gerbils tested
in the two-prod discrimination version of the paradigm displayed no burying behavior at all (Treit et al., 1980) they
may well have done so under different conditions. For example,
Silverman (1978) found that hamsters, as well as rats and mice,
covered a duct carrying cigarette smoke into the test chamber
with their feces.

Whillans and Shettleworth (1981) showed that the failure of
hamsters to bury a shock prod was not attributable to their
inability to learn the association between the prod and shock
prod in a single trial or to their inability to perform the
burying response. Following Wilkie et al.'s (1979) suggestion
that defensive burying, at least in some situations, might be
an "inherited altruistic reaction", they suggested that burying
did not evolve in hamsters because they are an asocial species.

Although such a proposal remains highly speculative, it illustrates an approach to the study of the origins of the burying response that could prove fruitful.

> Future comparative work might profitably be directed at understanding the patterns of occurrence of the various functional classes of "Burying" across different species and relating these patterns to differences in social organization and such aspects of the ecological niche as foods and predators. (Whillans and Shettleworth, 1981, pp. 361-362)

Ground Squirrels. A series of studies of encounters between California ground squirrels and rattlesnakes by Owings and his colleagues has provided the strongest evidence that burying behavior can serve a defensive function in the wild (e.g., Owings and Coss, 1977; Coss and Owings, 1978). During these encounters, the ground squirrels repeatedly sprayed sand at the snakes with a forelimb pushing response comparable to that used by rats to spray bedding material at sources of aversive stimulation. Because this behavior frequently caused the rattlesnakes to assume defensive postures or to withdraw from the encounter, there can be little doubt that it afforded the ground squirrel some protection from predation. The response of ground squirrels to a snake entering its darkened burrow was particularly interesting.

> The squirrels usually moved away from the entry point in reaction to the disturbance of introducing the stimulus animal... the squirrels began to appear extremely tense and cautious. They adopted elongate postures while oriented in the snake's direction, and moved slowly toward the snake, intermittently pausing and startling backward.... When rounding corners during the preliminary investigation, the squirrels very slowly extended the head around the corner first, and might retract the head and extend it again several times before actually moving into the next alley, if the alley was empty. If, however, the snake was in the next alley, and especially if the snake was moving, the squirrel often jumped back and kicked sand around the wall into the next alley. Such sand kicking appeared to function effectively to verify the presence of a snake, because the snake usually reacted by hissing, or striking and hissing. This usually caused the squirrel to retreat quickly.... Prior to returning to the snake, many squirrels accumulated a pile of sand, i.e., recruited sand, which they shoved ahead

> of them while approaching the snake... During a few
> trials, repeated trips of this sort brought enough
> sand to block the burrow completely. Once a burrow
> plug was constructed from floor to ceiling, the
> squirrels packed it by butting it with their heads.
> After the burrow was plugged, the squirrels appeared
> to move more confidently, as though they felt safer
> from the snake. (Coss and Owings, 1978, pp. 425-
> 426)

Although such observations leave little doubt that burying behavior can serve an important defensive function for rodents,
they do not rule out the possibility that the same behavior
can serve other functions in other contexts. For example, running can clearly be viewed as a defensive behavior in some
situations, and yet it can undeniably serve a variety of other
functions. In this vein, Poling, Cleary and Monaghan (1981)
observed that rats would sometimes bury nonthreatening objects
such as food pellets and glass marbles, and accordingly concluded that in some contexts burying might be a component of
hoarding behavior. Similarly, Pinel, Gorzalka and Ladak
(1981) have questioned whether the burial of dead conspecifics
by rats is most appropriately viewed as a defensive response.
And Montoya, Sutherland and Whishaw (1981) have made a similar
point concerning the burial by rats of cadaverine-soaked food
pellets.

It should also be emphasized that there is no evidence that the
burying behavior of ground squirrels observed in real or simu-
lated natural environments is the result of associative learning. In fact, naive laboratory-reared ground squirrels sprayed
sand at snakes even though they had never previously encountered them (Owings and Coss, 1977). The role of learning in the
defensive burying of rodents in their natural environment has
yet to be systematically investigated.

Support for a Cognitive View of Conditioned Defensive Burying

Although none of the studies of conditioned defensive burying
has been conducted for the expressed purpose of doing so,
several have provided particularly compelling evidence of its
cognitive basis. These studies are the focus of the present
section.

Learning Without Responding

Demonstrations of latent learning or "behaviorally-silent
associations" (Weisman and Dodd, 1979) provide strong support
for a cognitive approach to animal learning. In the classic

latent learning experiment (e.g., Tolman and Honzik, 1930), a rat was allowed to explore a maze that contained no food. When food was subsequently introduced to the maze, there was a sudden increase in the speed and accuracy of maze running. This learning has been termed latent because there was no observable behavioral change at the time the learning occurred, i.e., during the initial period of exploration. Such demonstrations of latent learning are clearly consistent with cognitive approaches to animal learning, which view learning as an acquisition of information, and challenge the S-R view of learning as a change in behavior.

Conditioned defensive burying studies are similar to latent learning experiments because the behavioral change does not occur during the acquisition phase of the experiment. Rather, animals seem simply to acquire the information that the prod is the source of the shock during the learning trial. It is not until the test that burying behavior is observed for the first time.

A control experiment included in the original series of studies by Pinel and Treit (1978) provides a particularly clear demonstration that learning occurs in the absence of the burying response. In this study, the experimental rats were shocked when they touched a prod inserted through the wall of their wire mesh, home cages. Burying was, of course, impossible in this context because there was no bedding, but when the shocked subjects were confronted with the prod in the standard test environment 1 min later, they buried it; whereas, unshocked control subjects or subjects shocked through a grid floor in a separate chamber did not.

Retention of Conditioned Defensive Burying

The fact that sufficient information about the source of an aversive stimulus is extracted from a single prod-shock pairing to enable the rat to deal effectively with such threats suggests that the rat processes information regarding sources of shock very rapidly. In this way, conditioned defensive burying resembles taste aversion learning, which often occurs after a single taste-poisoning episode (e.g., Grote and Brown, 1971). Although such rapid processing of information in both paradigms is apparently an evolutionary adaptation, there is one important difference between the two situations. Whereas rats often fail to associate poisoning with place or visual cues or do so only after several trials (e.g., Rozin, 1969), rats seem to extract both visual and place information regarding shock sources in a single trial (Pinel, Treit and Wilkie, 1980). Presumably there

is little adaptive advantage to be gained by rats from relating
gastrointestinal discomfort with nongustatory stimuli.

Memory is a necessary component of any cognitive approach to
learning (cf. Bolles, 1976). Information acquired, for example,
during the pairing of prod and shock must be remembered in some
form if the rat subsequently is to bury the prod. A basic
question concerning memory is its durability. We might expect
that information that is rapidly processed will also be retained
very well. This is the case with conditioned taste aversion
(e.g., Dragoin, Hughes, Devine and Bentley, 1973) and, as two
studies have shown, it is also the case with conditioned burying.

In the first of these studies (Pinel and Treit, 1978), experimental rats were shocked by the single prod that had been
mounted on the wall after the habituation phase of the experiment. Each rat was removed immediately from the test chamber
and tested for 15 min at one of five intervals: 10 sec, 5 min,
5 hr, 3 days, or 20 days. Although the amount of burying
declined monotonically over the 20-day period, the shocked rats
in the 20-day group still buried significantly more than did
unshocked control subjects tested at the same interval (see
Figure 5). Thus, rats appear to remember a single prod-shock
pairing for at least 20 days.

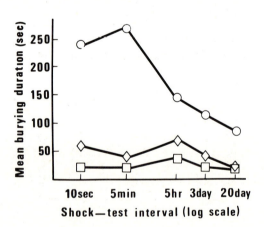

Figure 5. Mean duration of prod burying during a 15-
    min test at various intervals after a single prod
    shock (circles), after touching the prod but receiving
    no shock (diamonds), or after spending a few seconds
    in the chamber with neither prod nor shock (squares).
    (Pinel and Treit, 1978)

The second study of the effects of the duration of the learning-test interval on conditioned defensive burying was a study of discriminated burying. Pinel, Treit and Wilkie (1980) assessed the ability of rats to selectively bury one of two comparable prods (white or black) mounted on opposite walls of the test chamber at various intervals after the conditioning trial. Habituated subjects were shocked by one of the prods, removed immediately from the chamber, and returned 30 sec, 5 min, 1 hr, 8 hr, 24 hr, or 7 days later for the usual 15-min test. The rats buried the shock prod significantly more than they did the control prod at all but the 7-day interval (see Figure 6). Thus, even though the two prods were identical except for brightness, information necessary for discriminative burying was retained for more than 24 hr. The fact that the subjects in the 7-day group displayed substantial amounts of burying behavior but failed to direct it selectively at the shock prod, suggests that they still remembered the shock and some features of its source but had forgotten the distinguishing features of the shock prod.

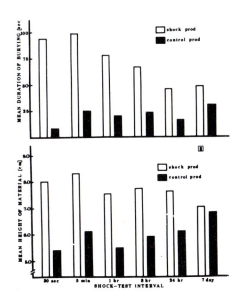

Figure 6. Mean duration of burying directed at the shock and control prods (top panel) and the mean height of the bedding at shock and control prods (bottom panel) at various intervals after conditioning. (Pinel, Treit and Wilkie, 1980)

An important point should be made about the use of conditioned
defensive burying, discriminated or otherwise, to investigate
animal memorial processes. Within the context of a cognitive
approach, memory is a capacity to subsequently change performance rather than a change in performance per se. Thus, although
the occurrence of burying behavior provides reasonable evidence
that the subjects remember the shock and some features of its
source, the lack of burying does not necessarily mean that there
is no memory. According to the cognitive view, animals can re-
tain information without it being necessarily reflected in their
ongoing behavior (cf. Wilkie and Masson, 1976; Wilkie and
Spetch, 1981).

Flexibility of the Burying Response

Further evidence that conditioned defensive burying reflects
the acquisition of information about sources of aversive stimulation rather than the simple strengthening of elicitation of
a particular response pattern is provided by demonstrations
that the burying response is extremely flexible. Considerably
different response patterns can be used by rats to bury aversive
objects, depending on the environmental conditions.

The flexibility of the burying response has not been obvious
in most studies of conditioned defensive burying. On the contrary, the topography has appeared to be remarkably constant
from rat to rat, and even from experiment to experiment. In
each case, rats spray bedding material over the test object
with the forelimb spraying response first described by Hudson
(1950). However, Pinel and Treit (1979) hypothesized that the
invariance of the burying response was less a reflection of the
inflexibility of the response than of the fact that most studies
of conditioned defensive burying had employed the same general
protocol. Conditioned defensive burying had previously been
studied only in situations in which a homogeneous bedding
material (e.g., wood shavings or San-i-cel) was the only
material available for burying the test object.

To test this hypothesis, Pinel and Treit (1979) tested rats
immediately after they had been shocked by one of two prods
(white or black) mounted on opposite walls of a test chamber
containing bedding, sand, or 1.0x1.6x2.4 cm wooden blocks. All
but one of the 30 subjects moved burying material toward one of
the prods, and in every case it was directed primarily at the
shock prod rather than at the control prod mounted on the
opposite wall (see Figure 7).

In the bedding and sand conditions, the rats buried the shock
prod using the same forelimb pushing response observed in other

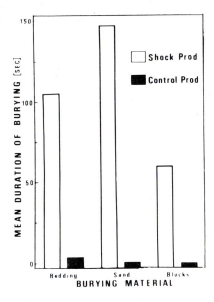

Figure 7. Mean duration of burying the shock and control
    prods with different burying materials. (Pinel
    and Treit, 1979)

experiments, but the burying response of the rats in the blocks
condition was substantially different. Although the rats sometimes used the forelimb response to move blocks toward the
shock prod, they ultimately picked up the blocks with their
teeth and either placed or threw them over the prod.

In a second study in this series, all of the blocks were placed
in the half of the chamber containing the control prod so that
there were no blocks in the immediate vicinity of the shock
prod. Nevertheless, all of the subjects in this condition
engaged in burying behavior, i.e., they were observed to spend
at least some time (M=107 sec) moving blocks toward the shock
prod.

These experiments show that the conditioned defensive burying
response is sufficiently flexible to function as an adaptive
response in the rat's natural environment, where suitable
materials for covering aversive objects sometimes may be
cumbersome and/or inconveniently located. Furthermore, they
also show that the single prod-shock pairing does not elicit
a particular response pattern. The prod shock seems to

identify for the rat the objects that should be buried, and the rat accomplishes this in a fashion suited to the environmental conditions in which that object is subsequently encountered. Thus, when sand or bedding is the only available burying material, the duration of forelimb spraying is good measure of burying, but when other materials are available, it is necessary to adopt a more flexible operational definition of the duration of burying, i.e. the amount of time spent moving materials toward the test objects regardless of the topography of the response employed to accomplish it.

These studies raise an interesting point concerning Bolles' (1970) suggestion that SSDRs are innate. Bolles argued that if an animal is to survive it must know how to react appropriately to dangerous stimuli the first time they are encountered; if it does not respond appropriately the first time, it may never get a second chance. The point raised by the finding that the burying pattern is modified to accommodate differences in the nature of available burying materials is that if burying is in fact innate, it may not be a particular response pattern that is inherited. Instead the rodents of some species may be born with a general tendency to bury aversive objects, and they accomplish this in a way that is consistent with the conditions of the encounter.

Burying in the Absence of the Aversive Object

In order to be classed as a defensive behavior, a behavior must be shown to occur in response to novel and/or dangerous objects or animals. Moreover, the response must at least potentially afford some protection to the animal or to its conspecifics. Pinel and Treit (1978) termed burying defensive because the results of their studies suggested that it satisfied both of these criteria. First, they showed that burying was not an inevitable response to shock; in their experiments, burying occurred only in the presence of the object that had been the source of the shock. Secondly, they showed that burying was always directed at the shock source when there was a comparable control object in the test chamber. Because the burying behavior was directed at the shock source, they argued that burying would reasonably be expected to reduce the probability of subsequent contact with it, either by driving it off as in the case of a predator (cf. Owings and Coss, 1977) or by physically covering it as in the case of inanimate objects.

The report by Pinel, Treit and Wilkie (1980) that shocked rats will under some conditions, direct their burying at sites that include neither the aversive object nor objects similar to it seemed to bring this conclusion into question. However, on

closer inspection this observation confirmed the defensive
function of the burying response as well as providing evidence
of its cognitive basis.

The observation in question occurred in experimental conditions
in which rats were immediately removed from the chamber after a
single prod shock and were returned 1 min later for the standard
15-min test. During the 1-min shock-test interval, the prod was
removed from hole in the wall of the chamber through which it
had been originally mounted and the same prod (n=10) or one
similar to it (n=10) was mounted in a like manner on the
opposite wall. Of the 10 subjects in each of these two conditions, 10 and 8, respectively, directed bedding material at
the prod in its new position. However, after spraying bedding
at the prod, 8 of these 20 subjects turned and directed bedding
at the hole through which the prod had originally been inserted.

It is difficult to see the adaptive value of burying areas of
environment (i.e., the empty prod hole) which no longer contain
the source of an aversive stimulation. However, a recent unpublished observation by one of us (Pinel) implicates an "expectancy" explanation of this hole burying behavior that seems
reasonable from an evolutionary perspective. Rats were shocked
by a prod lying on the surface of the bedding in a chamber containing a small hole in each end wall. The prod was initially
adjacent to one of the holes, but as soon as the shock was
administered, it was pulled by the wires across the box and out
through the hole in the opposite wall. Although none of the
10 subjects spent any time directing bedding towards the end of
the box in which it had been shocked, all 10 rats completely
sealed the hole through which the prod had exited. In other
words, it appears that rats do not always spray bedding at
actual sources of aversive stimulation; they will also accumulate bedding in a position that would be expected to block the
return of such objects to the test chamber. This "constructive"
behavior on the part of the rat is reminiscent of the results
of some of the early "insight" experiments (e.g., Maier, 1931).

Hudson (1950) provided an anecdotal description of a similar
behavior. In the relevant experiment, rats were shocked by the
panel-mounted metal food holder, which was immediately withdrawn through a hole in the side of the cage.

> The aperture in the side of the cage through which
> the pattern was withdrawn was closed by a shield
> that was attached to the slide. This shield
> covered the hole only while the pattern was in the
> cage. When the latter was withdrawn after shock,

a trap door dropped down to close the hole. In two
of the cages the hole was not completely covered by
the trap door, a small slot remaining about 1/4 of
an inch in width. The hole, discovered by eight
animals, then became the object of avoidance, and
the subsequent activity of the rat devoted to piling
shavings in that corner, this behavior alternating
with frequent approaches and withdrawals (Hudson,
1950, p. 123).

Backward Conditioning of Defensive Burying

It is reasonable to expect that animals would have the ability
to associate an aversive stimulus with a novel object that is
not detected until after the stimulus. An animal that did not
sight its attacker until after an abortive attack surely would
not submit to another attack (i.e., a forward pairing of the
predator and pain) before responding defensively. It is therefore surprising that the view emphatically espoused in most
textbook discussions of Pavlovian conditioning is that excitatory conditioning does not occur unless the conditioned stimulus
precedes the unconditioned stimulus.

After reviewing the backward conditioning literature, Spetch,
Wilkie and Pinel (1981) concluded that "the failure to recognize backward conditioning as a legitimate phenomenon seems
to reflect theoretical biases rather than a paucity of empirical evidence" (p. 163). Because the existence of excitatory
backward conditioning is inconsistent with the assumption,
common to many S-R theories, that a conditioned stimulus acquires excitatory properties because of its predictive relation
with the unconditioned stimulus, the well controlled demonstrations of excitatory backward conditioning that have been
published during the last decade (e.g., Kieth-Lucus and Guttman,
1975; Mahoney and Ayres, 1976; Wagner and Terry, 1975) have
received little attention. Even Pavlov (1927/1960) observed
excitatory backward conditioning effects and referred to his
frequently cited earlier claim that such effects do not occur
as "a brilliant illustration of the danger of too hasty generalizations" (p. 394).

Spetch, Terlecki, Pinel, Wilkie and Treit (1982) recently reported the excitatory backward conditioning of defensive burying in rats. In order to accomplish this, they modified the
standard paradigm so that the timing of the conditioned and unconditioned stimuli could be controlled by the experimenter.
Six shocks were administered at variable intervals through
wound clips attached to the rats' back, and on six different

occasions a retractable prod was inserted into the chamber for
5 sec in various temporal relations to shock delivery.

The burying behavior of rats in two conditioning groups was
compared with that of rats in three control groups. Rats in the
backward conditioning group received six backward pairings of
shock and prod, and the rats in the forward conditioning group
received six forward pairings. The forward group was included
to ensure that any burying that might occur following the backward shock-prod pairings was excitatory (i.e., a response quali-
tatively similar to that produced by forward pairings). The
three control groups served as controls for the possible nonassociative effects of exposure to the shock and prod. Rats in
the unpaired control group received six presentations of the
shock and prod in an unpaired manner. Rats in the prod-only
control group received six presentations of the prod without any
shock presentations. This group served as a control for unconditioned burying. Although little unconditioned burying
typically is directed at a stationary prod, it seemed possible
that a prod that moved in and out of the chamber might elicit
more burying. And lastly, a shock-only control group was included as a control for any effects of shock itself on subsequent burying of the prod.

The results of this study, presented in Figure 8, demonstrate
that the amount of burying produced by six backward pairings
of the prod and shock substantially exceeded that produced by
each of the three control conditions. Moreover, the amount of
burying produced by the six backward pairings was comparable to
that produced by the same number of forward pairings.

Although such demonstrations of excitatory backward conditioning are incompatible with most S-R theories of classical conditioning, they can be understood readily from a cognitive
perspective. When an animal experiences a sudden noxious
stimulation in the absence of any obvious source, it may explore
its immediate environment and "attribute" the aversive stimulation to any novel stimulus that it finds, particularly if it
is a moving object that is close by. Such an "attributional"
process may have evolved as an adaptation to attacks by con-
specifices or predators. An animal that can associate sudden,
aversive stimulation with a subsequently discovered novel
stimulus would gain clear reproductive advantage. Such a view
is consistent with the fact that most successful demonstrations
of backward conditioning have employed aversive unconditioned
stimuli.

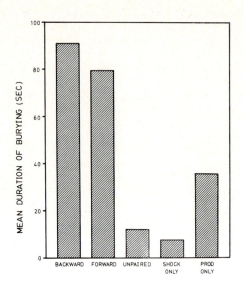

Figure 8. Mean duration of burying observed following
six backward or six forward conditioning trials, or
following three different control procedures.
(Spetch, Terlecki, Pinel, Wilkie and Treit, 1982)

Conclusion

One of the most notable features of conditioned defensive burying is that it occurs in such a variety of conditions after
only a <u>single</u> learning trial. In experiment after experiment,
almost every rat receiving a shock from a prod buried that
prod, whereas unshocked control rats displayed little if any
of this behavior. And almost every rat that we have tested in
the two-prod discrimination version of the conditioned defensive
burying paradigm has concentrated its burying almost exclusively
on the shock prod. What accounts for this rapid acquisition?

On the basis of evidence that has been discussed already, it is
clear that the consideration of this question within an S-R
framework is not likely to prove as enlightening as an analysis
from biological and cognitive perspectives. It has become increasingly clear during the course of its investigation that it
is advantageous to view the conditioned defensive burying of
rats in relation to the survival requirements of their natural
environment and that it is appropriate to view the learning
that takes place within the burying paradigm as acquisition of
information concerning the relation between environmental
stimuli.

From this combined perspective, there seem to be two different
factors that could account for the appearance of conditioned
defensive burying after a single prod-shock pairing. First,
the rapid acquisition might stem from the fact the conditioned
defensive burying paradigm taps an ability that is important
for the survival of the rat in its natural environment, i.e.,
the ability to associate rapidly painful exteroceptive stimulation with some concrete source. Second, acquisition is
undoubtedly facilitated by the fact that there is no need for
the rat to learn the response. The subject must simply identify
the source of the aversive stimulation and then use an already
existing response to defend itself. We assume that the burying
response is akin to Bolles' SSDRs, innate reactions to aversive
stimulation. Burying, however, is not triggered by the aversive
stimulation per se, but by objects perceived to be sources of
aversive stimulation, and its topography can be modified to suit
the situation.

Our increased understanding of the burying phenomenon resulting
from the application of both cognitive and biological perspectives encourages us to believe that the same approach will
continue to prove fruitful in other areas of animal learning.
For example, the learning involved in such phenomena as taste
aversion conditioning (e.g., Domjan and Wilson, 1972), autoshaping (e.g., Browne, 1974), and bird song learning (e.g.,
Marler and Peters, 1981) clearly entails the acquisition of information (that subsequently affects behavior), rather than
acquisition of the responses themselves. And in each of these
examples phylogeny is apparently important. As a result of
adaptive exchanges with their environment, rats apparently have
a proclivity to associate illness preferentially with novel
food, pigeons and other species are prepared to respond to
signals of reinforcers with "biologically pre-organized appetitive behavior patterns" (Woodruff and Williams, 1976), and song
birds such as sparrows sing the typical song of their species to
protect territories and attract mates.

We do not believe that the contemporaneous development of the
biological and cognitive approaches to animal learning has been
entirely fortuitous. The increasing tendency to study forms of
learning that have relevance to an animal's behavior in its
natural environment, free of the constraints imposed by most
traditional test environments, has led to an increasing appreciation for the impressive capacity of animals to learn in some
situations. These studies have served to demonstrate the speed,
complexity, and flexibility of forms learning that have been
the product of evolutionary pressure, and in so doing they have
encouraged the abandonment of simplistic S-R views.

References

1   Arnaut, L. & Shettleworth, S. J.   The role of spatial and
    temporal contiguity in defensive burying in rats.   Animal
    Learning and Behavior, 1981, 9, 275-280.

2   Bolles, R. C.   Species-specific defense reactions and avoid-
    ance learning.   Psychological Review, 1970, 77, 32-48.

3   Bolles, R. C.   Learning, memory, and cognition.   In
    W. K. Estes (Ed.), Handbook of learning and cognition
    (Vol. 1).   Hillsdale, N. J.: Erlbaum, 1975.

4   Bolles, R. C.   Some relationships between learning and
    memory.   In D. L. Medin, W. A. Roberts & R. T. Davis (Eds.),
    Processes of animal memory.   Hillsdale, N. J.: Erlbaum,
    1976.

5   Browne, M. P.   Autoshaping and the role of primary rein-
    forcement and overt movements in the acquisition of stimulus-
    stimulus relations.   Unpublished doctoral dissertation,
    Indiana University, 1974.

6   Coss, R. G. & Owings, D. H.   Snake-directed behavior by
    snake naive and experienced California ground squirrels in
    a simulated burrow.   Z. Tierpsychol., 1978, 48, 421-435.

7   Davis, S. F. & Rossheim, S. A.   Defensive burying as a
    function of insulin-induced hypoglycemia and type of
    aversive stimulation.   Bulletin of the Psychonomic Society,
    1980, 16, 229-231.

8   Domjan, M. & Wilson, N. E.   Contribution of ingestive be-
    haviors to taste aversion learning.   Journal of Comparative
    and Physiological Psychology, 1972, 80, 403-412.

9   Dragoin, W. B., Hughes, G., Devine, M. & Bentley, J.
    Long-term retention of conditioned taste aversions: Ef-
    fects of gustatory interference.   Psychological Reports,
    1973, 33, 511-514.

10 Estes, W. K.   The state of the field: General problems and
    issues of theory and metatheory.   In W. K. Estes (Ed.),
    Handbook of learning and cognition (Vol. 1).   Hillsdale,
    N. J.: Erlbaum, 1975.

11 Garcia, J., Rusiniak, K. W. & Brett, L. P. Conditioning
    food-illness aversions in wild animals: Caveant Canonici.
    In H. Davis & H. M. B. Hurwitz (Eds.), Operant-Pavlovian
    interactions.   Hillsdale, N. J.: Erlbaum, 1977.

12 Grote, F. W. & Brown, R. T. Rapid learning of passive avoidance by weaning rats: Conditioned taste aversion. Psychonomic Science, 1971, 25, 163-164.

13 Hudson, B. B. One-trial learning in the domestic rat.
Genetic Psychology Monographs, 1950, 41, 99-145.

14 Keith-Lucas, T. & Guttman, N. Robust single-trial delayed
backward-conditioning. Journal of Comparative and
Physiological Psychology, 1975, 88, 468-476.

15 Mahoney, W. J. & Ayres, J. J. B. One-trial simultaneous and
backward fear conditioning as reflected in conditioned
suppression of licking in rats. Animal Learning and Behavior, 1976, 357-362.

16 Maier, N. R. F. Reasoning and learning. Psychological
Review, 1931, 38, 332-346.

17 Marler, P. & Peters, S. Sparrows learn adult song and more
from memory. Science, 1981, 213, 780-782.

18 McKim, W. A. & Lett, B. T. Spontaneous and shock-induced
burying in two strains of rats. Behavioral and Neural
Biology, 1979, 26, 76-80.

19 Modaresi, H. A. Defensive behavior of the rat in a shock-
prod situation: Effects of the subject's location preference. Animal Learning and Behavior, 1982, 10, 97-102.

20 Montoya, C. P., Sutherland, R. J. & Whishaw, I. Q. Cadaverine and burying in the laboratory rat. Bulletin of the
Psychonomic Society, 1981, 18, 118-120.

21 Mowrer, O. H. On the dual nature of learning: A reinter-
pretation of "conditioning" and "problem solving".
Harvard Educational Review, 1947, 17, 102-148.

22 Owings, D. H. & Coss, R. G. Snake mobbing by California
ground squirrels: Adaptive variation and ontogeny.
Behaviour, 1977, 62, 50-69.

23 Pavlov, I. P. Conditioned reflexes: An investigation of
the physiological activity of the cerebral cortex
(G. B. Anrep, Ed. and trans). New York: Dover, 1960.
(Originally published, 1927).

24 Peacock, E. J. & Wong, P. T. P. Defensive burying in the
rat: A behavioral field analysis. Animal Learning and
Behavior, 1982, 10, 103-107.

25   Pinel, J. P. J. & Chorover, S. L.   Inhibition of arousal of
     epilepsy induced by Chlorambucil in rats.   Nature, 1972,
     236, 232-234.

26   Pinel, J. P. J., Gorzalka, B. B. & Ladak, F.   Cadaverine
     and putrescine initiate the burial of dead conspecifics by
     rats.   Physiology and Behavior, 1981, 27, 819-824.

27   Pinel, J. P. J., Hoyer, E. & Terlecki, L. J.   Defensive
     burying and approach-avoidance behavior in the rat.   Bul-
     letin of the Psychonomic Society, 1980, 16, 349-352.

28   Pinel, J. P. J. & Treit, D.   Burying as a defensive response
     in rats.   Journal of Comparative and Physiological Psycho-
     logy, 1978, 92, 708-712.

29   Pinel, J. P. J. & Treit, D.   Conditioned defensive burying
     in rats: Availability of burying materials.   Animal Learn-
     ing and Behavior, 1979, 7, 392-396.

30   Pinel, J. P. J. & Treit, D.   The conditioned defensive bury-
     ing paradigm and behavioral neuroscience.   In T. E. Robinson
     (Ed.), Behavioral approaches in brain research.   Oxford:
     Oxford University Press, in press.

31   Pinel, J. P. J., Treit, D., Ladak, F. & MacLennan, A. J.
     Conditioned defensive burying in rats free to escape.
     Animal Learning and Behavior, 1980, 8, 447-451.

32   Pinel, J. P. J., Treit, D. & Wilkie, D. M.   Stimulus control
     of defensive burying in the rat.   Learning and Motivation,
     1980, 11, 150-163.

33   Poling, A., Cleary, J. & Monaghan, M.   Burying by rats in
     response to aversive and nonaversive stimuli.   Journal of
     the Experimental Analysis of Behavior, 1981, 35, 31-44.

34   Rozin, P.   Central or peripheral mediation of learning with
     long CS-US intervals in the feeding system.   Journal of
     Comparative and Physiological Psychology, 1969, 67, 421-429.

35   Silverman, A. P.   Rodent's defense against cigarette smoke.
     Animal Behaviour, 1978, 26, 1279-1281.

36   Spetch, M. L., Wilkie, D. M. & Pinel, J. P. J.   Backward
     conditioning: A reevaluation of the empirical evidence.
     Psychological Bulletin, 1981, 89, 163-175.

37 Spetch, M. L., Terlecki, L. J., Pinel, P. J. P., Wilkie, D. M. & Treit, D. M. Excitatory backward conditioning of defensive burying in rats. Bulletin of the Psychonomic Society, 1982, 19, 111-114.

38 Terlecki, L. J., Pincl, J. P. J. & Treit, D. Conditioned and unconditioned defensive burying in the rat. Learning and Motivation, 1979, 10, 337-350.

39 Tolman, E. C. & Honzik, C. H. Introduction and removal of reward, and maze performance in rats. University of California Publications in Psychology, 1930, 4, 257-275.

40 Treit, D., Pinel, J. P. J. & Terlecki, L. J. Shock intensity and conditioned defensive burying in rats. Bulletin of the Psychonomic Society, 1980, 16, 5-7.

41 Treit, D. & Pinel, J. P. J. & Fibiger, H. C. Conditioned defensive burying: A new paradigm for the study of anxiolytic agents. Pharmacology Biochemistry and Behavior, 1981, 619-626.

42 Treit, D., Terlecki, L. J. & Pinel, J. P. J. Conditioned defensive burying: Organismic variables. Bulletin of the Psychonomic Society, 1980, 16, 451-454.

43 Wagner, A. R. & Terry, W. S. Backward conditioning to a CS following an expected vs. a surprising UCS. Animal Learning and Behavior, 1975, 3, 370-374.

44 Weisman, R. G. & Dodd, P. W. D. The study of association: Methodology and basic phenomena. In A. Dickenson & R. A. Boakes (Ed.), Mechanisms of learning and motivation: A memorial volume to Jerzy Konorski. Hillsdale, N. J.: Erlbaum, 1979.

45 Whillans, K. V. & Shettleworth, S. J. Defensive burying in rats and hamsters. Animal Learning and Behavior, 1981, 9, 357-362.

46 Wilkie, D. M., MacLennan, A. J. & Pinel, J. P. J. Rat defensive behavior: Burying noxious food. Journal of the Experimental Analysis of Behavior, 1979, 31, 299-306.

47 Wilkie, D. M. & Masson, M. E. Attention in the pigeon: A reevaluation. Journal of the Experimental Analysis of Behavior, 1976, 26, 207-212.

48 Woodruff, G. & Williams, D. R. The associative relation underlying autoshaping in the pigeon. Journal of the Experimental Analysis of Behavior, 1976, 26, 1-13.

ANIMAL COGNITION AND BEHAVIOR
Roger L. Mellgren, editor
© North-Holland Publishing Company, 1983

BIOLOGICAL CONSTRAINTS AND THE PURSUIT OF GENERAL THEORIES OF LEARNING

Michael Domjan

University of Texas at Austin

Students of animal learning are confronted with a dilemma. No one doubts that learning occurs in natural habitats and can have important adaptive functions. However, learning processes cannot be investigated outside the laboratory. Learning involves a change in behavior or behavioral potentiality produced by experience. Since experience takes time, learning is manifest in alterations of behavior over time. But not all time-dependent changes in behavior reflect learning. To be sure that one is observing the effects of learning and not the effects of other processes like maturation or fatigue, one has to conduct laboratory experiments in which the causes of a change in behavior are clearly identified. Thus, the nature of learning processes demands that we investigate them in the laboratory, but in so doing we cannot be certain that we are investigating phenomena of relevance outside the laboratory.

One way out of the dilemma is to investigate general characteristics of learning that exist in all learning situations. Such features might be discovered in circumstances that are neutral with respect to the innate behavioral repertoire of the particular animal being observed. The assumption that general characteristics of learning exist and can be discovered by studying learning in arbitrary situations has been one of the cornerstones of research on animal learning in the twentieth century. Investigators have made extensive observations about learning in a few laboratory situations hoping that relationships discovered in these studies would turn out to be general characteristics of learning.

The pursuit of general mechanisms of learning through the intensive investigation of a limited number of arbitrary laboratory preparations came under strong attack in the early 1970's in discussions about biological constraints on classical and instrumental conditioning (Bolles, 1970; Hinde and Stevenson-Hinde, 1973; Rozin and Kalat, 1971; Seligman, 1970; Shettleworth, 1972). These discussions were stimulated by numerous findings about learning that were inconsistent with general views of learning widely accepted at the time. Examples of biological constraints also appear to illustrate adaptive specializations in learning. Discussions about biological constraints on classical and instrumental conditioning suggested that there was a need for a new and revitalized framework for the study of animal learning integrated with considerations of ecological adaptation and evolutionary history.

Biological constraints soon became treated as a fundamental issue in the study of learning. Introductory psychology and other textbooks discussed it as a major problem. A revolution in the study of learning seemed to be in the making. Numerous investigators were drawn to the study of phenomena identified as biological constraints on classical and instrumental conditioning. Convention programs became filled with papers about taste-aversion learning, autoshaping, constraints on avoidance learning, and other such matters. The expectation was that students of learning would soon abandon arbitrary laboratory situations in favor of a more biologically-oriented approach sensitive to adaptive specializations in learning.

The biological revolution in the study of animal learning that seemed imminent in the early 1970's has not happened. Many new facts about learning have been discovered and many new interpretations have been proposed (see Rescorla and Holland, 1982). In addition, researchers have become less concerned about limiting their study to biologically-arbitrary forms of learning. However, the fundamental goal of discovering general principles of learning remains, and investigators continue to study learning in a small number of laboratory preparations. In the first part of this chapter, I will discuss possible reasons why biological constraints on classical and instrumental conditioning have not revolutionized the study of learning. I will then illustrate that the search for general principles is inescapable in the analysis of learning phenomena, even phenomena that appear to be adaptive specializations. In the third and last part of the chapter I will discuss how studies of animal learning can be better integrated with biological factors without abandoning the search for general principles.

Reasons for the Failure of Biological Constraints to Revolutionize the Study of Animal Learning

Discussions about biological constraints on classical and instrumental conditioning did not revolutionize the study of animal learning because the funcamental ideas involved did not have sufficient positive intellectual content to provide

a successful alternative to the general process approach (see
Domjan and Galef, 1982; Johnston, 1981). Biological constraint concepts did not lead to distinctive strategies for
the discovery of new learning phenomena or suggest ways to
systematize knowledge about biological constraints. In addition, phenomena identified as biological constraints on
classical and instrumental conditioning were easily assimilated
into revisions of general principles of learning, thereby eliminating many of them from deserving special attention.

What is a Biological Constraint?

A constraint on a phenomenon is a limitation or boundary condition for its occurrence. Scientific investigations invariably lead to the identification of such boundary conditions.
Studies of learning have documented numerous constraints, including, for example, the interval between conditioned and
unconditioned stimuli (CS and US), the intensity and novelty
of the CS and US, and the extent to which the unconditioned
stimulus is surprising. Such constraints constitute our
basic knowledge from which general theories of learning have
been constructed. However, they are not what are considered
to be biological constraints on learning.

The above comments suggest that at least two types of constraints on learning exist. However, conventional uses of
the term "biological" do not help in distinguishing between
them. It is unclear in what sense traditionally-known
boundary conditions on learning are "nonbiological". The
term "biological" implies something having to do with evolutionary considerations and adaptive function. Limitations on
learning brought about by factors such as the CS-US interval
and CS familiarity, for example, reliably occur in numerous
situations in which classical conditioning is observed.
Furthermore, such boundary conditions have functional utility
in the sense that they limit the learning of associations
between events that are causally independent in the environment (Dickinson, 1980). Therefore, one can argue that
traditionally-known limitations on learning have as much
adaptive function and are probably as much the products of
evolution as are examples of learning that have been considered
to illustrate biological constraints.

Because conventional uses of the term "biological" do not help
distinguish between biological constraints and other limitations on learning, one has to look for other criteria in writ-
ings about biological constraint phenomena. The primary im-
petus for discussions about biological constraints on classical and instrumental conditioning was that certain learning

phenomena were contrary to principles of learning widely held
at the time, and being an exception to the rule constituted a
working definition of biological constraints. The emphasis
on the fact that constraint phenomena were contrary to known
principles of behavior is evident in the title, "The misbehavior of organisms", used by Breland and Breland (1961) for
their classic discussion of biological constraints on instrumental conditioning. This working definition is also evident
in many textbook accounts of learning that present a discussion of biological constraints as exceptions to conventional
principles of classical and instrumental conditioning, which
are described first (e.g., Rachlin, 1976; Schwartz, 1978).

Some authors (e.g., Hinde, 1973) were not specific about what
aspects of traditional learning theory were violated by instances of biological constraints. In contrast, others (e.g.,
Seligman, 1970; Shettleworth, 1972) saw constraints on learning as violations of the principle of equipotentiality. This
principle states that mechanisms of learning are independent
of the cues, responses, and reinforcers involved in a particular instance of learning and was regarded as a fundamental
assumption of general-process learning theories. (For a
review of the impact of research on biological constraints
on the equipotentiality principle, see Domjan, 1982).

Another feature common to biological constraint phenomena is
that they illustrate situation specificity of learning. Instances of learning that violate the equipotentiality principle do so because they occur only in situations that contain
the required combination of cues, responses, and reinforcers.
All of the discussions about biological constraints on learning in the early 1970's involved descriptions of situation
specificity of learning. Bolles (1970), for example, discussed differences in avoidance learning as a function of the
instrumental response required in various avoidance conditioning situations. Rozin and Kalat (1971) discussed learning
in food selection situations as distinct from learning in
other circumstances. Seligman (1970) and Shettleworth (1972)
discussed situation-specific instances of learning as violations of the equipotentiality principle.

A third characteristic that was often attributed to instances
of biological constraints was adaptive specialization. Many
phenomena discussed as instances of biological constraints
appear to be adaptive specializations that help animals cope
with particular challenges to survival. Rozin and Kalat
(1971), for example, discussed several aspects of learning in
the feeding system (such as long-delay learning) as adaptive
specializations for food selection. Bolles (1970) discussed
the adaptive significance of the rapid acquisition of species-

specific defense responses in avoidance situations. The idea
that through evolution animals have acquired specialized learning abilities that facilitate certain kinds of learning and
hinder others was fundamental to many discussion about biological constraints.

Biological Constraints as an Approach to the Study of Learning

Implications for Discovery of New Phenomena. Progress in any
area of science depends on discovery and analysis of relevant
new phenomena. Biological constraint concepts did not present
a successful alternative to traditional approaches to the study
of learning in part because they did not provide distinctive
and productive strategies for the discovery of new phenomena.
If the identifying feature of biological constraints is that
they violate accepted principles of behavior, then discovery
of new instances requires the empirical challenge of general
process theories. To find an exception to the rule, we must
test various implications of the rule in the hope that one of
our tests will reveal an unpredicted outcome. However, such
activity makes the search for instances of biological constraints no different from other studies of learning. Investigators primarily interested in general process learning theory
also test various predictions in an effort to discover instances that violate the theory. Conceptions of biological
constraints as exceptions to the rule do not require departure
from this traditional approach.

The situation-specificity criterion for biological constraints
also fails to provide a distinctive method for the discovery
of new phenomena. This criterion is not helpful because it
does not prescribe how one should choose circumstances to
compare. To find a new biological constraint, should we compare learning to find a mate with learning to avoid poisonous
food, or should we compare learning to find food with learning
to find water? The choice is largely arbitrary, and this makes
studies of situation specificity arbitrary comparisons of
learning rather than biologically-oriented experiments. Consider, for example, the famous selective aversion conditioning
effect first demonstrated by Garcia and Koelling. When rats
are made sick, they quickly learn an aversion to taste but not
to auditory/visual cues, whereas when they receive foot-shock
they quickly learn an aversion to auditory/visual cues but not
to taste (Garcia and Koelling, 1966; Miller and Domjan, 1981).
This was one of the prominent findings that stimulated thinking about biological constraints and encouraged arguments about
the value of biological approaches to the study of learning.
However, nothing about the natural history of rats suggests
that comparison of fear conditioning and poison-avoidance

learning will yield important insights into behavior. Ironically, the best rationale for such experiments rests with general process learning theory. Since the general-process approach assumes that learning proceeds the same way in all situations, the arbitrariness of situations chosen for comparison is acceptable in tests of general process learning theory.

The adaptive specializations criterion for biological constraints on learning was as unsuccessful as the other conceptions in stimulating discovery of new phenomena. How the concept of adaptive specializations might be used in a search for new instances of biological constraints was not spelled out clearly. In addition, certain aspects of the adaptive specializations concept are similar to the exceptions-to-the-rule and situation-specificity criteria for biological constraints. Because adaptive specializations are assumed to occur against a background of general mechanisms of learning, adaptive specializations are exceptions to such general processes. They are also instances of situation-specific learning. To distinguish adaptive specializations from general adaptive mechanisms of learning, one has to show that they occur in some situations but not in others.

Implications for Mechanisms of Biological Constraint Effects

Another reason why discussions about biological constraints failed to revolutionize the study of animal learning is that ideas fundamental to biological constraints did not provide insights into the mechanisms of constraint phenomena. Identification of biological constraints as exceptions to general process theory does not insure that constraint phenomena will have any other common attributes. Particular instances of learning may be contrary to traditional theories for many different reasons. Situation specificity of learning may also occur for numerous reasons. Therefore, identification of constraints on learning in terms of situation specificity similarly fails to provide insights into underlying mechanisms.

Uses of the adaptive specializations concept were no more helpful in understanding biological constraints because many claims of adaptive function were post hoc. For example, the fact that rats learn aversions to poisonous food even if postingestional toxic effects are delayed for several hours was often discussed as an adaptive specialization of the feeding system. However, no direct evidence was presented that rats encounter palatable but toxic potential food in their natural habitats, that such toxins have effects delayed by many minutes or hours, or that

the resultant illness is readily associated with ingestion.
For all we know, rats undisturbed by man usually become ill
as a result of bacterial, viral, or parasitic infection, and
the formation of food aversions on the basis of such illness
may be, on average, counter-productive.[2]

Implications for General Process Theories of Learning. Because biological constraint concepts did not provide insights
into the mechanisms of constraint effects, investigations of
constraints on classical and instrumental conditioning ended
up being guided by the only available conceptual framework--
general process learning theory. This made studies of biological constraint phenomena like taste-aversion learning, autoshaping, and constraints on instrumental conditioning part of
the investigation of general process theory, with the result
that many constraint phenomena became integrated into revisions
of general process theory (see Domjan, 1982, for a detailed
discussion). Assimilation of biological constraints on classical and instrumental conditioning into revisions of general
process theory has contributed to their demise as special
instances of learning and diffused the biological revolution
in the study of learning that had been imminent.

Identification of biological constraints as exceptions to
general learning mechanisms makes them especially susceptible
to incorporation into revisions of the general principles. By
appropriately changing the general principles, the exceptional
nature of biological constraints can be easily eliminated.
Situation-specific instances of learning can be also easily incorporated in revisions of general process theory. Learning
situations basically differ in terms of the cues, responses,
and reinforcers they contain and how these are arranged. The
basic elements of general process learning theory are also
cues, responses, and reinforcers. Learning theory essentially
describes how learning differs as a function of these elements.
If general process theory cannot accommodate a new instance of
situation specificity, this shortcoming can be remedied by
appropriate modifications of, or additions to, the general
principles. For example, in response to evidence that associations are more easily formed between certain combinations of
conditioned and unconditioned stimuli than between other combinations (e.g., Garcia and Koelling, 1966), some have proposed the addition of a stimulus relevance or belongingness
principle to general process learning theory (e.g., Revusky,
1971). Others have suggested that a stimulus similarity
principle should be added to general process theory as a
factor in Pavlovian conditioning (e.g., Rescorla and Furrow,
1977; Testa and Ternes, 1977).

Because of changes in general conceptions of learning, certain
phenomena that were originally viewed as biological constraints
on classical or instrumental conditioning are now treated as
paradigmatic cases of learning. The phenomenon of autoshaping
provides a good illustration. Autoshaping refers to the fact
that animals come to approach and sometimes contact signals
for food and water even if approach and contact responses are
not necessary for delivery of the reinforcer (Hearst and
Jenkins, 1974). Autoshaping was considered to be a biological
constraint on learning in all of the original reviews of constraint phenomena. However, it is now used as a standard
technique for the investigation of Pavlovian conditioning
(e.g., see LoCurto, Terrace and Gibbon, 1981, especially
Parts II and III). This transformation from an anomaly to a
paradigmatic case occured through a reformulation of accepted
beliefs about conditioning. For example, research on autoshaping helped to undermine the idea that only visceral and
glandular responses can be modified by Pavlovian procedures.
Once it was accepted that classical conditioning can produce
changes in skeletal responses, the acquisition of approach
and contact responses to signals for food and water was no
longer seen as anomalous.

Taste-aversion learning is another phenomenon that was originally viewed as a biological constraint on learning but is now
treated by many as a paradigmatic case of Pavlovian conditioning (e.g., Logue, 1979). This transformation occurred in part
because of demonstrations that taste-aversion learning is
similar in many ways to other instances of Pavlovian conditioning. In addition, certain aspects of taste-aversion learning
that were originally viewed as unique (long-delay learning,
for example) have since been demonstrated in other situations
(see below).

Other instances of biological constraints on classical and
instrumental conditioning have not made the transformation
from anomaly to paradigmatic case but are also consistent with
revised general views of learning. Research suggests that misbehavior in depositing tokens for positive reinforcement, constraints on the topography of positively reinforced behavior,
constraints on punishment, constraints on avoidance learning,
and potentiation in Pavlovian conditioning may all be products
of general mechanisms of learning (see Domjan, 1982, for a
review).

The Inevitability of General Theories of Learning

In the above discussion I outlined some of the reasons why
biological constraints failed to create a revolution and
supplant the general process approach to the study of animal

learning. In considering further the impact of biological
constraints on classical and instrumental conditioning, two
alternatives arise. One possibility is that traditional focus
on general mechanisms in the study of animal learning can be
replaced, but biological constraint concepts were inadequate
to produce such a radical change. This view was recently
elaborated by Johnston (1981). Another possibility is that
a search for general mechanisms is so much a part of how
science progresses that it cannot be avoided in the scientific
investigation of animal learning. Recent research on biological constraint phenomena supports the second of these alternatives.

Scientific study of learning (and other) phenomena generally
involves formulation of explanations, derivation of the implications of each explanation, and empirical test of these implications. Results consistent with predicted outcomes provide
support for a proposed explanation or hypothetical mechanism.
Hypotheses that are closely based on the phenomenon to be
explained stand the best chance of being confirmed in experimental tests. However, the more one relies on the target
phenomenon in formulating a hypothetical mechanism, the
greater is the post hoc nature of the proposed explanation.
The best strategy for avoiding this tendency toward circularity
is to evaluate a hypothetical mechanism outside the situation
it was originally designed to explain.

Support for a proposed mechanism is considerably increased by
confirmatory evidence from new circumstances. This inevitably
draws scientific investigations toward explanations that are
relevant to more than the original phenomenon of interest.
By definition, the broader the applicability of a hypothetical
mechanism, the greater is its generality. Thus, a search for
explanations that can be confirmed by evidence from new
circumstances involves searching for general mechanisms.

Research on many biological constraints on classical and
instrumental conditioning has involved searching for underlying
general mechanisms of learning. Since much of this work has
been reviewed in detail elsewhere (Domjan, 1982), I will
describe only one example by way of illustration: long-delay
taste-aversion learning. Long-delay taste-aversion learning
refers to the fact that rats and other animals can learn
aversions to a novel-tasting substance followed by aversive
visceral consequences even if the aversive stimulation occurs
several hours after exposure to the taste (e.g., Revusky and
Garcia, 1970). Hypotheses formulated to explain long-delay
learning have been based on special features of taste-aversion

conditioning procedures. However, in many cases, efforts to substantiate a proposed mechanism have involved demonstrating how a presumed critical factor may produce long-delay learning outside the feeding system.

From Concurrent Interference to the Marking Hypothesis

One of the first mechanisms proposed to explain long-delay taste-aversion learning was the concurrent interference theory (Revusky, 1971, 1977). This theory sought to explain long-delay learning as a byproduct of stimulus relevance or selective associations. As we noted earlier, when rats are made sick they quickly learn aversion to taste but not to auditory/visual cues, whereas when they receive foot-shock they quickly learn aversion to auditory/visual cues but not to taste (Garcia and Koelling, 1966; Miller and Domjan, 1981). Concurrent interference theory assumes that long-delay learning is possible provided that the stimuli animals encounter during the interval between the target CS and the delayed US do not become associated with the US. In the typical long-delay taste-aversion experiment, animals experience various auditory and visual cues during the taste-toxicosis interval. Because these intervening events are not readily associable with toxicosis, they are assumed to provide no interference for the taste-toxicosis association, thereby permitting long-delay taste-aversion learning.

The concurrent interference theory was formulated using information from taste-aversion learning experiments, and initial tests of the theory focused on predictions about the boundary conditions of long-delay taste-aversion learning. The theory predicts, for example, that long-delay learning of a target taste will be disrupted by introducing other novel flavors during the delay interval. It also predicts that the degree of interference will depend on how familiar subjects are with the intervening flavors, because familiar tastes are less likely to become associated with toxicosis than novel tastes. Predictions such as these have been confirmed (Revusky, 1971), although there was some dispute whether the magnitude of the obtained effects was sufficient to consider concurrent interference as the only mechanism of long-delay learning (Kalat and Rozin, 1971).

The concurrent interference theory predicts long-delay learning in all sutiatuions where exposure to conditionable stimuli during the delay interval is minimized. Because support for the theory would be greatly enhanced by confirmed predictions of long-delay learning outside the feeding system, investigators sought to discover such examples. Lett (1973, 1974,

1975 and 1979) studied long-delay learning in T-maze situations.
She assumed that if animals receive a reinforcer in the T-maze,
only stimuli experienced in the maze would become associated
with the reinforcer. Based on this assumption, she removed
the subjects from the maze during the delay interval. In one
set of experiments, rats were reinforced either for going
right or for going left in a T-maze. After making the correct
choice, the animals were picked up and returned to the home
cage for the delay interval, at the end of which they were
placed in the start area of the maze to receive the reinforcer.
Using this procedure Lett reported successful conditioning
of a spatial discrimination with delay intervals of as long
as 60 min (Lett, 1973, 1975). In addition, as predicted by
the concurrent interference theory, she found that delayed
spatial discrimination learning was disrupted if subjects
were allowed to remain in the T-maze during the delay interval
(Lett, 1975, Experiment 3). Lett also reported successful
visual discrimination learning with 1-min delayed reinforcement in the T-maze (Lett, 1974). However, this aspect of her
work was later brought into question (see Roberts, 1976, 1977;
and Lett, 1977).

Demonstrations of long-delay spatial discrimination learning
in the T-maze provide considerable support for concurrent
interference theory because they confirm implications of the
theory outside the situation the theory was originally designed to explain (long-delay taste-aversion learning). The
empirical findings of long-delay spatial discrimination learning have been replicated in subsequent work, but research has
shown that the absence of concurrent interference is probably
not responsible for the effects. Lieberman, McIntosh and
Thomas (1979) showed that the critical procedural variable
for long-delay spatial discrimination learning in the T-maze
is picking up the subjects (or providing some other salient
stimulation) rather than removing them from the maze during
the delay interval. Animals that are picked up or given a
loud burst of noise or light after making a correct choice
learn the spatial discrimination regardless of where they
spend the delay interval. In contrast, subjects that do not
receive such stimulation do not learn with delayed reward
even if they spend the delay interval outside the choice maze.
Lieberman et al. (1979) proposed a marking hypothesis to explain these results. According to the marking hypothesis,
presentation of a salient stimulus immediately after an event
to be conditioned enhances the memory of that event so that
it is more readily available for association with a delayed
unconditioned stimulus or reinforcer.

The marking hypothesis was proposed to explain the results
of delayed reinforcement experiments conducted in the T-maze.

However, it is not necessarily limited to maze-learning situations. To apply the hypothesis to other situations, one must identify marking stimuli that might enhance retention. How could the marking hypothesis explain long-delay taste-aversion learning? One possibility is that the complex integrated series of responses and sensory stimuli involved in normal ingestion of a flavored solution provides the marking function. Taste-aversion conditioning procedures typically require subjects to drink a distinctively flavored solution. Therefore, exposure to the taste stimulus occurs in the context of responses involved in approaching a drinking spout, licking, and swallowing. These responses are in turn accompanied by a host of sensory stimuli provided by the texture, temperature, viscosity, and other features of the solution. The responses and stimuli involved in ingestion may make the flavor of the solution sufficiently memorable to allow association with delayed toxicosis. Consistent with this view, Domjan and Wilson (1972) found that rats that receive exposure to a flavored solution rinsed over the tongue in the absence of approach and drinking responses learn weaker aversions to the taste than animals that are allowed to drink the flavored solution in the normal fashion. (For a more detailed discussion of the role of ingestive behavior in food-aversion learning, see Domjan, 1980).

Affective Versus Instrumental Response Learning

Another effort to explain long-delay taste-aversion learning has focused on the fact that taste-aversion experiments involve conditioning an affective response (an aversion to the food) and use of a simple response measure (suppression of intake) that directly reflects what is learned. If these are critical features for long-delay learning, then long-delay learning should be evident in nonfeeding situations that also involve conditioning of affective responses and measurement of these conditioned responses with simple and direct indices of learning. D'Amato and his colleagues tested this prediction using spatial preference conditioning in cebus monkeys and rats (see D'Amato, Safarjan and Salmon, 1981). In one experiment (Safarjan and D'Amato, 1981), rats were pretested for preference for the right or left side of a T-maze having identical solid black choice arms. On the conditioning day, a striped pattern was placed on the floor, sides, and top of the nonpreferred arm of the maze, and independent groups were confined there for 10, 20 or 40 min. They were then placed in their home cages and moved to a designated area of the experimental room, where they spent a delay period of 30 or 120 min. At the end of the delay interval, the subjects were placed in a wastebasket to receive sucrose reward. Control

subjects received similar treatments except that for them reward was absent in the wastebasket. Two days later, each
subject received a preference tests in which unimpeded access
to the two maze arms was provided for 5 min, and the striped
pattern CS was applied to the surfaces of the previously-
nonpreferred arm.

During the postconditioning test, experimental subjects showed
a higher preference for the CS arm of the maze than control
subjects but only if they had been exposed to the CS for more
than 10 min during conditioning. There was also evidence that
effects of increasing the duration of CS exposure during conditioning could offset effects of increasing the delay of
reinforcement. With 20-min CS exposure, spatial preference
learning occurred with 30-min delayed reinforcement but not
with 120-min delayed reinforcement. With 40-min CS exposure,
spatial preference learning occurred with both 30-min and 120-
min delayed reinforcement. The difference between experimental
and control subjects was in part due to decreased preference
for the CS arm of the maze among control subjects as a function
of increasing exposure to the CS on the conditioning day. However, there was also some evidence that experimental subjects
increased their preference for the CS arm between the preconditioning and postconditioning tests.

A second experiment showed that conditioning of affective
responses is less susceptible to disruption by delay of reinforcement than conditioning of instrumental responses (D'Amato
and Safarjan, 1981). Rats were injected with lithium chloride
1 or 30 min after entering a visually-distinctive side of a
T-maze. Instrumental behavior was measured by which side the
subject entered first when it was subsequently given access
to both arms. Affective preference was measured by which side
the subject spent more time in during an unrestricted 5-min
choice test. Learning of the instrumental choice response was
severely disrupted when the lithium injection was delayed for
30 min during conditioning trials. In contrast, equal spatial
aversions were acquired with 1- and 30-min delayed reinforcement.

Relation of Response, Cue, and Reinforcer

Another interpretation of long-delay taste-aversion learning
attributes the phenomenon to a special relationship among the
response used to measure learning, the conditioned stimulus,
and the reinforcer or unconditioned stimulus (e.g., Testa and
Ternes, 1977). In poison taste-aversion conditioning, all of
these are centered on, or attributes of, the same object, a
flavored solution. The drinking behavior used to measure

learning is directed toward the flavored solution, CS taste is
provided by the flavored solution, and the solution may also
contain the toxin unconditioned stimulus. If the critical
factor for long-delay learning is having the indicant response,
CS, and US centered on the same object, then long-delay learning should occur in nonfeeding situations that share these
characteristics. This prediction was evaluated by Sullivan
(1979) in an experiment on shock-aversion conditioning of small
objects (such as a plastic pot cleaner and a piece of leather)
in rats. During conditioning trials, animals were given access
to a particular object for 5 min. The object was then removed
while the animal remained in the experimental chamber for a 35-
min delay interval, at the end of which several shocks were
delivered through the grid floor. On alternate trials, the
animals received equal access to a different object without
shock. The amount of time subjects spent touching the shock
object decreased across trials, whereas the amount of time they
spent touching the safe object increased. In addition, subjects showed a preference for handling the safe object in a
postconditioning preference test.

Other research has shown more definitively that having the
indicant response and conditioned stimulus centered on the
same object facilitates long-delay aversion learning. Galef
and Dalrymple (1981) demonstrated that poison-avoidance learning to a visual stimulus in rats is facilitated if the visual
cue is a distinctive feature of the food subjects ingest as
opposed to being a feature of the dish or chamber in which the
food is located. In a related study, Logue (1980) showed that
pigeons are more likely to learn an aversion to a red visual
stimulus paired with toxicosis if the red color is contained
in ingested water, or is provided by a light that illuminates
the water spout, than if the red color is located on a response
key the pigeons have to peck to obtain water.

Implications of Research on Long-Delay Learning

Research on long-delay learning has not progressed sufficiently
to allow firm conclusions about the mechanisms involved. Some
of the critical experiments have not been replicated, some
hypothetical mechanisms have not been tested in taste-aversion
conditioning, and some other promising explanations have received no experimental attention. Nevertheless, the research
illustrates the course of scientific investigation of a seem-
ingly-specialized form of learning. Several variables have
been discussed as critical for long-delay taste-aversion learning (e.g., absence of concurrent interference, conditioning of
affective rather than instrumental responses, and a special
relationship between response, CS, and US). Because the
strongest support for a hypothetical mechanism is provided by

confirmatory evidence outside the situation the hypothesis was originally designed to explain, much effort has been devoted to the discovery of long-delay learning of noningestive responses. Once a particular variable became the focus of investigation, experiments were conducted to show that the variable could lead to long-delay learning outside the feeding system. These efforts led to demonstrations of long-delay learning in new situations, and sometimes provided hypotheses (e.g., the marking hypothesis) deserving further experimentation in taste-aversion learning. The success of these efforts suggests that general mechanisms exist that happen to be strongly activated by typical long-delay taste-aversion conditioning procedures, but these mechanisms can also operate in other circumstances.

Biological and General Process Aspects of Animal Learning

We have seen that ideas about biological constraints were formulated in a way that prevented them from creating a successful alternative to the general process approach to the study of animal learning, and experimental analyses fo even seemingly specialized forms of learning inevitably involved a search for general mechanisms. These developments suggest that the general process approach cannot be abandoned in the study of animal learning. Therefore, attempts, to add a biological dimension to the study of animal learning have to focus on ways to complement rather than replace general process investigations. Two ways of accomplishing this have been recently suggested. One proposal emphasizes study of the adaptive functions of learning; the other emphasizes comparative investigations of learning.

Biological Functions of Learning

General process investigations of animal learning focus on how learning takes place--how stimulus, response, and post-experience variables control the learning process. A different kind of question concerns why learning takes place, and what are the biological functions of learning. The function of a behavior involves its net contribution to inclusive fitness. The function of learning depends on how well the products of learning facilitate transmission of genes. Thus, investigation of the functions of learning requires determining what animals learn and how the learned behavior is related to reproductive success. Since a given behavioral outcome may be produced by a variety of underlying mechanisms, study of the functions of learning is orthogonal to the study of learning mechanisms.

Several authors have recently advocated investigating the
functions of learning as a way to increase the biological content of research on animal learning (Hollis, 1981; Shettleworth,
1981). They have noted that direct study of the functions of a
behavior in terms of impact on reproductive fitness is often
very difficult, if not impossible. However, adaptive function
may be assessed in terms of the contribution of a behavior to
a nonreproductive process that has obvious or demonstrated
adaptive significance. For example, the function of different
ways of learning to find food may be assessed in terms of the
efficiency of each strategy (quantity of nutrients obtained
for the effort expended), on the assumption that efficiency of
foraging has adaptive significance.

Adaptive function is one of the fundamental concerns of biology
(Tinbergen, 1963). Therefore, studies of the adaptive function
of learning would help to integrate research on animal learning with biologically-oriented approaches to the study of animal behavior. Shettleworth (1981) has suggested that consideration of the functions of learning may also contribute
to investigations primarily concerned with elucidating learning mechanisms. For example, the functions of a particular
instance of learning may suggest underlying mechanisms responsible for the learning. Consideration of the functions of
learning may also help differentiate between several hypothetical mechanisms if these mechanisms predict behavioral
outcomes having different adaptive functions or different
degrees of functional utility. Functional considerations may
also provide insights into forms of learning that are multiply
determined. Multiple underlying mechanisms may evolve, for
example, to insure that a particular change in behavior will
be acquired despite unpredictable or varying environmental
contingencies.

Comparative Investigations of Learning

A second approach to adding a complementary biological
orientation to general process investigations of learning
involves studying comparative aspects of animal learning.
Modern investigations of animal learning developed from
comparative research on learning at the turn of the twentieth
century (Kline, 1898; Small, 1900 a,b). Domjan and Galef
(1982) recently advocated a renewed emphasis on comparative
research in animal learning as a way to introduce biological
considerations into the study of animal learning. In the
first part of this chapter, I suggested that concepts of
biological constraints on learning were unsuccessful in
adding a biological dimension to the study of animal learning
because they did not suggest a distinctive method for the

discovery of new phenomena, did not provide ways to systematize
knowledge about biological constraints, and permitted assimilation of constraint phenomena into revisions of general
principles of learning. An ecologically-oriented comparative
approach avoids these problems.

The general process approach to the study of learning generally
ignores species differences and focuses instead on how learning is determined by cues, responses, reinforcers, and their
combinations. In contrast, species differences constitute
the critical variable in a comparative investigation. This
provides comparative investigations with a distinctive strategy
for the discovery of new phenomena and makes it difficult to
incorporate evidence of species differences in learning into
revisions of general process theory by a rearrangement of
traditional parameters. Which species are chosen for comparison depends on other goals of the research, and this in
turn determines how the results are to be systematized.
Species widely differing in both ecology and phylogeny may be
investigated to reveal the generality of a learning phenomenon
or process in the animal kingdom. By investigating ecologically disparate but taxonomically similar species, information
can be obtained about the relationship between learning and
ecological variables. In addition, ecological factors producing convergent adaptations in learning phenomena can be
discovered by studying ecologically similar but taxonomically
disparate species (Campbell, 1976; Hailman, 1976). Species
differences in learning related to ecological differences can
be systematized using concepts from behavioral ecology.

So far, much of comparative research on animal learning has
been addressed to the question generality of learning phenomena and processes (e.g., Bitterman, 1975). However, ecologically-oriented comparative investigations of other aspects of
animal behavior are available. Tschanz and Hirsbrunner-Scarf
(1975), for example, compared the behavior of two species of
birds, guillemots and razorbills. The two species are of the
same family, Alcidae, and differ in few morphological, reproductive, and ethological traits. Both species breed on the
cliffs of coasts and islands, and in both species male-female
pairs produce a single egg, which parents alternate incubating
without building a nest. Razorbills are spatially spread out
and nest in caves and crevices. In contrast, guillemots stay
together in crowded breeding groups, and have eggs on open
ledges that are sometimes narrow and slope toward the sea.
This makes guillemot eggs more vulnerable to predation,
parental carelessness, or carelessness of conspecifics.

Differences in the ecology of the two species are associated
with a variety of differences in behavior. Guillemot parents
leave the egg only at times of extreme disturbance, and one
parent never leaves the nest before the other can immediately
take over. In contrast, razor bills are more likely to
leave the nest unattended. The chicks also evidence ecologi-
cally-related differences in behavior. For example, guillemot
chicks prefer to stay close to a vertical surface whereas
razorbill chicks do not show such a preference and are more
likely to wander close to the brink of a ledge.

Tschanz and Hirsbrunner-Scharf (1975) tested the adaptive
significance of these and other species differences in a cross-
fostering experiment conducted in the field. The hatching success of cross-fostering eggs was very similar for the two
species (73% and 80%). However, a much greater proportion of
cross-fostered razorbill chicks died in the first 22 days of
life (22 of 41) than guillemot chicks (3 of 39). Razorbill
chicks (which are normally reared in dry caves and crevices)
were at much greater risk being raised by guillemot foster
parents on cliff ledges than were guillemot chicks raised by
razorbill foster parents.

The type of comparative research described above may be also
profitably applied to the study of various forms of learning.
The ecological factors that have to be considered depend on
the form of learning under investigation. For example,
poison-avoidance learning may be studied from a comparative
perspective with closely-related species that differ in the
types and range of foods they eat. Generalist feeders would
be expected to learn food aversions more rapidly than
specialist feeders because generalists are more likely to
sample poisonous edibles (Daly, Rauschenberg and Behrends, in
press). In contrast, for an ecologically-oriented comparative
study of performance in a radial maze, species should be
selected that differ in their foraging patterns, since the
radial maze in an analogue of foraging situations (e.g.,
Olton, Handelman and Walker, 1981). Species adapted to
foraging in sparse patches by moving frequently from one patch
to another would be expected to perform differently in the
radial maze from species adapted to foraging in rich patches
by remaining in one patch for long periods.

Species differences related to ecological variables can be used
to reach conclusions about the adaptive function of a particular form of learning or behavior. Appropriately-conducted
comparative investigations are one source of information about
adaptive function (Hailman, 1976; Hodos and Cambell, 1969).
The specialized juvenile and parental behavior of guillemots,

for example, may be viewed as adaptations to nesting in colo-
nies on cliff ledges (Tschanz and Hirsbrunner-Scharf, 1975).
Therefore, the comparative approach shares elements of approaches emphasizing the functions of learning. However, the
comparative approach is more explicit in relating aspects of
learning to ecological predictions about adaptive function.
In conducting an ecologically-oriented comparative study, one
has to identify the relevant ecological differences between
species in advance and then strive to relate observed differences in learning to these ecological factors.

Concluding Comments

Research on animal learning has traditionally proceeded on the
assumption that general mechanisms of learning can be discovered through the intensive investigation of a few laboratory
preparations that are arbitrary with respect to the innate behavioral repertoire of the animal being observed. The traditional pursuit of general learning mechanisms was seriously
called into question during the early 1970's in discussions
about biological constraints on classical and instrumental conditioning. These discussions highlighted findings that appeared to be specialized adaptations and that were inconsistent
with extant general views of learning. In its extreme form,
this type of criticism suggested that the pursuit of general
principles of learning is an ill-fated venture. However,
ideas about biological constraints did not revolutionize the
study of animal learning because they did not suggest a distinctive strategy for the discovery of new phenomena, did not
provide ways to systematize knowledge about biological constraints, and permitted assimilation of constraints on classical and instrumental conditioning into revisions of general
process views of learning. In addition, subsequent empirical
analyses of phenomena identified as biological constraints
showed that a search for general mechanisms is so much a part
of the scientific method that it cannot be avoided in the study
of animal learning. Strongest proof that a particular variable
is responsible for a biological constraint effect like long-
delay learning is provided by evidence that the variable can
produce similar effects in new situations. Because of this,
many experiments have been conducted showing that biological
constraints on classical and instrumental conditioning are not
mediated by specialized or unique processes and can occur in
diverse situations. These developments suggest that to be
successful, biological approaches to the study of animal learning should stirve to complement rather than replace investigations of general mechanisms. Considerations of the adaptive

functions of learning and studies of species differences in
learning related to ecological variables have been proposed
to achieve this goal.

Footnotes

[1]Preparation of this chapter was supported by a Faculty Research Assignment from the University of Texas Research Institute. I am also indebted to Bennett G. Galef, Jr., for stimulating discussions that helped me clarify some of the ideas presented here.

[2]I am indebted to Bennett G. Galef, Jr., for this example.

References

1 Bitterman, M. E. The comparative analysis of learning.
   Science, 1975, 188, 699-709.

2 Bolles, R. C. Species-specific defense reactions and
   avoidance learning. Psychological Review, 1970, 77, 32-
   48.

3 Breland, K. & Breland, M. The misbehavior of organisms.
   American Psychologist, 1961, 16, 681-684.

4 Campbell, C. B. G. What animals should we compare? In
   R. B. Masterton, W. Hodos & H. Jerison (Eds.), Evolution,
   brain and behavior: Persistent Problems. Hillsdale, NJ:
   Lawrence Erlbaum, 1976.

5 Daly, M., Rauschenberger, J. & Behrends, P. Food aversion
   learning in kangaroo rats: A specialist-generalist comparison. Animal Learning & Behavior, in press.

6 D'Amato, M. R. & Safarjan, W. R. Differential effects of
   delay of reinforcement on acquisition of affective and
   instrumental responses. Animal Learning & Behavior, 1981,
   9, 209-215.

7 D'Amato, M. R., Safarjan, W. R. & Salmon, D. Long-delay
   conditioning and instrumental learning: Some new findings.
   In N. E. Spear & R. R. Miller (Eds.), Information processing in animals. Hillsdale, N.J.: Lawrence Erlbaum,
   1981.

8 Dickinson, A. Contemporary animal learning. Cambridge,
   England: Cambridge University Press, 1980.

9 Domjan, M. Ingestional aversion learning: Unique and
   general processes. In J. S. Rosenblatt, R. A. Hinde,
   C. Beer & M. C. Busnel (Eds.), Advances in the study of
   behavior. Vol. 11. New York: Academic Press, 1980.

10 Domjan, M. Biological constraints on learning 10 years
    later: Implications for general process learning theory.
    Submitted, 1982.

11 Domjan, M. & Galef, B. G., Jr. Biological constraints on
    animal learning: Retrospect and prospect. Submitted,
    1982.

12 Domjan, M. & Wilson, N. E. Contribution of ingestive behaviors to taste-aversion learning in the rat. Journal of
Comparative and Physiological Psychology, 1972, 80, 403-
412.

13 Galef, B. G., Jr. & Dalrymple, A. J. Toxicosis-based
aversions to visual cues in rats: A test of the Testa and
Ternes hypothesis. Animal Learning & Behavior, 1981,
9, 332-334.

14(Garcia, J. & Koelling, R. A. Relation of cue to consequence in avoidance learning. Psychonomic Science, 1966,
4, 123-124.

15 Hearst, E. & Jenkins, H. M. Sign-tracking: The stimulus-
reinforcer relation and directed action. Austin, TX:
Psychonomic Society, 1974.

16 Hinde, R. A. Constraints on learning: An introduction to
the problem. In R. A. Hinde & J. Stevenson-Hinde (Eds.),
Constraints on learning. London: Academic Press, 1973.

17 Hinde, R. A. & Stevenson-Hinde, J. (Eds.), Constraints
on learning. London: Academic Press, 1973.

18 Hodos, W. & Campbell, C. B. G. Scala naturae: Why there
is no theory in comparative psychology. Psychological
Review, 1969, 76, 337-350.

19 Hollis, K. L. Pavlovian conditioning of signal-centered
action patterns and autonomic behavior: A biological
analysis of function. In J. S. Rosenblatt, R. A. Hinde,
C. Beer & M. -C. Busnel (Eds.), Advances in the study of
behavior, Vol. 12. New York: Academic Press, 1981.

20 Johnston, T. D. Contrasting approaches to a theory of
learning. The Behavioral and Brain Sciences, 1981, 4,
125-173.

21 Kalat, J. W. & Rozin, P. Role of interference in taste-
aversion learning. Journal of Comparative and Physiological Psychology, 1971, 77, 53-58.

22 Kline, L. W. Methods in animal psychology. American
Journal of Psychology, 1898, 10, 256-279.

23 Lett, B. T. Delayed reward learning: Disproof of the
traditional theory. Learning and Motivation, 1973,
4, 237-246.

24   Lett, B. T.   Visual discrimination learning with a 1-min
     delay of reward.   Learning & Motivation, 1974, 5, 174-181.

25   Lett, B. T.   Long delay learning in the T-maze.   Learning
     and Motivation, 1975, 6, 80-90.

26   Lett, B. T.   Regarding Roberts' reported failure to obtain
     visual discrimination learning with delayed reward.   Learn-
     ing and Motivation, 1977, 8, 136-139.

27   Lett, B. T.   Long-delay learning: Implications for learn-
     ing and memory theory.   In N. S. Sutherland (Ed.),
     Tutorial essays in psychology.   Vol. 2.   Hillsdale, N.J.:
     Lawrence Erlbaum, 1979.

28   Lieberman, D. A., McIntosh, D. C. & Thomas, G. V.   Learn-
     ing when reward is delayed: A marking hypothesis.
     Journal of Experimental Psychology: Animal Behavior
     Processes, 1979, 5, 224-242.

29   LoCurto, C. M., Terrace, H. S. & Gibbon, J. (Eds.), Auto-
     shaping and conditioning theory.   New York: Academic
     Press, 1981.

30   Logue, A. W.   Taste aversion and the generality of the
     laws of learning.   Psychological Bulletin, 1979, 86, 276-
     296.

31   Logue, A. W.   Visual cues for illness-induced aversion in
     the pigeon.   Behavioral and Neural Biology, 1980, 28,
     372-377.

32   Miller, V. & Domjan, M.   Specificity of cue to consequence
     in aversion learning in the rat: Control for US-induced
     differential orientations.   Animal Learning & Behavior,
     1981, 9, 339-345.

33   Olton, D. S., Handelman, G. E. & Walker, J. A.   Spatial
     memory and food searching strategies.   In A. C. Kamil and
     T. D. Sargent (Eds.), Foraging behavior.   New York:
     Garland STPM Press, 1981.

34   Rachlin, H.   Behavior and Learning.   San Francisco:
     W. H. Freeman, 1976.

35   Rescorla, R. A. & Furrow, D. R.   Stimulus similarity as a
     determinant of Pavlovian conditioning.   Journal of Experi-
     mental Psychology: Animal Behavior Processes, 1977, 3,
     203-215.

36 Rescorla, R. A. & Holland, P. C.   Behavioral studies of associative learning in animals.   Annual Review of Psychology, 1982, 33, 265-308.

37 Revusky, S.   The role of interference in association over a delay.   In W. K. Honig & P. H. R. James (Eds.), Animal memory.   New York: Academic Press, 1971.

38 Revusky, S.   The concurrent interference approach to delay learning.   In L. M. Barker, M. R. Best & M. Domjan (Eds.), Learning mechanisms in food selection.   Waco, TX: Baylor University Press, 1977.

39 Revusky, S. H. & Garcia, J.   Learned associations over long delays.   In G. H. Bower (Ed.), The psychology of learning and motivation.   Vol. 4.   New York: Academic Press, 1970.

40 Roberts, W. A.   Failure to replicate visual discrimination learning: A reply to Lett.   Learning and Motivation, 1977, 8, 140-144.

41 Rozin, P. & Kalat, J. W.   Specific hungers and poison avoidance as adaptive specializations of learning. Psychological Review, 1971, 78, 459-486.

42 Safarjan, W. R. & D'Amato, M. R.   One-trial, long-delay, conditioned preference in rats.   The Psychological Record, 1981, 31, 413-426.

43 Schwartz, B.   Psychology of learning and behavior.   New York: W. W. Norton, 1978.

44 Seligman, M. E. P.   On the generality of the laws of learning.   Psychological Review, 1970, 77, 406-418.

45 Shettleworth, S. J.   Constraints on learning.   In D. S. Lehrman, R. A. Hinde & E. Shaw (Eds.), Advances in the study of behavior.   Vol. 4.   New York: Academic Press, 1972.

46 Shettleworth, S. J.   Function and mechanism in learning. In M. Zeiler & P. H. Harzem (Eds.), Advances in analysis of behavior, Vol. 3.   New York: Wiley, 1981.

47 Small, W. S.   An experimental study of the mental processes of the rat.   American Journal of Psychology, 1900, 11, 133-165. (a)

48 Small, W. S. An experimental study of the mental processes of the rat II. American Journal of Psychology, 1900, 12, 206-239. (b)

49 Sullivan, L. Long-delay learning with exteroceptive cue and exteroceptive reinforcement in rats. Australian Journal of Psychology, 1979, 31, 21-32.

50 Tschanz, B. & Hirsbrunner-Scharf, M. Adaptation to colony life on cliff ledges: A comparative study of guillemot and razorbill chicks. In G. Baerends, C. Beer & A. Manning (Eds.), Function and evolution of behavior. Oxford: Clarendon Press, 1975.

51 Testa, T. J. & Ternes, J. W. Specificity of conditioning mechanisms in the modification of food preferences. In L. M. Barker, M. R. Best & M. Domjan (Eds.), Learning mechanisms in food selection. Waco, TX: Baylor University Press, 1977.

52 Tinbergen, N. On aims and methods of ethology. Zeitschrift Fur Tierpsychologie, 1963, 20, 410-433.

ANIMAL COGNITION AND BEHAVIOR
Roger L. Mellgren, editor
© *North-Holland Publishing Company, 1983*

PROPERTIES AND FUNCTION OF AN INTERNAL CLOCK

Seth Roberts

University of California, Berkeley

The term time discrimination (shortened to timing) has no exact
meaning, but the common examples of it share two properties:
(a) The probability of some learned response changes with the
time since some event; and (b) the function relating response
probability and time changes with the time of the motivating
event (e.g., food or shock). As an example, Figure 1 (from
Roberts and Church, 1978) shows the result of training rats with
two signals that were followed by food at different times.
After a random intertrial interval, the 30-sec signal (light)
or the 60-sec signal (sound) began. After the appropriate time
--30 sec for light, 60 sec for sound--the next response (lever
press) was rewarded with food and the signal was turned off.
On the first day of exposure to this procedure, neither condition for a time discrimination was clearly met: (a) Response
ratio (response rate at the given time divided by overall response rate) was almost constant with time, and (b) the 30-sec
and 60-sec functions were similar. On the last day, both conditions were met: (a) Response ratio increased substantially with
time, and (b) the 30-sec and 60-sec functions were very different.

Russell Church, Mark Holder, Villu Maricq and I have done experiments to determine how rats discriminate time (properties);
to our surprise, these experiments have also suggested why rats
discriminate time (function). This chapter summarizes our
work, emphasizing our conclusions about the internal clock that
is used. In addition, this chapter tries to place our work in
context--by describing the generality of animal timing, by
comparing our explanation with earlier explanations of animal
timing, and by pointing out connections between this work and
work outside of animal timing.

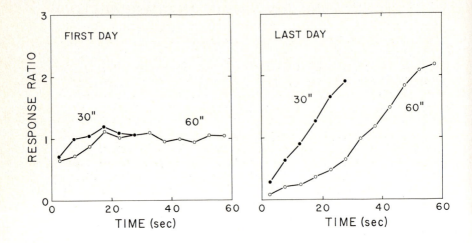

Figure 1. The two features of a time discrimination, present on the last day (right panel) but not on the first
day (left panel). Figure 10 of Roberts and Church,
1978. The left panel shows results from the first day
of fixed-interval training; the right panel shows the
results from the last day, 30 days later. "Time" is
time since the interval began. Response ratio is local
response rate divided by overall response rate; for
example, the response ratio for the interval Sec 5-10
is response rate during Sec 5-10 divided by the response
rate for the whole interval, Sec 0-30 or Sec 0-60.
Different curves are the results from different signals;
the label of each curve is the procedure of the signal,
FI 30 Sec or FI 60 sec. Each point is a mean over 12
rats. (Copyright 1978 by the American Psychological
Association.)

The Generality of Animal Timing

The inhibition-of-delay results of Pavlov (1926/1960) were the first examples of animal time discrimination. One of Pavlov's examples, originally published in 1980, used as the conditioned stimulus (CS) the sound of a whistle; it was followed after 3 min by the unconditioned stimulus (US), acid in the dog's mouth. During the first minute of the whistle there were 0 drops of saliva; during the second minute, about 5 drops; during the third minute, about 9 drops. This satisfies the first condition. In addition, Pavlov found that the pause before salivating was proportional to the length of the CS; this satisfies the second condition. The introduction by Skinner of fixed-interval schedules of reinforcement made it much easier to observe timing, especially in rats. Skinner was training rats in the first Skinner boxes. He began by rewarding every response, but changed to rewarding responses only once a minute (a fixed-interval schedule) in order to make his supply of pellets last longer (Skinner, 1956). After training with this procedure, a cumulative record of the number of responses showed the now-familiar fixed-interval scallops--response rate was low at the beginning of the interval and much higher near the end.

A figure in Skinner (1956) showed three cumulative records with fixed-interval scallops, one record from a rat, one from a pigeon, and one from a monkey. The three records were similar, and Skinner stated, "Pigeon, rat, monkey, which is which? It doesn't matter" (Skinner, 1956, p. 40). He was right--the ability to discriminate time is possessed by a wide range of vertebrates, and the accuracy of discrimination does not vary substantially. Not only dogs, pigeons, rats, and monkeys, but also mice (Sprott and Symons, 1974), opossums (Cone and Cone, 1970), rabbits (Rubin and Brown, 1969), racoons (King, Schaeffer and Pierson, 1974), bats (Schumake and Caudill, 1974), crows (Powell, 1973), and goldfish (Rozin, 1965) have discriminated time, usually with a fixed-interval schedule. On the other hand, bees apparently can not discriminate duration (Grossman, 1973), although they can discriminate other dimensions (e.g., color and smell: Klosterhalfen, Fischer and Bitterman, 1978) including time of day (Koltermann, 1974). Richelle and Lejeune (1980) reviewed results from 28 species scattered throughout the vertebrates; three of them (a bird, a fish, and a primate) did not clearly show timing.

The range of procedures that can produce timing is also wide. The duration that is discriminated can be: (a) the time from an experimenter-controlled event, such as food or shock (e.g., LaBarbera and Church, 1974); (b) the time from a response, as with differential-reinforcement-of-low-rate schedules

(Kramer and Rilling, 1970) and Sidman avoidance schedules
(e.g., Libby and Church, 1974); (c) the duration of an experimenter-controlled stimulus, such as a light (e.g., Libby
and Church, 1974); (d) the duration of a response, such as a
key peck (Ziriax and Silberberg, 1978) or lever press (Platt,
Kuch and Bitgood, 1973). The durations can be as short as tens
of milliseconds (Ziriax and Silberberg, 1978, with pigeons
or hundreds of milliseconds (Killeen, 1981, with pigeons;
Millenson, Kehoe and Gormezano, 1977, with rabbits) or as long
as tens of minutes (Pavlov, 1926/1960, pp. 41-2, with dogs)
or tens of hours (Dews, 1965, and Eckerman, Note 1, with
pigeons). With rats there seem to be two clocks available.
One seems to have a range roughly from 0 to 5 min (Richardson
and Loughead, 1974; Sherman, 1959), so that, for example,
accuracy is normal with FI 1 min and FI 3 min but much worse
with FI 10 min. Another seems to have a range roughly from
20 to 28 hrs (Boulos, Rosenwasser and Terman, 1980).

Some of the responses with which the discrimination can be
expressed are salivation, a lever press, a key peck, the cross-
ing of a shuttle box, a chain pull (Carlson, Wielkiewicz and
Modjeski, 1972), a choice response in a maze (Cowles and Finan,
1941), the closing of a nictating membrane (Millenson, Kehoe
and Gormezano, 1977), an attack (Azrin and Hutchinson, 1967),
general activity (Killeen, 1975), and scrabbling and gnawing
(Anderson and Shettleworth, 1977). Reinforcement can be food,
shock, water, blood (Shumake and Caudill, 1974), brain stimulation (Cantor and Wilson, 1981), or drugs such as cocaine
(Balster and Schuster, 1973). The reinforcer can be response-
contingent or response-independent. For a longer review of
the timing literature, see Richelle and Lejeune (1980).

Explanations of Animal Timing

Any mechanistic explanation of timing must assume at least
two internal mechanisms or operations: (a) A clock (something
that changes with time in a regular way) that measures the
relevant duration. Because time changes performance in a
regular way, time must change something inside the animal in a
regular way. (b) A memory (something allowing storage and
retrieval) that stores and compares against storage the output
of the clock. Because the time of the reinforcer (on one
trial) changes performance (on later trials), the time of the
reinforcer must be measured and stored (on one trial) and
accessed (on later trials). In this chapter, I focus on the
clock and pay little attention to the memory. There has been
some controversy about what is and is not an explanation (e.g.,
Brown, 1965; Skinner, 1950). To simply say "a rat has an
internal clock" is neither an explanation of timing nor a
description of timing, but something in between the two for

which no word exists. It resembles a description because it describes something hidden; it differs from an explanation because it is not testable, it is more like an incomplete sentence than anything else. The sentence is completed, and becomes an explanation, when a theorist proposes a property of the clock--e.g., "the rat has an internal clock that times both light and sound."

Somewhere between ten and twenty explanations of animal timing have been proposed. To give an idea of their range:

1. To explain his inhibition-of-delay results, Pavlov (1926/ 1960, p. 104) noted that "we get accustomed to stimuli of smell, sound or illumination"--their effective intensity changes (decreases) with time. Thus one intensity of the CS (at its end) is reinforced, but other intensities of the CS (e.g., at its beginning) are not. So the time discrimination is really an intensity discrimination. Logan (1979) proposed a more general version of this idea to explain fixed-interval performance. According to Logan, the brief event at the start of the interval--e.g., a food pellet--produces a variety of memories that last different amounts of time. The short-lived memories do not reach the next reward and do not become associated with time since food because the ratio of long-lived memories (associated with reward) to short-lived memories (not associated with reward) increases with time since food.

2. Dews (1962) proposed that the fixed-interval scallop arises because the effectiveness of reward decreases with delay and responses early in the interval are rewarded with longer delays of reward than responses later in the interval. This assumes, of course, that responses early in the interval are somehow distinct from responses later in the interval.

3. Killeen (1975) used equations derived from control theory to fit the response-rate functions from a number of procedures where food is given at fixed intervals. The emphasis was more on the range of procedures used and the goodness of fit than on the mechanism that would produce the equations.

4. Gibbon (1977) developed a detailed and quantitative theory to describe results from fixed-interval procedures, choice procedures, and a few others. The emphasis was mainly on how changes in the durations to be discriminated changed the accuracy of discrimination.

Some other recent theories are Staddon (1974), Ambler (1976),
and Keller (1980).

All of these theories assume that something internal changes
with time in a regular way. In Pavlov's theory, the clock (or
its output) is the effective intensity of the CS, which decreases with time. In Logan's theory, the clock is the changing memory of food, increasingly composed of its long-lived
components. In Dews' theory, the response changes with time,
but not necessarily in a regular way; the required regularity
is the progressive weakening of the effects of reward--not very
different, in essence, from Logan's theory. The mechanistic
derivation of Killeen's equations assumes that, with time, the
animal passes through a progression of states. The probability
of passing from one state to the next is constant with time,
but of course the probability of being in one of the later
states rises with time. Gibbon simply assumes that the animal
has available an accurate estimate of the amount of time that
has passed.

Our work differs from this earlier work in a number of ways.
Perhaps the most important difference is that the theory is
less detailed. The psychological process producing the clock
is not named (in contrast to Pavlov, Logan and Dews); and
assumptions about the shapes of distributions are not made
(in contrast to Killeen and Gibbon). In general, the approach
is to add to the assumption of a clock as little as possible,
only enough to produce an experimental test. If it seems that
we assume an internal clock and other theories do not, this is
only because less has been added. A second important difference is that the data used to support the theory usually come
from procedures designed for that purpose--procedures designed
to choose between theoretical alternatives. Earlier theories
have been based on procedures originated for other purposes
(e.g., fixed-interval schedules). A third difference is that
the clock metaphor is taken more seriously. A number of
experiments (e.g., Roberts and Church, 1978) have been based
on the differences between man-made clocks. For example, a
stopwatch times up from zero while a kitchen timer times down
from a preset value; we asked if the rat's clock timed up or
down. Experimental comparisons of animal clocks and man-made
clocks had not previously been made.

Properties of the Clock

This section describes the properties of the clock that our
research has found.

Distinct

The clock is distinct from, or independent of, the other mental
operations used in time discrimination. It is distinct in
two ways: (a) the clock can be changed without changing other
operations; and (b) other operations can be changed without
changing the clock. This is the clock's most fundamental
property; it has been taken for granted by almost every discussion of "the clock" and by almost every detailed theory of
timing. For example, Church and Deluty (1977) explained their
results by assuming a clock, a criterion, and a response rule,
each of which was distinct from the other two. Roberts and
Church (1978) assumed that the clock could be changed without
changing the translation of clock time to response rate. The
theory of Gibbon (1977) has a number of parameters that are
assumed to change without changing the clock. Although none
of these writers had any evidence that the clock was distinct,
it was a natural assumption to make. Distinctness is so
common in our view of the world that it is hard to imagine
a clock without it; it is in the nature of things that they
are distinct from other things.

The simplest evidence that the clock is distinct comes from
the application of logic that I will call the independent-
measures method; it is similar in form and goals to Sternberg's
(1969) additive-factor method. The method is essentially
three assumptions and a prediction. The assumptions are:

Distinct operations. There are (at least) two distinct
operations, A and B, that control the observed response.

Selective influence. Factor F changes A and not B, and
Factor G changes B and not A ("factor" meaning "experimental factor," or "independent variable").

Selective measurement. Measure M changes when A changes
and does not change when A does not change; Measure
N changes when B changes and does not change when B
does not change.

The prediction: If the assumptions are true, Factor F should
change Measure M but not Measure N, and Factor G should change
Measure N but not Measure M. The predicted result is sometimes called "double dissociation" (Teuber, 1955), but
"independent measures" seems more descriptive. A result of
independent measures suggests that all three assumptions are
correct. Warrington and Weiskrantz (1973) used this logic
to infer separate short- and long-term memories; Blackmore
and Campbell (1969) used it to infer spatial-frequency analysis

by the visual system. In animal learning, one of its few uses
is by Maki (1979), who argued that discrimination learning and
delayed matching-to-sample involve different memories.

This logic applies to an experiment (unpublished) where rats
were trained to make two discriminations simultaneously--a
response discrimination and a time discrimination. The subjects were six rats, who worked in standard lever boxes.
There were two measured responses: lever presses, of course,
and also pulls on a flexible metal bead chain that hung from
the ceiling of the chamber. The discriminative stimuli were
sound (white noise) and light (a houselight). Responses were
rewarded with a food pellet. There was one 3-hr session each
day.

After the rats learned to make both responses, a response
discrimination was trained: During one signal (light or
sound), only lever responses were rewarded; during the other
signal, only chain responses were rewarded. The lever signal
was light for half the rats, sound for the rest. The two
signals alternated with periods of darkness and silence when
neither response was rewarded. This procedure lasted 5 days.
Then a time discrimination was added to the response discrimination. After an intertrial interval of random length, one
of the two signals started. After 60 seconds, the next response appropriate to the signal was rewarded. When the response was rewarded, the signal went off. This procedure
lasted 10 days. The results described below are from the
last 5 days, when the time discrimination had stopped improving.

Figure 2 shows the time discrimination. The top panel shows
that response ratio was not affected by the modality of the
signal. In a context where time changed response ratio by
2.4, the largest absolute difference between light and sound
at any time was .11, and the median absolute difference (over
the 12 times) was .02. The middle panel shows the effect of
the meaning of the signal (whether lever or chain responses
were rewarded). Again, the two functions are very similar;
the median absolute difference is .03. The bottom panel shows
the effect of response (chain or lever) averaged over the two
signals. Again, the two functions are very close; the median
absolute difference is .02. The averages in the bottom panel
do not include Rat 6, which showed consistent differences
between the two responses, as large as .57.

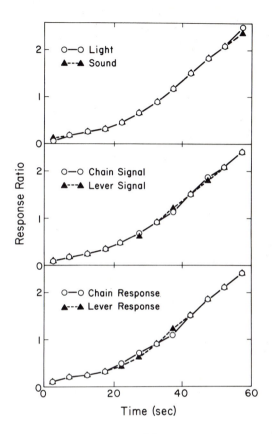

Figure 2.   Time has a large effect on response ratio;
   modality (top panel), signal (middle panel), and
   response (lower panel) have very small effects.
   (The data are from the last 5 days of time-discrimination training. Response ratio is defined in the
   caption of Figure 1. In the top two panels, the two
   responses are not distinguished; e.g., a rate of 3
   chain resp/min and 5 lever resp/min is considered
   8 resp/min. In the bottom panel, the two signals
   are not distinguished. Each point in the upper two
   penels is a mean over six rats. Each point in the
   bottom panel is a mean over 5 rats; Rat 6 was omitted
   for reasons described in the text. For Rats 1, 2, and
   6, the lever signal was light; for the rest, the
   lever signal was sound.)

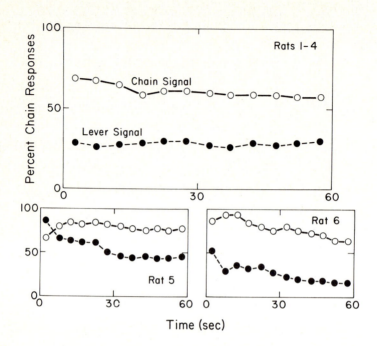

Figure 3. Signal has a large effect on the percentage of
   chain responses; time has a small effect. (The data
   are from the last 5 days of time-discrimination training. Percentage of chain responses refers to the
   percentage of chain responses relative to all responses at a given time; for example if the chain
   rate was 3 resp/min and the lever rate was 7 resp/min
   during the interval sec 10-15, the percentage of
   chain responses would be 30%.)

Figure 3 shows the response discrimination. It was nearly
constant throughout the interval for four rats, but clearly
varied for the other two. Over the last 30 sec of the interval, it was constant for both signals and all rats except
Rat 6.

Here, then, are two independent measures. With a few exceptions, response ratio was changed by time and not by the
meaning of the signal, while the percentage of chain responses
was changed by the meaning of the signal and not by time.
This suggests that the clock (the operation changed by time)
was distinct from the operation changed by the meaning of the
signal. Because response ratio was sensitive to changes in

the clock but did not change with modality or response, we
can also conclude that the clock was not changed by modality
or response.

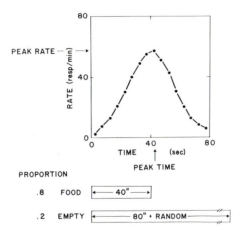

Figure 4. The basic elements of the peak procedure
    and some typical results. (Figure 1 of Roberts,
    1981. Time is measured from the start of a
    signal, such as light. Copyright 1981 by the
    American Psychological Association.)

Because the clock is distinct, we can hope to "isolate" it;
that is, find a measure such that (a) a change in the measure
implies a change in the clock and/or (b) no change in the
measure implies no change in the clock. The results just
described suggest that response ratio has these properties
for some manipulations (the assumption of selective measurement); however, I guess that with a wider range of manipulations, some of them would change response ratio without
changing the clock. For example, it seemed likely that some
manipulations could change accuracy without changing the clock,
yet any change in accuracy would change response ratio. To
find a better measure, I used the peak procedure, a modification of a procedure used by Catania (1970). The basic
elements of the peak procedure are shown in Figure 4. It is
similar to a discrete-trials fixed-interval procedure with
intermingled long extinction trials. There are two types of
trial, randomly mixed: (a) Food trials, on which the first
response after a fixed time--in Figure 4, 40 sec--is rewarded
with food; the trial then ends. (b) Empty trials, on which no
food is given. The trial lasts a relatively long time (say,

160 sec) and ends independently of responding. Figure 4
also shows some typical results. After training with this
procedure, response rate within a trial reaches a maximum at
about the time that food is given. The main measures of performance are peak time, the time of the maximum response rate
measured from the start of the trial, and peak rate, the value
of the maximum. I hoped that peak time would isolate the
clock. Underlying this hope was the idea that the peak in
response rate happened when the rat expected food--when
elapsed time, as measured by its clock, matched the time when
food had been given on earlier trials. Changing the clock
would thus change peak time; for example, speeding up the
clock would decrease peak time. And with the time of food
constant, a change in peak time would imply a change in the
clock. The work to check these ideas had essentially two
parts. The first part asked if peak time isolated something;
i.e., if it was sensitive to changes in one operation that
controlled responses and not to changes in another operation
that controlled responses. The second part asked if the
operation isolated was the clock.

Peak time isolates something. Figure 5 shows evidence that
peak time isolated something (Experiment 1 of Roberts, 1981).
The subjects and apparatus of this experiment were much like
those in the experiment just described. The experiment of
Figure 5 had two phases. During both phases, there were two
signals (light and sound) with different treatments. In the
first phase, food trials with one signal had food primed at
Sec 20 (i.e., 20 sec after the start of the trial); food
trials with the other signal had food primed at Sec 40. For
both signals, food trials were 80% of all trials. In the
second phase, food trials with both signals had food primed
at Sec 20. The difference between the signals was that for
one signal (high food), food trials were 80% of all trials;
for other signal (low food), they were 20% of all trials.
The main result, shown in Figure 5, was that peak time and
peak rate were independent measures. The time of food changed
peak time but not peak rate, while the probability of food
changed peak rate but not peak time. Roberts and Holder
(Note 2) also found that changing the probability of food did
not change peak time. Based on the independent-measures
logic (the assumption of selective measurement), we can conclude the peak time isolated something.

Peak time isolates the clock. In general, the support for
this statement is that conclusions about the clock based on
peak time--assuming that peak time isolates the clock--agree
with conclusions about the clock based on other evidence
(Maricq, Roberts and Church, 1981; Roberts, 1981, 1982;

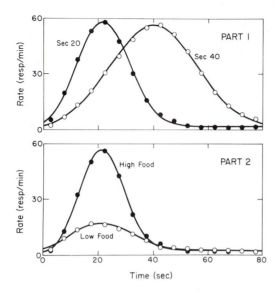

Figure 5.   Peak time and peak rate are independent.
(Figure 2 of Roberts, 1981.   Each point is a mean
over 10 rats.   The two curves in each panel are the
results from different signals.   During Part 1,
food was given at different times with the two
signals.   During Part 2, food was given with different probabilities during the two signals.   For
more details, see the text.   Copyright, 1981 by
the American Psychological Association.)

Roberts and Holder, Note 2).   It is clear enough that peak
time is sensitive to the operation that changes with the
duration of the signal, which by definition is the clock.
Consider, for example, what would happen if trials with light
were preceded by a 20-sec interval of light.   The 20-sec
interval of light, just before the trial, would decrease peak
time by about 20 sec, and this shows that peak time is sensitive to the operation measuring the duration of the light.
However, it is possible that peak time is sensitive to other
operations as well, and this is why it is important that conclusions about the clock based on peak time agree with conclusions based on other evidence.   The most important agreement has been with conclusions from choice procedures, where
rats are trained to make one response after one duration and
another response after a second duration.   For example,

Roberts (1981, Experiment 3) found that feeding the rats before the session increased peak time, suggesting that prefeeding slowed down the clock. Support for this conclusion
came from an experiment by Yagi (1962), who found a similar
result using rats trained to choose between the arms of a T-
maze based on their duration of confinement before the choice.
The procedure of Yagi and the peak procedure both required the
rat to use its clock, but they differed in other ways, such
as the input operations (how the rat responded). However,
both the peak procedure and choice procedures also require
use of a long-term memory for durations; to conclude that a
change in peak time was due to a change in the clock rather
than a change in this memory required other evidence (Roberts,
1981).

The experiments of Roberts (1981) also suggested two problems
with the assumption that peak time isolates the clock. First,
although changing the time of food changed peak time, it
apparently did not change the clock (Roberts, 1981, Experiment
5). This was not surprising, and it s-emed to be best dealt
with simply by limiting the statement: When the time of food
is constant, peak time isolates the clock. Second, two experiments (Roberts, 1981, Experiments 3 and 4) suggested that
under some circumstances a change in the clock would produce
an opposite change in memory; the net effect would be no
change in peak time, even though the clock was changed. However, these circumstances seemed easy to avoid. Roberts (1981)
discusses these problems in detail.

As the following sections show, the operation isolated by peak
time has many clock-like properties--in particular, many stopwatch-like properties. This is another sort of justification
for the statement that peak time isolates the clock.

Linear Scale

The sclae of the clock seems to be linear with physical time,
in the sense that equal differences in seconds produce equal
differences along the scale. This conclusion is based on the
symmetry of the response-rate functions from the peak procedure (Figure 5). Equal distances in seconds from the peak
of the response-rate function--comparing a move toward earlier
times with a move toward later times--produce roughly equal
changes in response rates; we assume that the behavioral
equality reflects an equality on the underlying scale. Another way to make the argument is to note that the response-
rate functions have a simple shape (symmetric) on a simple
scale (linear with physical time). While any single-peaked
function will be symmetric on some scale, and any function

will have some shape on a linear scale, the conjunction of the two simplicities is remarkable. If the rat's clock does not have a linear scale, it is hard to explain. The symmetry result has been found by others using a different procedure and shorter times (Church and Gibbon, 1982). The slight asymmetry in the function (it starts at zero resp/min, but flattens out above zero resp/min) can be explained in various ways that are compatible with a linear scale (Church and Gibbon, 1982; Roberts, 1981). A more serious problem is that Millenson, Kehoe, and Gormez (1977) found a timing function that was symmetrical on a log scale (where the distance between 10 sec and 20 sec equals the distince between 20 sec and 40 sec) rather than on a linear scale. They were studying the rabbit's nictating-membrane response using durations of a few hundred milliseconds; the best resolution of the difference in results is probably that a different clock was involved.

Earlier scaling with animals has usually taken the underlying scale to be the one on which overall discriminability is constant (e.g., Church and Deluty, 1977). If we apply this criterion to the data of Figure 5 (upper panel), we would conclude that the underlying scale is logarithmic, because the response-rate function with a peak near Sec 40 is about twice the width of the function with a peak near Sec 20. A logarithmic scaling of the time axis would give both functions the same shape and width, a criterion for scaling used by Shepard (1965). With exactly the same data, then, we can argue either for a linear scale or a logarithmic scale, and one of the arguments must be wrong. A simple reason for thinking that the equal-discriminability argument is wrong (or at least unattractive) is that it leads to the conclusion that a stopwatch does not have a linear scale. A stopwatch inevitably measures longer times with more error (on a linear scale) than shorter times; it will measure all times with equal error only on a scale transformed away from linearity toward logarithms. Other reasons for doubting the equal-discriminability argument are that work not involving symmetry supports a linear scale (Gibbon and Church, 1981), and that earlier data taken to support a logarithmic scale is compatible with a linear scale (Gibbon, 1981). It is probably wrong to use the equal-discriminability criterion to scale continua that have natural zeros, such as time and length.

Accurate

Another conclusion from the symmetry of the response-rate functions of Figure 5 is that the clock is relatively accurate--that the error in the measurement of duration is a small part

of the total error in performance ("error" meaning variance).
The argument behind this conclusion is given in full by Roberts
(1981); it is outlined below.

If all aspects of the discrimination were done without error,
there would be a sharp rise near the time of food and a sharp
decline thereafter. In fact, the response-rate functions have
considerable spread. What produces the spread? Other work
(Roberts, 1981; Roberts and Church, 1978) suggests that performance in this task is based on a comparison of the current
time (measured from the start of the trial) to the memories of
the times when responses had been rewarded on earlier trials.
so the spread in the functions has at least three possible
sources: (a) error in measuring the current time (clock error);
(b) error in remembering the time of reward; and (c) error in
comparing the two. Whatever the size of the clock error, we
expect it to increase with longer times, perhaps doubling when
the measured duration is doubled. For example, if the clock
measures 20 sec with a standard deviation of 2 sec, we expect
it to measure 40 sec with a larger standard deviation, perhaps
4 sec. If the clock error is large relative to other error,
longer times should be measured with noticeable less accuracy
than shorter times; in particular, the discrimination after
the peak. In fact, the two sides of the function are about
equally steep, implying that the clock error is about the same
on both sides. Because the clock error is probably small at
short times (e.g., 5 sec will be measured with a small standard
deviation), it is probably small throughout.

This may be the most surprising conclusion from this work.
Earlier theoretical work on the accuracy of timing both
human and animal, had always assumed that the clock was the
main source of error (e.g., Church, Getty and Lerner, 1976;
Creelman, 1962; Getty, 1976; Kinchla, 1972). A similar
assumption--that the important error is in the sensory signal
--is made by most theoretical applications of signal-detection
theory. For timing, apparently, this assumption is false.

Can be Stopped

Like a stopwatch, the clock can be stopped and restarted at
the time it was stopped. Some of the evidence for this is
shown in Figure 6 (Roberts, 1981). Rats were trained with the
peak procedure with light as the signal. After they learned
the discrimination, I compared their performance on two types
of empty trials: (a) baseline trials, where the signal comes
on and goes off without interruption; and (b) blackout trials,
with a 10-sec blackout starting at Sec 10. Figure 6 shows
the results. The blackout simply shifted the response-rate
function rightward by roughly the length of the blackout.

Five-sec blackouts shifted the function about half as much as 10-sec blackouts. This is what would be expected if the clock stopped during the blackout, and it is hard to explain in other ways. More evidence that the clock can be stopped, based on different durations and two other procedures, comes from Roberts and Church (1978) and Church (1978).

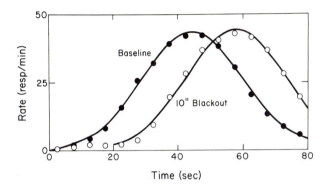

Figure 6. The effect of a blackout. (Figure 5 of Roberts, 1981. The signal timed was light; the blackout started 10 sec after the beginning of the trial and lasted 10 sec. Each point is the mean over 10 rats. Copyright 1981 by the American Psychological Association.)

Internal Pacemaker

Most man-made clocks are essentially counters. They may count, for example, the swings of a pendulum (a grandfather clock) or the reversals in the direction of house current (a wall clock). The source of the events counted is the clock's pacemaker. The pacemaker may be near the counter (a stopwatch) or far away (a wall clock). If the rat's clock is a counter, its pacemaker(s) might be inside the rat (e.g., a cell that fired at regular intervals), outside the rat (e.g., the stimulus energy), or both (if the clock counted events from more than one source).

The results of two manipulations suggest that the clock has an internal pacemaker. First, Roberts (1981), using the peak procedure, found that feeding the rats just before the experimental session increased peak time. Assuming that this reflected a change in the clock, it might have been due to (a) an increase in clock latency--the time between the start of the signal and the start of the clock--or (b) a decrease in clock rate. A change in latency would "shift" the response-rate function (as in Figure 6) while a change in rate would "stretch" it (as in the upper panel of Figure 5). In fact, prefeeding stretched rather than shifted the response-rate function, suggesting that it changed clock rate. Second, Maricq, Roberts and Church (1981) found that methamphetamine injections decreased peak time. This suggested that methamphetamine changed the clock but by itself does not indicate if the drug changed latency or rate. To distinguish these two possibilities we used a choice procedure. After the signal (a blackout), the rats had a choice of two levers. The correct (rewarded) response depended on the duration of the blackout; for example, if the blackout had been short (e.g., 1 sec), responses on the right lever were rewarded, while if the blackout had been long (e.g., 4 sec), responses on the left lever were rewarded. After blackouts of intermediate duration (between 1 and 4 sec), neither choice was rewarded; the intermediate durations were used to estimate the duration that the rat would classify as "long" and "short" equally often (the indifference point). There were three groups: One learned to discriminate 1 and 4 sec, another 2 and 8 sec, and the third 3 and 12 sec. After the discrimination was learned, we injected the drug. Based on our earlier work, we expected the drug to decrease the indifference point. If methamphetamine changed only the latency of the clock, then the indifference point should decrease equally for all three groups. But if methamphetamine changed the rate of the clock, the decrease should be largest in the 3-vs.-12-sec group. Figure 7 shows that this is what happened. The change in the indifference point was roughly proportional to the duration being measured. We concluded that methamphetamine changed the rate of the clock.

Because manipulations that produce only internal changes (prefeeding, methamphetamine) can change the rate of the clock, the clock apparently has an internal pacemaker. Note, however, that the clock may also have an external pacemaker-- e.g., its rate may depend on features of the signal being timed, such as its intensity or modality.

Times Up

Some man-made clocks, such as stopwatches, time "up" from zero; others, such as kitchen timers, time "down" toward zero. The essential difference is whether it is the starting point (timing up) or the ending point (timing down) that is the same for different durations.

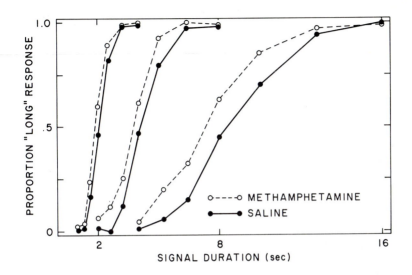

Figure 7. The effect of methamphetamine (Figure 7 of
Maricq, Roberts and Church, 1981. Each pair of curves
is from a different group of rats. Unfilled circles
show results from sessions with methamphetamine;
filled circles show results from sessions with
saline. Each point is the median over 8 or 10 rats.
Copyright 1981 by the American Psychological Association.)

The rat's clock apparently times up. The simplest evidence
for this comes from choice experiments like the one just
described, where the rat is trained to press one lever after
a short duration and the other lever after a long duration.
These experiments have used retractable levers. The levers
are retracted during the intertrial interval and during the
presentation of the duration; after the duration, they are
extended, and the rat chooses-essentially classifying the
duration as "long" or "short." In this situation, no duration
at all--a 0-sec signal--is usually classified as "short"
(Church, 1980; Roberts and Holder, Note 2), and this suggests
that the clock times up. This conclusion assumes that if the
clock times down, it is not set to the appropriate duration
until the signal begins; however, other work, not making this
assumption, also suggests that the clock times up (Roberts and
Church, 1978, Experiment 3, using a fixed-interval procedure;
Roberts, 1981, Experiment 5, using the peak procedure).

Cross-modal

As Figure 1 (right panel) shows, rats can discriminate the durations of both lights and sounds. The two modalities might be timed by different clocks or by the same clock; apparently they are timed by the same clock.

The best evidence for this comes from an experiment by Roberts (1982). Rats were trained with the peak procedure using both trials of light and trials of sound, so that they developed time discriminations in both modalities. Then transfer trials were added; these consisted of a short (10 or 20 sec) signal of one modality followed by a long (e.g., 180 sec) signal of the other modality. The first signal defines the preset interval; the second signal, the measurement interval. The main empirical question was the effect of preset intervals on peak time during the measurement interval (measured from the start of the measurement interval). If light and sound are timed by the same clock, then preset intervals should decrease peak time (because the clock will not be at zero when the measurement interval begins). If light and sound are timed by different clocks, there is no reason to expect preset intervals to change peak time.

Figure 8 shows that preset intervals decreased peak time during the measurement interval. The size of the decrease roughly corresponded to the length of the preset interval: The decrease was about 4 sec with the 10-sec preset interval and about 14 sec with the 20-sec preset interval. The fact that preset intervals decreased peak time suggests that the same clock times light and sound; the fact that the transfer was less than perfect (e.g., a 10-sec preset interval decreased peak time by less than 10 sec) suggests that the clock can distinguish light and sound. There is apparently a single clock that receives separate inputs from different modalities. The results of Figure 8 have been repeated in other experiments (Roberts, 1982; Roberts and Holder, Note 5); the conclusion that there is a cross-modal clock is also supported by results not involving the peak procedure (Meck and Church, 1982; Roberts, 1982).

In a way, the experiment of Figure 8 unites two earlier lines of work. On the one hand, there had been about ten previous findings, with five different dimensions, of cross-modal transfer of training with animals (see Roberts, 1982, for references). As an example of this work, Over and Mackintosh (1969) trained rats to respond fast or slowly depending on the intensity of a light. For some rats, a high intensity meant "respond fast"; for others, it meant "respond slowly." When

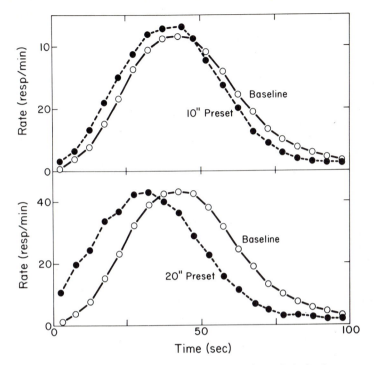

Figure 8. Preset intervals of sound (or light) decrease
    peak time during light (or sound). (Figure 9 of Roberts,
    1982. If the preset interval was sound, the measure-
    ment interval was light, and vice versa. The baseline
    function comes from all trials without preset intervals.
    Each point is a mean over six rats. Copyright 1982
    by the American Psychological Association.)

exposed to sound, the rats to some extent treated stimuli in
the new modality as if they were stimuli in the old modality;
for example, rats trained to respond fast to a high-intensity
light responded faster to a high-intensity sound than rats
trained to respond slowly to a high-intensity light. This
kind of result shows that non-verbal animals can perform cross-
modal abstraction that directs learned behavior. (This is
"abstraction" because one quality of the stimulus--its modality
--is being ignored while another quality--its intensity--is
being attended to.) However, the abstraction might have been
due to associative learning. For example, tactile-to-visual
transfer of shape discriminations (e.g., Cowey and Weiskrantz,
1975, with monkeys) is probably the result of earlier experience with co-occuring visual shapes and tactile shapes. On
the other hand, in the physiological literature there had been
about five previous reports (see Roberts, 1982), based on

single-cell recording, of cross-modal cells sensitive to
location. For example, Morell (1972) found cells in the
visual cortex of cats that responded only to stimuli in a
particular horizontal direction, but to both lights and sounds.
This sort of result suggests that there is cross-modal abstraction not due to associative learning; however, the
abstraction does not necessarily direct learned behavior--it
might be used for orienting, for example. The timing work
described above is an example of cross-modal abstraction that
both directs learned behavior and cannot plausibly be due to
associative learning.

Path-independent

Figure 9 shows the response-rate functions of Figure 8 plotted
so that their peak times coincide; the three functions are now
very similar. The same result has been found in other situations (Roberts, 1981, Experiment 2; Roberts, 1982, Experiment
4). One way of describing this result is to say that response
rate was path-independent, in the sense that response-rate
functions shifted to be equal at one time were equal at all
later times. The term path-independent was, of course, originally used to describe mathematical learning theories that
assumed that learning curves horizontally shifted to be equal
on one trial would be equal on all later trials. The independence of path shown by response rate presumably reflects
the independence of path of the underlying clock: What the
clock will read after, say, 10 sec, depends only on the current
reading, and not on how the current reading was reached.

All man-made clocks (e.g., stopwatches, wall clocks) are path-
independent, and it may be hard to imagine other possibilities.
Another way to describe the property of path independence is
to say that the state of the clock is one-dimensional: it
can be fully described by a single number, in the sense that
by knowing only a single number (its current reading) we can
predict what it will read, say, 10 sec from now. But if we
consider clocks in general--anything that changes with time
in a regular way--we can see that not all clocks have this
property. Consider a man driving home from work. We plot
distance travelled from work versus physical time since leaving
work, and we see that distance travelled is a clock--it changes
with time in a regular way. But the clock has at least two
dimensions because future progress depends not only on current
position but also on the time of day; for example, there will
be less progress during rush hour. To predict where the man
will be 10 min from now requires knowing not only where he is
now, but also the time of day. Functions from two trips

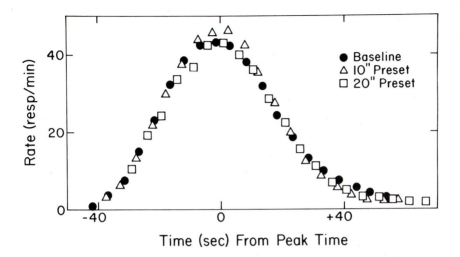

Figure 9. The path independence of response rate.
(Figure 10 of Roberts, 1982. This shows the
functions of Figure 8 of this paper shifted so that
their peak times coincide. Copyright 1982 by the
American Psychological Association.)

horizontally shifted to be equal at one time will not always
be equal at all later times. The rat could have a two-
dimensional, path-dependent clock if the output of the clock
that guides behavior, Clock A, was the result of combining
the output of two other clocks, Clocks B and C.

Times Different Intervals and Times Different Intervals Using the Same Rate

Some man-made clocks, such as eggtimers, can only time a
single interval, but most can time a range of intervals. Some
of the clocks that time a range of intervals use different
rates with different intervals. For example, there is a lab
clock whose rate is set according to the interval to be timed;
it times 30 sec at twice the rate that it times 60 sec. Most
computer clocks measure similar intervals using the same rate
but measure very different intervals (e.g., 1 sec and 1 hour)
using different rates. A stopwatch, however, measures all
intervals using the same rate.

The rat apparently times different intervals using the same
clock and the same rate. Some of the evidence for this comes

From an experiment by Roberts (1981) using the peak procedure.
As in the cross-modal experiment described above, there were
both trials with light and trials with sound. Unlike the
cross-modal experiment, the two signals differed in the time
of food: During one signal, the first response after Sec 20
was rewarded, during the other signal, the first response
after Sec 40. After the discrimination was learned, shift
trials were added: They began with the Sec-20 signal but
shifted to the Sec-40 signal at Sec 5, 10, or 15 (i.e., after
5, 10 or 15 sec of the Sec-20 signal). Shift trials were the
same length as empty trials, and they ended without food.

On shift trials, when will the rats expect food? There are
three cases to consider: (a) The Sec-20 and Sec-40 signals
are timed by different clocks; (b) the Sec-20 and Sec-40
signals are timed by the same clock at different rates; and
(c) the Sec-20 and Sec-40 signals are timed by the same clock
at the same rate. If the two signals are timed by different
clocks--case (a)--then the duration of the Sec-20 signal should
have no effect on when rats expect food during the Sec-40
signal; the peak in response rate should happen about 40 sec
after the shift regardless of the time of the shift. Measured
from the start of the Sec-20 signal, peak time should be about
45 sec (shift at Sec 5), 50 sec (Sec 10), or 55 sec (Sec 15).
If the two signals are timed by the same clock at different
rates--case (b)--the most plausible arrangement is that the
rate of the clock is proportional to the interval to be timed
(the shorter the interval, the faster the clock). Then the
clock essentially measures the proportion of the interval
completed, and later shifts will produce earlier peaks.
Suppose the shift is at Sec 5. At the time of the shift the
clock will essentially read "25% done," and 30 sec (75% of
40 sec) of the Sec-40 signal will be required to finish the
interval. Or suppose the shift is at Sec 15. At the time of
the shift the clock will essentially read "75% done," and 10
sec will be required to finish the interval. Measured from
the start of the Sec-20 signal, peak time should be about 35
sec (shift at Sec 5), 30 sec (Sec 10), or 25 sec (Sec 15).
Finally, it is possible that the two signals are timed by the
same clock at the same rate--case (c). In this case, the clock
essentially measures absolute time, and the time of the shift
will not change peak time. Suppose the shift is at Sec 5. At
the time of the shift the clock will essentially read "5 sec
done," and 35 sec of the Sec-40 signal will be required to
finish the interval. Measured from the start of the Sec-20
signal, peak time should be about 40 sec (shift at Sec 5),
40 sec (Sec 10), or 40 sec (Sec 15).

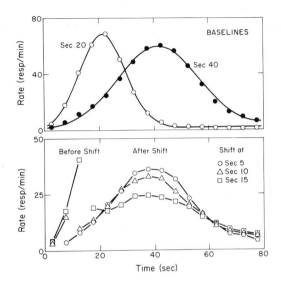

Figure 10. The effect of changing from a Sec-20 signal
to a Sec-40 signal early in the trial. (Figure 15 of
Roberts, 1981. The upper panel shows response rates
on trials with only one signal present throughout the
trial. Different curves show the results from different signals; the label for each curve is the time
that food was primed during the signal. The lower
panel shows response rates on trials that began with
the Sec-20 signal but shifted to the Sec-40 signal
at Sec 5, 10, or 15. Each point is a mean over 10
rats. Copyright 1981 by the American Psychological
Association.)

Figure 10 shows the results. The upper panel shows the discrimination between the two signals, and the lower panel shows
the effect of shifts. Measured from the start of the Sec-20
signal, peak time was near 40 sec for all three shifts. This
suggests that the two intervals were measured by the same
clock, and that the clock timed both intervals at the same
rate. Roberts and Church (1978, Experiment 3) reached the
same conclusions using a fixed-interval procedure.

Times Selectively

As we move from sensory processes to cognitive processes, there
is a considerable increase in selectivity. We see more events
than we notice, and we notice more events than we think about.

The clock might time all stimuli above some sensory threshold, or it might time only some of them. In fact, the clock seems to be selective; apparently, it only times stimuli that signal important events.

To learn what stimuli are timed, we took advantage of cross-modal transfer. After rats were trained to time stimuli from one modality (light or sound), we asked if they timed stimuli from the other modality. An example is an experiment by Roberts and Holder (Note 2) that used a choice procedure. During an initial training phase, rats were trained to press one lever ("short") after a 3-sec signal and the other lever ("long") after a 12-sec signal. We then added treatments and tests of the untrained modality. For concreteness, suppose the trained modality was light and the untrained modality was sound. Sound was treated in three ways: (a) First, it was presented alone, for random durations. (b) Second, it was paired with food. The sound lasted 20 sec and ended with a food pellet. (c) Finally, it was extinguished. The sound lasted 20 sec and ended without a food pellet. The effect of the treatments was measured with test trials. There were two kinds: (a) 0-sec tests. No stimulus was presented, and then the levers were extended into the box. (b) 12-sec tests. The sound was presented for 12 sec, and then the levers were extended into the box. The two tests differ in the duration of the sound. If the rats are timing the sound, they should respond differently on the two types of test; if they are not timing the sound, they will not respond differently.

Figure 11 shows the results. During the first treatment, when sound was presented alone, the rats treated the 0-sec and 12-sec sounds similarly. During the second treatment, when sound was paired with food, the 12-sec sound was treated as longer than the 0-sec sound. During the third treatment, when sound was extinguished, the 0-sec and 12-sec sounds were again treated similarly. During all three treatments, there was no change in the rats' response to light (not shown). These results suggest that the rats timed sound during the second treatment, but not during the first or third treatments. We have found similar results using a 3-sec test instead of a 0-sec test; we have also found similar results with the peak procedure (Roberts and Holder, Note 2). Generalizing from these results, we conclude that (a) rats time stimuli that signal important events and (b) they do not time stimuli that do not signal important events. (We do not know if they time novel stimuli.) Our experiments have only used food as the reinforcer, but we generalize to "important events" because

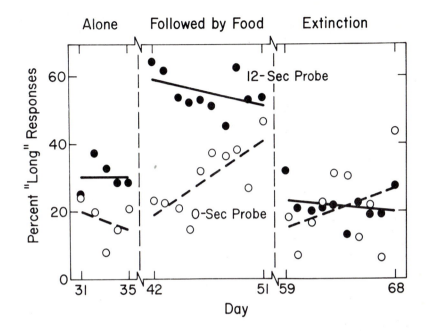

Figure 11. A stimulus paired with food is timed, but a
  stimulus presented alone or extinguished is not timed.
  (From Roberts and Holder, Note 2. Each point is a
  mean over 8 rats.)

timing occurs when shock is the reinforcer (e.g., LaBarbera
and Church, 1974; Libby and Church, 1974); the term "important
events" is intentionally vague.

The Function of the Clock

Daniel (1976), writing about the design of experiments and the
need for contact with actual problems, commented: "Many
mathematical statisticians are under the illusion that they
and their graduate students are writing for a future which
they forsee without benefit of detailed knowledge of the present. A tiny proportion of their work may be remembered 20
years from now" (p. 7). I have the same doubts about laboratory research that is inspired by other laboratory research.
When we began this work, the contact with a reality outside
the lab was not clear. We knew of no examples of timing in
natural situations; we did not even know where to look. We
could not cite a plausible, concrete way in which timing would
benefit an animal. I worried that the laboratory task was

entirely artificial, and that the clock was designed (evolved)
for a different purpose. As an example of what I feared, there
is the task of X-ray detection. Laboratory results show that
rats can detect X-rays, at some intensities, they do this with
their olfactory system (Garcia and Koelling, 1971). We can be
sure that animals in nature never detect X-rays, and that the
olfactory system was designed for another purpose. Since then,
however, three sets of facts have come along that show con-
clusively, I think, (a) that animals in natural situations
commonly measure time (duration), (b) that the clock was
designed for timing, and (c) that timing can be useful.

The first set of facts is the remarkable number of similarities
between the rat's clock and an ordinary stopwatch. The
similarities fall into two categories: (a) properties of a
stopwatch that distinguish it from other man-made clocks (e.g.,
that it can be stopped--not all clocks can be stopped); and
(b) properties of a stopwatch that it shares with other man-
made clocks (e.g., a linear scale--all man-made clocks that I
know of have a linear scale). Properties in the first category
(specific to stopwatches) shared by the rat's clock are: (a)
Can be stopped temporarily. Sundials, for example, cannot be
stopped. (b) Internal pacemaker. Electric clocks that require
wall current usually use its alternation in voltage to set
their rate; their pacemaker is at the power station. (c) Times
up. Kitchen timers, for example, time down. (d) Times intervals of different lengths. An egg timer can only time a
single interval. (e) Times intervals of different lengths
using the same rate. Many computer clocks have adjustable
rates. (f) Selectively started. Sundials are always running.
Properties in the second category (general to most man-made
clocks) shared by the rat's clock are: (a) Distinct. Most man-
made clocks can be moved, started, stopped, etc., without
changing other objects. (b) Linear scale. (c) Cross-modal.
Few man-made clocks are restricted to timing stimuli in only
one modality. (d) Path-independent. Our notion of time is
one-dimensional. Two other similarities are less certain:
(a) Accurate. When a person uses a stopwatch to discriminate
duration (e.g., at a track meet), the variance in measurement
probably comes much more from such operations as starting the
clock and recording the result than from the stopwatch. (b)
Can be reset quickly--i.e., it can be moved back to zero much
more quickly than it moves away from zero. An egg timer cannot be reset quickly unless the interval has finished. We
have not studied this property in the rat's clock; however, it
is suggested by the familiar observation that on an ordinary
fixed-interval schedule--where food delivery defines the beginning of each interval--there is a sudden change from a
high response rate just before food is given to a very low

response rate just afterward. If we assume that response rate reflects clock setting, then the change in rate reflects the resetting of the clock. Roberts and Church (1978) describe other evidence for this property.

It is unlikely that ten or more similarities are due to chance. The only explanation seems to be that the similar properties are the result of evolution to perform a similar function (Roberts and Church, 1978). For the stopwatch, of course, it was technological evolution; for the rat's clock, biological evolution. We know that stopwatches were designed to measure times on the order of seconds; therefore the rat's clock was designed to measure times on order of seconds. This is not the first time that a resemblance between biology and technology has been used to suggest that the biology is adaptive (although it may be the first case involving behavior). For example, Williams (1966) took the analogies "between bird wings and airship wings, between bridge suspensions and skeletal suspensions, between the vascularization of a leaf and the water supply of a city" (p. 10) as convincing evidence that the biological structure served the same purpose as the man-made structure.

The second set of facts is the evidence suggesting that the clock times selectively; in particular, that it times stimuli that signal important events and does not time stimuli that do not signal important events. This property suggests that animals commonly measure durations in natural situations because they must commonly encounter signals for important events, especially food.

Selective timing is also a guide to what the clock was designed to do. A stopwatch-like clock could be used by an animal in at least five ways: (a) to measure the time between a stimulus and an important event; (b) to measure the time between a response and an important event; (c) to measure the overall density of an important event (Gibbon and Balsam, 1981); (d) to measure the density of an important event during a stimulus (Gibbon and Balsam, 1981); and (e) to predict when an important event will happen. Erasers make pencils better for writing, but not for punching holes; the presence of an eraser therefore suggests that a pencil has been designed for writing, and not for punching holes. In a similar way, the selective-timing property makes the clock better for predicting when important events will happen (Use e) but not better for the other uses (a-d); thus it suggests that the clock was designed for predicting when important events will happen.

Consider the following sequence of events: (a) At 1:00 p.m.,
a light goes on. (b) At 1:05 p.m., a sound goes on. (c)
At 1:10 p.m., food is given and the clock reading is stored.
At 1:10, the clock could read either "0 min," "5 min," or
"10 min." If we wish to predict the time of food, which
duration would be best to store? There are four cases to
consider: (a) If light is a signal for food and sound is not,
then the clock should read "10 min"--that is the actual
duration between the start of the signal for food and the food.
(b) If sound is a signal for food and light is not, by the
same reasoning the clock should read "5 min." (c) If neither
light nor sound is a signal, then the clock should read "0
min." If the clock timed non-signals, it would lose time
resetting to zero when an actual signal started. (d) If both
light and sound are signals, then the clock should read either
"5 min" or "10 min"; it is hard to say which reading would be
more helpful. The selective-timing results suggest that the
rat's clock will in fact produce the desired readings. Other
possible clocks would not. For example, a clock that timed
all stimuli, and reset completely when a new stimulus started,
would read "5 min" in all cases. This match--between the performance of the clock and the needs of the task--suggests
that the clock was designed for the task.

The conclusion that the clock was designed to predict the time
of important events fits well with what we know about the
function of associative learning. New phenomena of the last
20 years, especially the blocking effect and the selective
nature of taste-aversion learning, make a strong case, I think,
that the function of associative learning is to predict the
future, especially the future likelihood of important events
such as food. Associative learning in rough outline seems to
be a mechanism for predicting the future, and subtleties such
as blocking and the ease with which tastes but not sounds become associated with illness make it better for predicting
the future. Dickinson (1980) makes a similar argument. However, associative learning, by itself, only allows the animal
to predict what important events will happen. By timing the
signals that associative learning has detected, the animal will
be able to predict when important events will happen.

The third set of facts comes from the study of foraging by
zoologists. When foraging in a patchy environment (where food
is not uniformly distributed), an animal must decide when to
leave one patch and find another. A recent theory about how
animals make this decision (independently proposed by many;
see Krebs, 1978, for a review) emphasizes the use of a
criterion giving-up time, the length of time an animal should
wait without finding food before leaving the patch. The

mechanism is not made explicit, but the underlying idea is that the animal starts its clock at zero when it enters the patch, and resets the clock to zero whenever it finds food. When the clock time is greater than the giving-up time, the animal leaves the patch. The giving-up time is set and changed based on the measurement of interfood intervals. When food becomes sparse (longer interfood intervals), the giving-up time is increased; when food becomes plentiful (shorter interfood intervals), the giving-up time is decreased. This will help the animal maximize its rate of food intake.

This theory shows a specific way in which timing can be useful, and the generality of the problem (finding food in a patchy environment) more or less corresponds with the generality of timing (throughout the vertebrates, at least). More important, however, it has led to the collection of observations that show timing in a natural setting. In a large garden, Davies (1977) watched spotted flycatchers (a small bird) waiting on a perch for flies. Now and then, the birds would leave one perch and go to another. Davies measured the probability of departing as a function of time without a capture attempt (since arriving at the perch or since the last capture attempt). Figure 12 shows the results. The rate of departure increased with the time since the last capture attempt, as foraging theory would predict. Note that the times in Figure 12 are within the range we have been discussing, and that the accuracy of the time discrimination (the line is straight, rather than concave up or down) is not very different from the accuracy of the time discrimination seen in the data of this chapter (e.g., Figure 1). The accuracy is also similar to some laboratory data with pigeons (e.g., Dews, 1970, Figure 2-2). This makes it more plausible that the same timing system is being used in the lab and in nature.

Figure 12 shows one property of laboratory time discrimina-tions--behavior changed with time in a regular way. The other property is that the function relating behavior to time depends on the time of the motivating event. With foraging, this corresponds to the finding that the rate of departure depends on the density of food. In the situation of Davies, the average wait before leaving was 30 sec. The density of food did not change; however, the mean time between captures was 18 sec, so an average wait of 30 sec before leaving was reasonable, and a computer simulation suggested that the average wait was roughly optimal (maximizing the rate of find-ing food). Other work with birds, in more artificial situ-ations, has shown directly that the average waiting time be-fore departure depends on the average density of food (Cowie, 1977; Krebs and Cowie, 1976; Krebs, Ryan and Charnov, 1974).

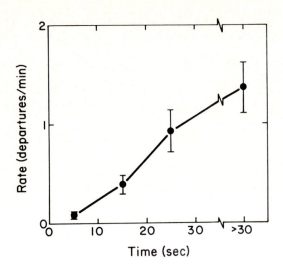

Figure 12. Departure rate changes with time. (Computed
   from data of Figure 21 of Davies, 1977. Before the
   first capture, time was measured from the bird's
   arrival at the perch; after the first capture, time
   was measured from the most recent capture. The
   vertical bars are standard errors based on the
   variance of a binomial probability.)

All three sets of facts suggest that timing is useful--that
an animal that can measure durations has an advantage over one
that cannot. The first set of facts, the convergent evolution
of the stopwatch and the rat's clock, would hardly have been
possible without the same selective force acting on both, and
that force must have been the usefulness of timing. The second
set of facts, the evidence for selective timing, suggests that
timing helps animals predict the future. Prediction of the
future increases control of the future, which is obviously
useful. The third set of facts, the foraging results, shows
a concrete way in which prediction of the future by timing is
useful: It allows the animal to forage more efficiently.

Questions about the "function" of a tool may refer (a) to the
goal of the designers of the tool (its intended function);
(b) to how the tool is used (its actual function); or (c)
especially in the life sciences, to the usefulness of the tool,
its adaptive value. In summary, our work and the work of
others suggest answers to all three questions about the
function of the clock. (a) The rat's clock is designed for
timing--in particular, for predicting the time of important

events. This is suggested both by the similarity between the
rat's clock and a stopwatch and by the fit between the properties of the clock and the properties needed to use timing
to predict the future. (b) Animals in natural situations use
the clock for measuring durations. This is suggested both by
the conclusion that the clock times all signals for important
events and by the observation of timing in a natural situation.
(c) By helping the animal predict the future, timing helps the
animal control the future. When foraging, for example, timing
can help the animal increase its rate of food intake.

Timing is usually treated in textbooks of animal learning as
no more than a by-product of fixed-interval schedules of reinforcement. However, the facts described above suggest that
timing plays a role in the life of the animal similar to the
role of associative learning, which is the main subject of
these textbooks. Both associative learning and timing are used
by the animal to predict the future. Associative learning predicts what; timing, in conjunction with associative learning,
predicts when.

Relation to Work in Other Areas

Human Timing

Work on duration measurement by humans has covered a much
larger range of durations than our work--from times on the
order of milliseconds (e.g., Massaro and Idson, 1976; Thomas
and Weaver, 1975) to times on the order of many minutes (e.g.,
Block, George and Reed, 1980; Ornstein, 1969). My comments
below are restricted to work involving times of 1-20 sec; the
measurement of longer and shorter times is too likely to involve clocks different from the one we have studied. In
animals, the clock used for measuring durations less than a
second may be different than the clock used with longer times;
the evidence is that, as mentioned earlier, Millenson, Kehoe
and Gormezano (1977), using durations of a few hundred milliseconds, found symmetry on a log scale of time, whereas our
work, using durations of seconds, has found symmetry on a
linear scale. In humans, work described below suggests that
the clock used for measuring durations less than 20 sec is
different than the clock used with longer times.

For the measurement of durations on the order of seconds, is
a human's clock the same as a rat's? We can compare the two
clocks on the following properties:

Internal pacemaker. The rate of the rat's clock is determined
internally, at least partially. The same seems to be true of
the human's clock. There are three lines of evidence: (a)

Drug effects. Adam, Rosner, Hosick and Clark (1971) found that
breathing anesthetic gases, such as cyclopropane, caused subjects to produce time intervals about 20% longer than usual.
(With the method of production, subjects are told the name of
a time interval--say, "10 sec"--and asked to mark it off.)
The increase was roughly a constant proportion of the interval
to be produced; this is easy to explain if the drugs changed
the rate of the clock, and hard to explain otherwise.
Frankenhaeuser (1959) found similar evidence that nitrous
oxide slowed down the clock by about 20%; she used the method
of estimation (e.g., the subject is asked to name the interval
between two clicks). Agreement between the results of estimation and production--e.g., a drug that causes 10% overestimation also causes 10% underproduction--suggests that the drug
changes the clock; this is because the clock is probably the
only operation that the two tasks have in common. Goldstone,
Boardman and Lhamon (1958) and Goldstone and Kirkham (1968)
found that dextroamphetamine decreased by about 15% the physical duration that subjects judged equal to 1 sec. This is
similar to what we found with methamphetamine (Maricq et al.,
1981), and presumably the mechanism is the same--the rate of
the clock was increased. Many other experiments have found
that a drug changes time estimation or production, but none of
the twenty or so I have looked at had enough evidence to
suggest that the rate of the clock had been changed. (b)
Stress effects. Subjects said that more time had passed during an interval when they expected a shock at the end of the
interval than when they did not expect a shock (Hare, 1968);
the overestimation was about 3 times larger with a 20-sec
interval than a 5-sec interval. Frankenhaeuser (1959) and
Langer, Wapner and Werner (1961) found similar results also
using the method of estimation; Falk and Bindra (1954) found
similar results using the method of production. (c) Temperature effects. Many experiments have found that increasing body
temperature causes overestimation and underproduction of durations, and lowering body temperature has the opposite effects.
The changes in body temperature have been produced by fever
(Hoagland, 1933), diathermy (Francois, 1927; Hoagland, 1933),
time of day (Francois, 1927, Pfaff, 1968), hot and cold air
(Fox, Bradbury, Hampton and Legg, 1967), and cold water
(Baddeley, 1966; Bell, 1975). Results of production and
estimation are similar (Pfaff, 1968), suggesting that the
clock is changed. Temperature effects are the evidence usually
used to suggest an internal pacemaker; actually, this line of
evidence is weaker than the first two because all of the
results could be explained by changes in any feature of the
clock, not just its rate. Better evidence would show that the
temperature effect increases linearly with the duration used.

Cross-modal. The rat has a clock that measures durations of both light and sound. Likewise, there is something cross-modal about the human timing system. The clearest evidence is the work of Warm, Stutz and Vassolo (1975), who found that training for accuracy in timing transferred from light to sound and from sound to light. In addition, Behar and Bevan (1961) found large cross-modal anchor effects in the judgment of duration; for example, a 10-sec light changed the reported durations of subsequent 1- to 5-sec sounds. Roberts (1982) summarizes other experimental evidence. Another sort of evidence, less convincing but more vivid, comes from language. Duration is a word attached to no particular modality, and adjectives of duration (brief, long) can be applied to stimuli from any modality.

The human evidence, unlike the rat evidence, does not allow us to say where the timing system for light and the timing system for sound converge. It might be at the clock, or it might be later. As an example, consider the cross-modal use of adjectives of duration. Let us assume that two physical events will be labeled with the same word if they produce to some extent the same mental event. Then adjectives of duration could be cross-modal for two reasons: (a) Learned associations. We may use brief with both lights and sounds because brief lights and sounds often occur together. Via learning, brief sounds recall brief lights and vice-versa. Co-occurrence is probably why sharp has both a visual and a tactual meaning (a knife can both look and feel sharp). (b) Unlearned wiring. Unlearned wiring is probably why we used red to describe both left- and right-eye stimuli. Because similar results (cross-modal effects in timing) in similar animals (rats and people) are probably due to the same mechanism, it is reasonable to conclude that adjectives of duration are cross-modal due to unlearned wiring--light and sound are timed by the same clock. Here is a case where a result in animal learning (the rat's clock is cross-modal) seems to shed light on a feature of human thought (our cross-modal concept of duration). Usually, clearcut applications of animal learning to human life are in less cognitive areas, such as food preferences or fear.

Linear scale. The rat's clock has a scale that is linear with physical time. The same seems to be true of the human's clock. Eisler (1976) reviewed about a hundred scaling studies, and concluded that clock time was a power function of physical time with an exponent of about .9. Allan (1979), making somewhat different assumptions, concluded that an exponent of 1.0 was most consistent with the literature. The difference between .9 and 1.0 is not important for our purposes because

the rat data are not that precise. The symmetry of the response-rate functions (e.g., Figure 3) suggests that the scale
of the rat's clock is much closer to linear (corresponding to
an exponent of 1) than to logarithmic (corresponding to an
exponent of 0-see Tukey, 1977, pp. 86-91); it does not allow
one to say if the exponent is closer to .9 or 1.0.

Range. The range of the rat's clock seems to be larger than
the range of the human's clock. In most of our work, we have
used procedures involving times on the order of 1-10 sec and
times on the order of 20-60 sec. The conclusions with both
ranges of times have been the same for four properties: (a)
can be stopped (Roberts and Church, 1978; Roberts, 1981);
(b) internal pacemaker (Maricq, Roberts and Church, 1981;
Roberts, 1981); (c) cross-modal (Roberts, 1982); and (d) times
selectively (Roberts and Holder, Note 3). This suggests that
rats use the same clock for measuring times in both ranges.
In contrast, there is evidence that humans use different
clocks with the two ranges. Richards (1973), studying the
patient H. M., found that his reproduction of durations was
normal (power function with exponent of 1.0) for times less
than 20 sec. and clearly abnormal (power function with exponent of .4) for times more than 20 sec. This suggests that
different clocks are used for the measurement of times less
than and greater than 20 sec. Because H. M.'s most obvious
deficit is the inability to form long-term memories, these
results also suggest that, in normals, the measurement of times
greater than 20 sec uses this ability; for example, we may
judge the duration of an interval by how much we remember from
it (e.g., Ornstein, 1969). This is probably different than
using a specialized clock. More evidence that humans measure
durations above and below 20 sec with different clocks comes
from Frankenhauser (1959). Her subjects read digits at what
the subject considered to be a 1-sec pace. The number of
digits read was determined by the experimenter. After reading
the required number of digits, the subject estimated how much
time had passed. The ratio of the estimated time to the actual
(physical) time varied with the actual time; the ratio started
at about .85 at 4 sec, declined to about .65 near 20 sec, and
was constant from 20 sec to 50 sec, the edge of the range.
The change in slope at 20 sec is evidence for a change in
mechanism.

As far as I know, there is no conclusive evidence that shows
if the human's clock has other properties of the rat's clock
that I have discussed above (e.g., distinctness). In general,
work on human timing has been much different than our work,
with much less emphasis on the underlying clock.

In summary, the human and rat clocks seem to be similar in
three ways (internal pacemaker, cross-modal, linear scale)
and different in one way (range). It is too early to say if
the underlying mechanism is the same. It is clear, though,
that data in one field bear on questions asked in the other
field.

Associative Learning

One view of our work is that it provides a new example of
associative learning: The situations that do and do not start
the clock are similar to the situations that do and do not
produce traditional examples of classical conditioning. A dog
does not salivate during a sound after the sound has been presented alone; it salivates during the sound after the sound
is followed by food; and it no longer salivates during the
sound after the sound is extinguished. A rat's clock does not
time a sound after the sound has been presented alone; it does
time the sound after the sound is followed by food; and it no
longer times the sound after the sound is extinguished (Roberts
and Holder, Note 2). Whenever a new preparation is found to
follow the rules of associative learning, the question arises:
Are the new results (e.g., with the clock) due to the same
mechanism as the older results (e.g., with salivation) on which
the rules are based? earlier examples? Taste-aversion learning was different in three or four ways (e.g., learning with
delays of hours between CS and US) from earlier examples of
classical conditioning, and the differences led some people
(e.g., Garcia, Hankins and Rusiniak, 1974) to conclude that
the underlying mechanisms were different. Our results differ
in at least one way from earlier examples of classical conditioning: The operations of associative learning (e.g., pairing, extinction) are apparently changing an earlier part of
the stimulus-response path. Earlier examples of classical
conditioning have all involved changes in the linkage of
stimuli to responses, and could be explained by changes in
associative memory. However, in this case what is changed is
the clock; the clock is "earlier" than associative memory in
the sense that the output of the clock goes to associative
memory--the output of the clock can be a CS. We have only begun to assess the analogy between timing and earlier examples
of associative learning, and it is too early to choose between
two possibilities: (a) The operations of associative learning
act on two different mechanisms, one before the clock, one
after. This parallels the position that taste-aversion learning and earlier examples of classical conditioning are due to
different mechanisms, i.e., that taste aversion is a new form
of learning. (b) The operations of associative learning act
on a single mechanism, and this mechanism is more complicated

than we thought--it must send output to operations from which
it receives input. This parallels the position that taste
aversion learning and earlier examples of classical conditioning are due to the same mechanism, but this mechanism is
more complicated than we thought--e.g., a CS may sometimes
be stored for hours.

Another view of our work emphasizes a more subtle similarity
to traditional associative learning. By a sort of triangu-
lation, conclusions about the function of the clock help us
decide the function of associative learning. Knowing only
one means to an end, it is hard to decide what is means and
what is end. A compass points in a certain direction; is it
attracted to the nearby tree, the distant mountain, or something else? To find out, we need another placement. In a
similar way, the function of a tool can be described with varying amounts of abstraction, and we need to study other tools to
choose the right amount. Are pencils for (a) making graphite
marks on paper or (b) making dark marks on paper? There are
no other tools that do (a); and there are pens, which do (b).
This suggests that (b) is a better answer than (a). The
existence of the more abstract goal of writing, which we might
deduce from video terminals, does not change the fact that
we have learned something about pencils by considering pens.
Is associative learning for (a) detecting cause and effect
(e.g., Dickinson, 1980, p. 9) or (b) predicting the future?
Because there is apparently no other mechanism designed to do
(a), and because the clock is apparently designed to do (b),
(b) is the better answer. The existence of the more abstract
goals of survival and reproduction, which we might deduce from
a hundred other adaptations, does not change this. This argument assumes that the mechanism responsible for associative
learning is somehow distinct from the clock, just as pencils
are distinct from pens. Support for this assumption comes
from the fact that bees show associative learning but not
timing.

Animal Cognition

Like other work in the field of animal cognition, this work
tries to study only part of what happens between stimulus
and response; but it does this in an unusual way. Three
methodological features of this work may be useful in other
situations.

First, this work provides evidence for the assumption that the
clock is distinct--i.e., that the clock can be changed without
changing other operations, and that other operations can be
changed without changing the clock. Assumptions of

distinctness are very common, but the supporting evidence is
usually weak or non-existent. For example, Rescorla (1978)
distinguishes between associative bonds and the representations
being bonded; in his words, between "knowledge about the relation" of two events and "the value of the event representations" (p. 39). The distinction is appealing, but, as
far as I know, there is no evidence for it. Rescorla (1978)
uses the notion of distinct event representations to interpret four diverse experiments.

Second, this work makes explicit the notion of "isolating" a
mental process and shows one way of doing it. Isolating a
mental process, call it P, means finding a measure of behavior such that (a) a change in the measure implies a change
in P; and/or (b) no change in the measure implies no change
in P. P might be a clock, short-term memory, the association
between two events, an event representation, stimulus en-
coding, the retrieval stage, etc. To learn about P it is
essential to isolate P. Suppose you do an experiment that
varies Factor F (e.g., shock density, contingency, pre-
exposure). The results will reveal nothing about P unless you
can decide if changing F changed P; to decide, you need to have
isolated P. Of course, experimenters often use results to
draw conclusions about a mental process; they implicitly assume
that they have isolated the process. For example, Grant (1981)
takes a variety of results involving percent correct in delayed
-symbolic-matching-to-sample experiments to suggest that
pigeons remember the test stimulus rather than the sample
stimulus during the delay interval. Implicit in Grant's conclusion about the contents of memory is the assumption that
percent correct isolates the contents of memory. One experiment, for example, measured the effect of repeating the sample,
and compared the effects of "physical" repetition (exacly the
same sample is presented twice) with "instructional" repetition (two different samples are shown, but both have the same
correct match, e.g., the green key, so that both mean "peck
green.") Repetition increased percent correct (two samples
were better than one), and this was taken to suggest that repetition changed the contents of memory; this assumes, of
course, that a change in percent correct implies a change
in the contents of memory. The two types of repetition produced the same increase, and Grant concluded from this that
the contents of memory were the same in the two cases; this
assumes, of course, that no change in percent correct implies
no change in the contents of memory. Without the assumption
that percent correct isolates the contents of memory, nothing
could have been concluded about the contents of memory; however, Grant gives no evidence for this assumption.

Third, this work uses a procedure (the peak procedure) chosen to <u>explain</u> the result of interest (timing); most procedures now used to study animal cognition seem chosen to <u>demonstrate</u> the result of interest (e.g., associative learning, short-term memory). To see this distinction, compare an ordinary fixed-interval schedule with the peak procedure. The fixed-interval schedule is better than the peak procedure for demonstrating timing; it is easier to implement and to describe. On the other hand, it is worse than the peak procedure for explaining timing; with a fixed-interval schedule, most changes in behavior could be equally well explained by a change in the clock or a change somewhere else (for an example, see Roberts, 1981, Experiment 4). The Olton and Samuelson (1976) eight-arm radial-maze procedure is another example of a procedure that is good for demonstration but, probably, poor for explanation. When the experimenter chooses the first few arms that the rat will visit (Beatty and Shavalia, 1980), it is apparently easy to demonstrate the existence of a high-capacity, long-lasting spatial memory. However, it is probably hard to use the procedure to learn much more about this memory. The usual measure of performance is percent correct. Because of ceiling effects (accuracy is usually high), it is unlikely that no change in percent correct implies no change in spatial memory. Because the task requires more than spatial memory (e.g., motivation, sensory input), it is unlikely that a change in percent correct implies a change in spatial memory. The interests of psychologists studying animal behavior seem to have shifted over the last ten years from the behavior in general to the mental components (e.g., short-term memory, associations) that produce the behavior (e.g., Hulse, Fowler and Honig, 1978); it was probably inevitable that the procedures used would lag behind.

McFarland (1981) defines animal cognition as the study of "mental processes that are presumed to be occurring within the animal, but which cannot be directly observed" (p. 71). I think that most people would agree with this definition (e.g., Skinner, 1950; Honig, 1978; but see Bolles, 1979, p. 192); however, I do not. The phrase "directly observed" seems to imply that there is a qualitative difference between the way we observe, say, behavior (directly) and the way we observe mental processes (not directly). We may say this, but it is contradicted by our actions. I measure response rate ("directly") by counting the number of times a switch is closed, and use my counts to make statements about what the animal has done. Implicit in these statements is the assumption that my counts isolate response rate--that a change in count implies a change in rate and/or no change in count implies no change in rate. It is a very plausible assumption, but it is testable (measure rate in some other way, e.g., by watching) and therefore could be wrong. I measure the clock ("not directly") by measuring peak time with the peak procedure. When I use peak time to make statements about the clock, I am assuming that peak time isolates the clock--that a change in peak time implies a change in the clock and/or no change in peak time implies no change in the clock. It is less plausible than the counts-to-rate assumption, but it is testable (measure the clock in some other way, e.g., with a choice procedure) and therefore could be right. Implicit in eighty years of conclusions about the mental processes of animals are eighty years of assuming that mental processes can indeed be isolated. A more accurate definition of animal cognition would refer to "mental processes . . . which cannot be easily observed."

Footnote

This work was supported by National Institutes of Health grant GM23247, Russell M. Church principal investigator, and National Science Foundation grant BNS 79-00829, Seth Roberts principal investigator. I thank Michael Brown, Robert Cook, Mark Holder, and John Watson for comments on the manuscript. Requests for reprints should be sent to Seth Roberts, Department of Psychology, University of California, Berkeley, California 94720.

References Notes

1. Eckerman, D. A. Performance fixed-interval reinforcer
   schedules of about a day. Paper presented at the meeting
   of the Psychonomic Society, November, 1977.

2. Roberts, S., & Holder, M. What starts an internal clock?
   Manuscript submitted for publication, 1982.

References

1 Adams, N., Rosner, B. S. Hosick, E. C. & Clark, D. L.
   Effect of anesthetic drugs on time production and
   alpha rhythm. Perception & Psychophysics, 1971,
   10, 133-136.

2 Allan, L. G. The perception of time. Perception & Psychophysics, 1979, 26, 340-354.

3 Ambler, S. A comparison of two models for performance
   under fixed interval schedules of reinforcement. Journal
   of Mathematical Psychology, 1976, 14, 53-71.

4 Anderson, M. C. & Shettleworth, S. J. Behavioral
   adaptation to fixed-interval and fixed-time food delivery in golden hamsters. Journal of the Experimental
   Analysis of Behavior, 1977, 25, 33-49.

5 Azrin, N. H. Conditioning of the aggressive behavior of
   pigeons by a fixed-interval schedule of reinforcement.
   Journal of the Experimental Analysis of Behavior, 1967,
   10, 395-402.

6 Baddeley, A. D. Time-estimation at reduced body-tempera-
   ture. American Journal of Psychology, 1966, 79, 475-
   479.

7 Balster, R. L. & Schuster, C. R. Fixed-interval schedule
   of cocaine reinforcement: Effect of dose and infusion
   duration. Journal of the Experimental Analysis of Behavior, 1973, 20, 119-129.

8 Beatty, W. W. & Shavalia, D. A. Spatial memory in rats:
   Time course of working memory and effect of anesthetics.
   Behavioral and Neural Biology, 1980, 28, 454-462.

9 Behar, I. & Bevan, W. The perceived duration of auditory
   and visual intervals: Cross-modal comparison and interaction. American Journal of Psychology, 1961, 74, 17-26.

10 Bell, C. R. Effects of lowered temperature on time
   estimation. Quarterly Journal of Experimental Psychology,
   1975, 27, 531-538.

11 Blakemore, C. & Campbell, F. W. On the existence of
   neurones in the human visual system selectively sensitive
   to the orientation and size of retinal images. Journal
   of Physiology, 1969, 203, 237-260.

12   Block, R. A., George, E. J. & Reed, M. A.   A watched
     pot sometimes boils: A study of duration experience.
     Acta Psychologia, 1980, 46, 81-94.

13   Bolles, R. C.   Learning theory (2nd Ed.).   New York: Holt,
     Rinehart, and Winston, 1979.

14   Boulos, Z., Rosenwasser, A. M. & Terman, M.   Feeding
     schedules and the circadian organization of behavior
     in the rat.   Behavioral Brain Research, 1980, 1, 39-65.

15   Brown, J. S.   Generalization and discrimination.   In
     D. I. Mostofsky (Ed.), Stimulus generalization.
     Stanford, Calif.: Stanford University Press, 1965.

16   Cantor, M. B. & Wilson, J. F.   Temporal uncertainty as
     an associative metric: Operant simulations of Pavlovian
     conditioning.   Journal of Experimental Psychology:
     General, 1981, 110, 232-268.

17   Carlson, J. G., Wielkiewicz, R. M. & Modjeski, R. B.   The
     Psychological Record, 1972, 22, 531-542.

18   Catania, A. C.   Reinforcement schedules and psychophysical
     judgments: A study of some temporal properties of be-
     havior.   In W. N. Schoenfeld (Ed.), The theory of rein-
     forcement schedules.   New York: Appleton-Century-Crofts,
     1970.

19   Church, R. M.   The internal clock.   In S. H. Hulse,
     H. Fowler & W. K. Honig (Eds.), Cognitive processes in
     animal behavior.   Hillsdale, N. J.:   Lawrence Erlbaum
     Associates, 1978.

20   Church, R. M.   Short-term memory for time intervals.
     Learning and Motivation, 1980, 11, 208-219.

21   Church, R. M. & Deluty, M. Z.   Bisection of temporal
     intervals.   Journal of Experimental Psychology: Animal
     Behavior Processes, 1977, 3, 216-228.

22   Church, R. M., Getty, D. J. & Lerner, N. D.   Duration
     discrimination by rats.   Journal of Experimental Psycho-
     logy: Animal Behavior Processes, 1976, 2, 303-312.

23   Church, R. M. & Gibbon, J.   Temporal generalization.
     Journal of Experimental Psychology: Animal Behavior
     Processes, 1982, 8, 165-186.

24   Cone, A. L. & Cone, D. M.   Operant conditioning of
     Virginia opossum.   Psychological Reports, 1970, 26, 83-86.

25   Cowey, A. & Weiskrantz, L.   Demonstration of cross-modal
     matching in rehesus monkeys, Macaca mulatta.   Neuropsycho-
     logia, 1975, 13, 117-120.

26   Cowie, R. J.   Optimal foraging in great tits (Paras major).
     Nature, 1977, 268, 137-139.

27   Cowles, J. T. & Finan, J. L.   An improved method for
     establishing temporal discrimination in white rats.   The
     Journal of Psychology, 1941, 11, 335-342.

28   Creelman, C. D.   Human discrimination of auditory duration.
     Journal of the Acoustical Society of American, 1962, 34,
     582-593.

29   Daniel, C.   Applications of statistics to industrial
     experimentation.   New York: Wiley, 1976.

30   Davies, N. B.   Prey selection and the search strategy of
     the spotted flycatcher (Muscicapa striata): A field study
     on optimal foraging.   Animal Behaviour, 1977, 25, 1016-
     1033.

31   Dews, P. B.   The effect of multiple S[delta] periods on
     responding on a fixed-interval schedule.   Journal of the
     Experimental Analysis of Behavior, 1962, 5, 369-374.

32   Dews, P. B.   The effect of multiple S[delta] periods on
     responding on a fixed-interval schedule: III.   Effect of
     changes in pattern of interruptions, parameters and
     stimuli.   Journal of the Experimental Analysis of Be-
     havior, 1965, 8, 427-435.

33   Dews, P. B.   The theory of fixed-interval responding.   In
     W. N. Schoenfeld, The theory of reinforcement schedules.
     New York: Appleton-Century-Crofts, 1970.

34   Dickinson, A.   Contemporary animal learning theory.
     Cambridge, England: Cambridge, 1980.

35   Eisler, H.   Experiments on subjective duration 1868-1975:
     A collection of power function exponents.   Psychological
     Bulletin, 1976, 83, 1154-1171.

36   Falk, J. L. & Bindra, B.   Judgment of time as a function
     of serial position and stress.   Journal of Experimental
     Psychology, 1954, 47, 279-282.

37 Fox, R. H., Bradbury, P. A., Hampton, I. F. G. & Legg, C. F. Time judgment and body temperature. Journal of Experimental Psychology, 1967, 75, 88-96.

38 Francois, M. Contribution a l'etude du sens du temps: La temperature interne comme facteur de variation de l'appreciation subjective des durees. Annee psychologie, 1927, 28, 186-204.

39 Frankenhaeuser, M. Estimation of times: An experimental study. Stockholm: Almqvist & Wiksell, 1959.

40 Garcia, J., Hankins, W. G. & Rusiniak, K. W. Behavioral regulation of the milieu interne in man and rat. Science, 1974, 185, 824-831.

41 Garcia, J. & Koelling, R. A. The use of ionizing rays as a mammalian olfactory stimulus. In L. M. Beidler (Ed.), Handbook of sensory physiology (Vol. 4). Berlin: Springer-Verlag, 1971.

42 Getty, D. J. Counting processes in human timing. Perception & Psychophysics, 1976, 20, 191-197.

43 Gibbon, J. Scalar expectancy theory and Weber's law in animal timing. Psychological Review, 1977, 84, 279-325.

44 Gibbon, J. On the form and location of the psychometric bisection function for time. Journal of Mathematical Psychology, 1981, 24, 58-87.

45 Gibbon, J. & Balsam, P. Spreading association in time. In C. M. Locurto, H. S. Terrace & J. Gibbon (Eds.), Autoshaping and conditioning theory. New York: Academic Press, 1981.

46 Gibbon, J. & Church, R. M. Time left: Linear versus logarithmic subjective time. Journal of Experimental Psychology: Animal Behavior Processes, 1981, 7, 87-108.

47 Goldstone, S., Broadman, W. K. & Lhamon, W. T. Effect of quinal baritone, dextrose-amphetamine, and placebo on apparent time. British Journal of Psychology, 1958, 49, 324-328.

48 Goldstone, S. & Kirkham, J. E. The effects of secobarbital and dextroamphetamine upon time judgment: Intersensory factory. Psychopharmacologia, 1968, 13, 65-73.

49 Grant, D. S. Short-term memory in the pigeon. In N. E. Spear & R. R. Miller (Eds.), Information processing in animals: Memory mechanisms. Hillsdale, N. J.: Lawrence Erlbaum Associates, 1981.

50 Grossman, K. E. Continuous, fixed-ratio, and fixed-interval reinforcement in honey bees. Journal of the Experimental Analysis of Behavior, 1973, 20, 105-109.

51 Hare, R. D. The estimation of short temporal intervals terminated by shock. Journal of Clinical Psychology, 1963, 19, 378-380.

52 Hoagland, H. The physiological control of judgments of duration: Evidence for a chemical clock. Journal of General Psychology, 1933, 9, 267-287.

53 Honig, W. K. On the conceptual nature of cognitive terms: An initial essay. In S. H. Hulse, H. Fowler & W. K. Honig (Eds.), Cognitive processes in animal behavior. Hillsdale, N. J.: Lawrence Erlbaum Associates, 1978.

54 Hulse, S. H., Fowler, H. & Honig, W. K. (Eds.). Cognitive processes in animal behavior. Hillsdale, N. J.: Lawrence Erlbaum Associates, 1976.

55 Keller, K. J. Inhibitory effects of reinforcement and a model of fixed-interval performance. Animal Learning & Behavior, 1980, 8, 102-109.

56 Killeen, P. On the temporal control of behavior. Psychological Review, 1975, 82, 89-115.

57 Killeen, P. R. Learning as causal inference. In M. L. Commons & J. A. Nevin (Eds.), Quantitative analyses of behavior: Discriminative properties of reinforcement schedules. Cambridge, Mass.: Ballinger, 1981.

58 Kinchla, J. Duration discrimination of acoustically defined intervals in the 1- to 8-sec range. Perception & Psychophysics, 1972, 12, 318-320.

59 King, G. D., Schaeffer, R. W. & Pierson, S. C. Reinforcement schedule preference of a racoon (Procyon lotor). Bulletin of the Psychonomic Society, 1974, 4, 97-99.

60 Klosterhalfen, S., Fischer, W. & Bitterman, M. E. Modification of attention in honey bees. Science, 1978, 201, 1241-1243.

61 Koltermann, R. Periodicity in the activity and learning
    performance of the honeybee. In L. B. Browne (Ed.),
    Experimental analysis of insect behavior. Berlin:
    Springer-Verlag, 1974.

62 Kramer, T. J. & Rilling, M. Differential reinforcement of
    low rates: A selective critique. Psychological Bulletin,
    1970, 74, 224-254.

63 Krebs, J. R. Optimal foraging: Decision rules for pre-
    dators. In J. R. Krebs & N. B. Davies (Eds.), Behavioral
    ecology: An evolutionary approach. Oxford: Blackwell
    Scientific Publications, 1978.

64 Krebs, J. R. & Cowie, R. J. Foraging strategies in birds.
    Ardea, 1976, 64, 98-116.

65 Krebs, J. R., Ryan, J. C. & Charnov, E. L. Hunting by
    expectation or optimal foraging: A study of patch use by
    chickadees. Animal Behavior, 1974, 22, 953-964.

66 LaBarbera, J. D. & Church, R. M. Magnitude of fear as a
    function of expected time to an aversive event. Animal
    Learning & Behavior, 1974, 2, 199-202.

67 Langer, J., Wapner, S. & Werner, H. The effect of danger
    upon the experience of time. American Journal of
    Psychology, 1961, 74, 94-97.

68 Libby, M. E. & Church, R. M. Timing of avoidance re-
    sponses by rats. Journal of the Experimental Analysis of
    Behavior, 1974, 22, 513-517.

69 Libby, M. E. & Church, R. M. Fear gradients as a function
    of the temporal interval between signal and aversive
    event in the rat. Journal of Comparative and Physiological
    Psychology, 1975, 88, 911-916.

70 Logan, F. A. Hybrid theory of operant conditioning.
    Psychological Review, 1979, 86, 507-541.

71 McFarland, D. The Oxford companion to animal behavior.
    Oxford: Oxford University Press, 1981.

72 Maki, W. S. Discrimination learning without short-term
    memory: Dissociation of memory processes in pigeons.
    Science, 1979. 204, 83-85.

73 Maricq, A. V., Roberts, S. & Church, R. M. Methamphetamine and time estimation. Journal of Experimental Psychology: Animal Behavior Processes, 1981, 7, 18-30.

74 Massaro, D. W. & Idson, W. L. Temporal course of perceived auditory duration. Perception & Psychophysics, 1976, 20, 331-352.

75 Meck, W. H. & Church, R. M. Abstraction of temporal attributes. Journal of Experimental Psychology: Animal Behavior Processes, 1982, 8, 226-243.

76 Millenson, J. R., Kehoe, E. J. & Gormezano, I. Classical conditioning of the rabbit's nictating membrane response under fixed and mixed CS-US intervals. Learning and Motivation, 1977, 8, 351-366.

77 Morrell, F. Visual system's view of acoustic space. Nature, 1972, 238, 44-46.

78 Olton, D. S., & Samuelson, R. J. Remembrance of places passes: Spatial memory in rats. Journal of Experimental Psychology: Animal Behavior Processes, 1976, 2, 97-116.

79 Ornstein, R. E. On the experience of time. Baltimore: Penguin, 1969.

80 Over, R. & Mackintosh, N. Cross-modal transfer of intensity discrimination by rats. Nature, 1969, 224, 918-919.

81 Pavlov, I. P. Conditioned reflexes (G. B. Anrep, Ed. and trans.). New York: Dover, 1960. (Originally published, 1926.)

82 Pfaff, D. Effects of temperature and time of day on time judgments. Journal of Experimental Psychology, 1968, 76, 419-422.

83 Platt, J. R., Kuch, D. O. & Bitgood, S. C. Rats' lever-press duration as psychophysical judgments of time. Journal of the Experimental Analysis of Behavior, 1973, 19, 239-250.

84 Powell, R. W. Time-based responding in pigeons and crows. Auk, 1973, 90, 803-808.

85 Rescorla, R. Some implications of a cognitive perspective
    on Pavlovian conditioning. In S. H. Hulse, H. Fowler
    & W. K. Konig (Eds.), Cognitive processes in animal be-
    havior. Hillsdale, N. J.: Lawrence Erlbaum Associates,
    1978.

86 Richards, W. Time reproductions by H. M. Acta Psycho-
    logica, 1973, 37, 279-282.

87 Richardson, W. K. & Loughead, T. E. Behavior under large
    values of the differential-reinforcement-of-low-rate
    schedule. Journal of the Experimental Analysis of Be-
    havior, 1974, 22, 121-129.

88 Richelle, M. & Lejeune, H. Time in animal behaviour.
    Oxford: Pergamon Press, 1980.

89 Roberts, S. Isolation of an internal clock. Journal of
    Experimental Psychology: Animal Behavior Processes,
    1981, 7, 242-268.

90 Roberts, S. Cross-modal use of an internal clock. Journal
    of Experimental Psychology: Animal Behavior Processes,
    1982, 8, 2-22.

91 Roberts, S. & Church, R. M. Control of an internal clock.
    Journal of Experimental Psychology: Animal Behavior
    Processes, 1978, 4, 318-337.

92 Rozin, P. Temperature independence of an arbitrary
    temporal discrimination in the goldfish. Science, 1965,
    149, 561-563.

93 Rubin, M. B. & Brown, H. J. The rabbit as subject in be-
    havioral research. Journal of the Experimental Analysis
    of Behavior, 1969, 12, 663-667.

94 Shepard, R. N. Approximation to uniform gradients of
    generalization by monotone transformations of scale.
    In D. I. Mostofsky (Ed.), Stimulus generalization.
    Stanford, Calif.: Stanford University Press, 1965.

95 Sherman, J. G. The temporal distribution of responses
    on fixed-interval schedules. Unpublished Ph.D.
    dissertation, Columbia University, 1959.

96 Shumake, S. A. & Caudill, C. J. Operant conditioning
    of licking in vampire bats, Desmodus rotundus. Behavioral
    Research Methods and Instrumentation, 1974, 6, 467-470.

97 Skinner, B. F. Are theories of learning necessary? Psychological Review, 1950, 57, 193-216.

98 Skinner, B. F. A case history in scientific method. American Psychologists, 1956, 2, 221-233. Reprinted in M. H. Siegel & H. P. Zeigler (Eds.), Psychological research: The inside story. New York: Harper & Row, 1976.

99 Sprott, R. L. & Symons, J. P. Oqerant performance in in-bred mice. Bulletin of the Psychonomic Society, 1974, 4, 46-48.

100 Staddon, J. E. R. Temporal control, attention, and memory. Psychological Review, 1974, 81, 375-391.

101 Sternberg, S. The discovery of processing stages: Extensions of Donders' method. Acta Psychologica, 1969, 30, 276-313.

102 Teuber, H. L. Physiological psychology. Annual Review of Psychology, 1955, 6, 267-296.

103 Thomas, E. A. C. & Weaver, W. B. Cognitive processing and time perception. Perception & Psychophysics, 1975, 17, 363-367.

104 Tukey, J. W. Exploratory data analysis. Reading, Mass.: Addison-Wesley, 1977.

105 Warm, J. S., Stutz, R. M. & Vassolo, P. Intermodal transfer in temporal discrimination. Perception & Psychophysics, 1975, 18, 281-286.

106 Warrington, E. K. & Weiskrantz, L. An analysis of short-term and long-term memory deficits in man. In J. A. Deutsch (Ed.), The physiological basis of memory. New York: Academic Press, 1973.

107 Williams, G. C. Adaptation and natural selection: A critique of some current evolutionary thought. Princeton, N. J.: Princeton, 1966.

108 Yagi, B. The effect of motivating conditions on "the estimation of time" in rats. Japanese Journal of Psychology, 1962, 33, 8-24.

109 Ziriax, J. M. & Silberberg, A.   Discrimination and
     emission of different key-peck durations in the pigeon.
     Journal of Experimental Psychology: Animal Behavior
     Processes, 1978, 4, 1-21.

ANIMAL COGNITION AND BEHAVIOR
Roger L. Mellgren, editor
© *North-Holland Publishing Company, 1983*

THE MALLEABILITY OF MEMORY IN ANIMALS

William C. Gordon

University of New Mexico

We often define a memory as an internal representation of some
event an organism has experienced in the past. In effect, this
definition suggests that memories are little more than pictures
or isomorphic replicas of our experiences as we have perceived
them. Yet, our own experience, as well as research in the area
of human memory, tells us that this is not the case. Human
memories are seldom precise representations or reliable replicas
of past events. Sometimes, we recall less than we previously
experienced because of some source of forgetting. At other
times, however, our memories may include even more information
than we originally experienced when these memories first were
formed. This latter phenomenon has led a number or researchers
to suggest that humans possess "constructive" memory processes.
In other words, humans actively construct memories by combining
the memory of a given event with related information that is
available in the memory store.

Empirical evidence for constructive processes in humans was
first introduced by Bartlett (1932). Since that time numerous
investigators have demonstrated that humans tend to distort
their memories for events and that these distortions often depend on what additional information already is contained in a
person's memory store when the target memory is formed (e.g.
Bransford and Franks, 1972; Hertel and Ellis, 1979). More
pertinent to the present discussion, however, is a recent series
of experiments conducted by Loftus and her associates (Loftus
and Zanni, 1975; Cole and Loftus, 1979; Loftus, 1979). Loftus
has demonstrated quite convincingly that information that is
presented to a subject at the time a memory is recalled, often
becomes incorporated into the retrieved memory. In other words,
humans tend to update or add to their memories whenever these
memories are retrieved in the context of new information. Although this phenomenon may be viewed as a special case of memory construction, these studies do indicate that the time of
memory retrieval may be an important time for the operation
of constructive processes.

Given these kinds of findings it is now quite common to view
human memories as malleable or susceptible to change long
after these memories have originally been formed. It is surprising, therefore, that the idea of memory malleability has
had so little impact on theories of memory processing in animals. To date only a few theoretical papers have suggested
that animal memories might be dynamic or alterable in the
absence of specific learning contingencies (e.g. Lewis, 1979).
And, until recently there has been virtually no experimental
evidence to support such a view directly.

In response to this lack of evidence our own recent research
has attempted to confront the issue of memory malleability in
animals. The question we have addressed in these studies is
essentially the same as that addressed by Loftus in her work
with human subjects. Specifically, we have sought to answer
the following question: When animals retrieve a training memory does new information present at the time of retrieval become added to that memory? The present chapter reviews the
results of these studies and discusses the implications of
these findings for theories of memory processing in animals.

Memory Processing Induced by Retrieval

Initially, our own interest in memory malleability grew out of
some experiments we and others were doing with cueing or reactivation treatments. Typically, these studies involved
training rats to perform some response; then, prior to retention testing, exposing the animals to a cueing treatment.
Such treatments normally consisted of confronting the animals
with some part of the training stimulus complex without exposing them to a complete relearning trial. In short, the
cueing treatments were designed to serve as retrieval cues for
the training memory, but were constructed to minimize the opportunity for additional learning about the response.

The results of these studies and the implications these results
have had for theories of memory processing have been detailed
elsewhere (c.f. Spear, 1978; Lewis, 1979; Gordon, 1981). However, one aspect of these data merits particular attention.
In a number of these cueing experiments striking similarities
between newly formed and recently cued memories were found.
For example, it had long been known that newly formed memories
seemed to be uniquely susceptible to the disruptive effects of
treatments such as electroconvulsive shock (ECS). However,
Misanin, Miller and Lewis (1967) demonstrated that while ECS
normally has little effect on a memory 24 hours after learning,
such a memory is susceptible to disruption if the memory is
cued just prior to ECS administration. In a like manner

recently cued memories now have been shown to be susceptible
to the disruptive effects of other treatment such as hypo-
thermia (e.g. Riccio and Ebner, 1981).

Similarly, it had been assumed that C.N.S. stimulants facilitate retention performance only when administered shortly after
a learning experience (c.f. Calhoun, 1971). However, drugs
such as strychnine sulphate have now been shown to facilitate
retention when administered shortly after an "old" memory is
cued (e.g. Gordon and Spear, 1973a; Gordon, 1977a). Other
facilitating treatments such as electrical stimulation of the
midbrain reticular formation also have been shown to be effective when given following a cueing treatment (De Vietti,
Conger and Kirkpatrick, 1977).

Finally, aside from being susceptible to many of the same
treatments, newly acquired and recently cued memories were
found to be similar in at least two additional ways. First,
several studies demonstrated that both new and recently cued
memories have the capacity for producing interference in proactive interference paradigms (e.g. Gordon, 1977b; Gordon,
Frankl and Hamberg, 1979; Gordon and Feldman, 1978). Secondly
it has been shown that retention performance tends to be excellent when an animal is tested shortly after either learning or cueing. And, retention performance tends to decrease
monotonically as the interval between either acquisition or
cueing and the retention is lengthened (Feldman and Gordon,
1979).

In an effort to account for these similarities, several researchers have suggested that cueing or retrieval may initiate
memory processing that is at least similar to the processing
which occurs at the time of learning. This view has been
stated most explicitly by Lewis (1979). According to Lewis,
when new information is acquired that information is processed
for a short time in an active memory store. During this time
the memory is elaborated or encoded before entering a long-
term or passive memory store where it resides until retrieval
occurs. Retrieval, Lewis hypothesizes, results in the transfer of a memory from the passive back to the active store.
And, when a retrieved memory enters the active store it is
subject to the same processing that new information normally
undergoes at the time of acquisition.

This view of memory processing accounts quite well for the
apparent similarities between new and recently retrieved
memories. In effect, it is proposed that shortly after either
learning or retrieval a memory undergoes processing in the
active memory store. Furthermore, this processing model con-

tains an important implication concerning memory malleability.
It suggests that the same processes responsible for the original formation of a memory also operate on memories at the time
of retrieval. This means that while memories may be formed
at the time of a learning experience they may be re-formed
whenever they are retrieved. The studies reported below were
designed to test this implication.

The Alleviation of Retention Deficits Resulting From Context Change

It is well known that substantial changes in context between
the time of training and testing usually result in significant
retention deficits in animals. Because of this effect we assume that when an animal forms a memory for a training experience that memory normally includes some representation of the
training context. For this reason, we chose to concentrate
on contextual information in our studies of memory malleability.
We hypothesized that when an animal retrieves a training memory in a novel context (i.e. a context dissimilar from the
training context) the novel context will become represented
as part of the training memory. If this is the case, we reasoned, then the retrieval context should come to control the
performance of the trained behavior just as the training context originally controlled that behavior. In other words, the
context in which retrieval occurs should begin to function as
if it had been the training context.

The first series of experiments testing this prediction
(Gordon, McCracken, Dess-Beech and Mowrer, 1981) involved
training rats to perform a one-way avoidance task and then
testing them for retention of the avoidance response 24 hours
later. Both training and testing occurred in an avoidance
apparatus containing a white start chamber, a black goal
chamber and a grid floor. During training each animal was
placed in the start chamber of the apparatus and, after 3 sec.
the door to the black chamber was opened. The opening of the
door was accompanied by the onset of a flashing light located
directly behind the white chamber. If an animal failed to
cross into the black chamber within 5 sec. after the door was
opened a gridshock was applied to the floor of the white
chamber and was continued until the animal escaped to the black
side. Responses made within 5 sec. resulted in the avoidance
of the gridshock. Following either an escape or a successful
avoidance response, the chamber door was closed and the animal was removed to a holding cage for a 30 sec. intertrial
interval. The criterion for active avoidance acquisition was
5 consecutive avoidances within 26 trials. Animals which
failed to meet this criterion were eliminated from the study.

Retention testing consisted of 5 trials which were identical to the training trials with one exception. During testing no shocks were administered regardless of an animal's latency to cross to the black chamber. Of primary interest were each animal's response latency on test trial 1 and each animal's mean latency over the 5 test trials. In both cases shorter latencies were assumed to reflect better retention of the avoidance response.

Since these studies were concerned with the effects of context on retention performance, training and testing often were carried out in different contexts. In these and subsequent studies the different contexts employed were actually different rooms. These rooms differed along a variety of dimensions including illumination, size, odor and noise level. The specific characteristics of these rooms are detailed in Table 1. It should be noted that each room contained its own avoidance apparatus and that all of these apparati were constructed identically.

The purpose of the first experiment was simply to replicate the well established finding that a context change between training and testing produces deficits in test performance. To this end half of our animals were trained to avoid in Room A while the remaining animals received training in Room B., half of the animals were tested in the same room in which training had occurred (A-A or B-B) while the other half were tested in a different room (A-B or B-A). The results of this manipulation can be seen in Figure 1 which presents the response latencies on test trial 1 as a function of the training-testing sequence. As this figure suggests, animals trained and tested in different rooms performed significantly worse than animals trained and tested in the same room. Furthermore, these same results were evident when the mean latency scores for all 5 test trials were analyzed. Importantly, these results suggested that the different rooms used in our study were sufficiently different to produce a context change deficit in performance.

The purpose of experiment 2 in this series was to test the hypothesis that the context present at the time of retrieval will begin to function as if it had been the training context. We reasoned that if this hypothesis is correct it should be possible to alleviate the performance deficit due to context change by giving the animals a cueing treatment in the test context prior to testing. According to our hypothesis cueing should induce retrieval of the training memory in the test context causing the test context to become represented in the training memory. This would mean that at the time of testing the test

Room	A	B	C
Dimensions(ft.)	16.0 x 7.5	16.0 x 11.0	13.5 x 7.5
Lighting	Dark .04 ft. lumens	Bright 1.0 ft. lumens	Dim .17 ft. lumens
Odor (Room Deodorizer)	Lemon	Rose	None
Noise Level	Fan 63-65 db	None 45-50 db	None 45-50 db
Gloves	Rough	Smooth	Textured
Holding Cage	Wire Mesh	Wooden	Plastic Cage

Table 1.   Characteristics of Experimental Treatment Rooms.

context would match the contextual information represented in the training memory and the usual retention deficit would not result.

Based on this reasoning the second experiment contained 5 treatment groups. In the first group (B-B) all animals were trained and tested in Room B. The second group (A-B) included animals that were trained to avoid in Room A, but were tested in Room B. These groups were included to replicate the results of experiment 1. In the third condition (A-B-B) animals were treated identically to animals in the A-B condition with one exception. The A-B-B animals were given a 15 sec. cueing treatment in Room B, 3.5 min. prior to testing. Cueing

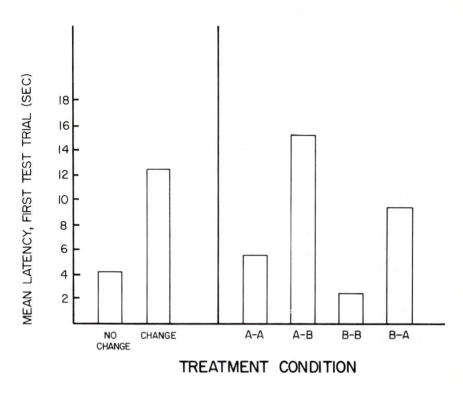

Figure 1.

consisted of placing the animals in a white chamber with a grid
floor. This chamber was constructed identically to the start
chamber of the avoidance apparatus. After cueing, these animals were confined in a holding cage in Room B for 3.5 min. and
then were tested in that room. A fourth condition (A-C-B) was
identical to the A-B-B condition except that for these animals
the cueing and subsequent confinement occurred in Room C adjacent to the test room. Finally, animals in group A-conf
were tested the same as the A-B-B animals except that these animals were never placed in the cueing chamber. These animals
spent both the cueing and confinement periods confined to a
holding cage in Room B.

The results of these treatments can be seen in Figure 2 which presents the response latencies on test trial 1 as a function of treatment condition. As in the first experiment B-B animals

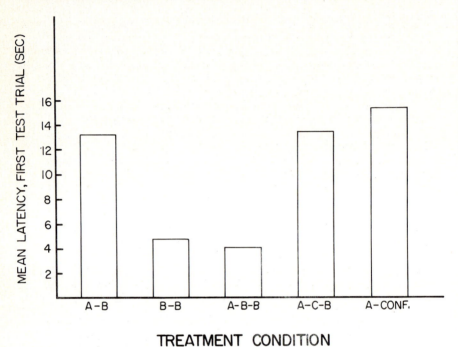

TREATMENT CONDITION

Figure 2.

had significantly shorter latencies and, thus, better retention of the avoidance response than the A-B animals. More important-ly, however, the A-B-B animals also exhibited significantly shorter latencies than did the A-B rats. This supports the notion that the performance deficit resulting from context change can be alleviated by cueing the animals in the novel test context (Room B) prior to testing there. That this re-duced deficit is not due solely to exposure to the test room prior to testing is shown by the failure of the A-conf group to perform well. Similarly, the poor performance of the A-C-B group suggests that the actual location of cueing and not just the procedure itself is important for reducing the retention deficit.

In an effort to strengthen these conclusions, one further experiment was done. This was necessary because at least two alternative explanations could have been applied to the results

of experiment 2. First, it is possible that the particular
cueing treatments used might have proactively affected retention performance. In other words, the same pattern of results might have been found even if the animals had received no
prior active avoidance training. Secondly, it is possible that
the effects of the A-B-B condition might have been due to the
fact that these were the only animals given a cueing treatment
as well as exposure to Room B prior to testing. In other words,
the same results might have been obtained regardless of the
location of cueing as long as cueing did occur and as long as
exposure to Room B was given before testing. The purpose of
experiment 3 was to test these alternate explanations.

This experiment involved six treatment conditions. Two of
these conditions A-B and A-B-B were identical to those used in
the previous study. In a third condition (A-X) animals were
trained in Room A and tested in Room B. Before testing, however, these animals first were confined in Room B for 3.5 min.,
then were cued in Room C, then spent the 3.5 min. cueing-test
interval confined in Room C. The remaining three conditions
(NC-B, NC-B-B and NC-X) were treated identically to the A-B,
A-B-B and A-X animals, respectively with one exception. Instead of receiving avoidance training these animals were given
non-contingent footshocks in Room A in a large conditioning
chamber.

Figure 3 presents the results of this study. As the figure
indicates the A-B-B animals exhibited significantly shorter
latencies than all other groups. No other groups differed
from each other. These results show again that the performance
deficit produced by context change can be reduced by cueing
in the test context prior to testing. They also demonstrate
1) that this effect obtains only when cueing itself occurs in
the test context (c.f. groups A-B-B and A-X), and 2) that this
effect results only when active avoidance training has preceded
the cueing treatment.

Taken together this group of studies provided strong support
for our initial prediction, i.e. that the context in which an
animal retrieves a memory should begin to function as if it had
been the training context. Furthermore, such findings clearly
are consistent with the hypothesis that contextual information
present at the time of retrieval can become represented as
part of the memory being retrieved. Our next series of studies
sought to demonstrate the plausibility of this hypothesis in
a different way.

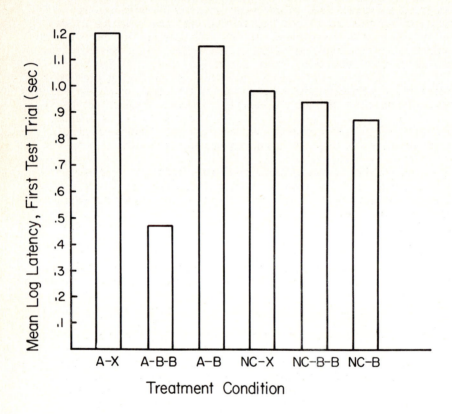

Figure 3.

Enhancing Pre-test Cueing Effects
By Cueing During Acquisition

It is very common to find that when animals are given a cueing
treatment prior to retention testing these animals exhibit
better test performance than do non-cued animals (c.f. Spear,
1978). However, pilot work in our laboratory showed that this
enhancement of test performance depends to a large degree on
the context in which pre-test cueing occurs. For example,
we had found that when rats received both training and testing
in one context (e.g. Room A) pre-test cueing significantly
facilitated test performance when it was administered in that
same context. However, when cueing occurred in a different

context (e.g. Room B) no facilitation was found relative to
non-cued animals. Our interpretation of this finding was
that cueing in a novel context is less effective in inducing
retrieval of the training memory than is cueing in the training
context and that the facilitating effects of pre-test cueing
depend on the inducement of retrieval.

Since the effects of pre-test cueing appeared to be dependent
on the similarity of the training and cueing contexts we decided to use this phenomenon to test further the notion of memory
malleability. Our reasoning was as follows. Pre-test cueing
effectiveness appears to depend on the cueing context matching
the context represented in the training memory. Thus, based
on our hypothesis concerning memory malleability, pre-test cueing in a non-training context should become effective if animals are given prior cueing treatments in that non-training
context. We based this prediction on the idea that prior cueing should result in the non-training context becoming represented in the training memory. Therefore, at the time of pre-
test cueing, the cueing context should match contextual information represented in the training memory and the cueing
treatment should become a more effective facilitator of performance.

Our first experiment (Wittrup and Gordon, experiment 1, in
press) formally confirmed the findings of our pilot work. In
this study all animals received active avoidance training in
Room A and then were tested for retention in Room A 72 hours
later. The training and testing procedures were the same as
those previously described with one exception. All animals
received 20 avoidance training trials regardless of their performance. To reach criterion each animal was required to make
at least 3 consecutive avoidances during training. The 3
treatment conditions differed in terms of the pre-test cueing
treatments. Group A-A received no cueing treatment. Group
A-A-A was given a 15 sec. exposure to the cueing chamber in
Room A and was then confined in Room A for the 3 min. cueing-
test interval. Group A-B-A was treated identically to the
A-A-A group except that pre-test cueing and confinement occurred in Room B.

As in our pilot studies the results of these manipulations were
clear. A-A-A animals performed significantly better on the
retention test than did A-A animals. This finding indicates
that pre-test cueing in the training context does facilitate
retention performance. On the other hand, group A-B-A did not
perform better than the A-A animals. In fact animals in the
A-B-A condition performed significantly worse than animals in
group A-A-A.

The purpose of the second experiment (Wittrup and Gordon, experiment 2, in press) was to attempt to enhance the effectiveness of pre-test cueing in Room B by giving animals prior cueing treatments in that room. Since the design of this study was somewhat complex the treatments are summarized in Table 2.

GROUP	TRAINING	PRE-TEST CUEING	TEST
A-A	RM. A	NONE	RM. A
A-B-A	RM. A	RM. B	RM. A
A(B)-B-A	RM. A (ACQ CUE RM. B)	RM. B	RM. A
A(C)-B-A	RM. A (ACQ CUE RM. C)	RM. B	RM. A
A(B)-C-A	RM. A (ACQ CUE RM. B)	RM. C	RM. A
A-CONF	RM. A (CONF. RM B)	RM. B	RM. A
(B)A-B-A	RM. A (PRE-TRAINING CUES) RM. B	RM. B	RM. A

Table 2. Summary of Treatment Conditions.

This experiment involved 7 treatment conditions. Two of the conditions, A-A and A-B-A were run identically to the corresponding conditions in experiment 1. Animals in the remaining 5 conditions received only 17 rather than 20 acquisition trials. In place of the 3 additional trials allotted to the A-A and A-B-A animals, each animal in these conditions received 3 "acquisition cueing treatments." These cueing treatments consisted of a 15 sec. exposure to the cueing chamber in a given room followed by a 15 sec. holding cage confinement in that same room. These cueing treatments replaced acquisition

trials 7, 13, and 19 for most animals in these conditions. In
group A(B)-B-A animals received the same treatment as the A-B-A
animals except that they received acquisition cueing in Room B
prior to the pre-test cue in that same room. The fourth group,
A(C)-B-A received the same treatment as the A(B)-B-A group
except that acquisition cueing occurred in Room C. Group
A(B)-C-A received acquisition cueing in Room B, but the pre-
test cue was given in Room C. The next group A(conf B)-B-A
was treated the same as the A(B)-B-A animals except that these
animals were simply confined in Room B for 30 sec. rather than
receiving both acquisition cueing and confinement in Room B.
The final group (B)A-B-A, was given its 3 acquisition cueing
treatments in Room B, but in this group the 3 cueing treatments
occurred in place of training trials 1, 2 and 3. Except for
this difference the (B)A-B-A and the A(B)-B-A animals were
treated identically.

The results of these treatments on test performance are represented in Figure 4. An analysis of these data revealed that

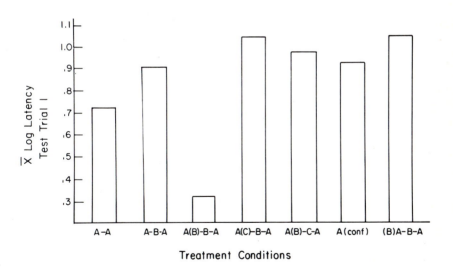

Figure 4.

animals in the A(B)-B-A condition performed significantly better than animals in the 6 remaining conditions and that none of
the other groups differed from each other. Identical findings
were evident when the mean test scores for all 5 trials were
considered. These results showed that while pre-test cueing
in a novel context normally does not facilitate retention per-
performance (c.f. groups A-A and A-B-A) such cueing can be
made effective by prior cueing treatments in that context (c.f.
group A(B)-B-A). For prior cueing to have this effect, both
prior cueing and pre-test cueing must occur in the same context (c.f. groups A(B)-C-A and A(C)-B-A). Furthermore, neither
prior exposure to the pre-test cueing context, nor prior cueing
in that context before a training memory is established are
sufficient to enhance the effectiveness of the pre-test cue
(c.f. groups A(conf B)-B-A and (B)A-B-A). As with the initial
series of results, these findings indicate once again that the
context in which an animal is cued subsequently functions as if
it had been the training context.

Cue-induced Increases in Resistance to Extinction

Since both of our initial sets of studies had been concerned
with retention of avoidance responding, we next attempted to
test our malleability hypothesis in an appetitively motivated
learning paradigm. According to at least some early studies,
runway training under varied stimulus conditions appears to be
more resistant to extinction than training under constant
conditions (c.f. Sutherland and Mackintosh, 1971). We reasoned
that if this were the case it should be possible, based on our
hypothesis, to increase resistance to extinction by cueing a
training memory in an alternate context as well as by actually
training in multiple contexts. In either case we would expect
the net effect to be that the training memory should include
representations from more than one context.

Thus, in our first study (Gordon and Wittrup, experiment 1, in
preparation) we attempted to confirm the notion that resistance
to extinction can be increased by training animals in more than
one context. The apparati used in these studies were 2 identically constructed, straight alleyways. Each alleyway was made
of plywood with a Plexiglas top, and each had a start box, a
4 ft. runway and a goal box which contained a small food cup.
One runway was located in Room A while the other was in Room B.

To begin the study each animal was placed on a restricted diet
and was reduced to 85% of normal body weight over a period of
approximately 7 days. For each of the next 5 days the animals
received a 15 min. exploration period in the Room A runway.
During these exploration periods food pellets were distributed

throughout the alleyway and the start and goal box doors remained open. Twenty-four hours after the last exploration period, training was begun. Each training trial consisted of placing the animal in the start box of an alleyway and after 3 sec. opening the door to the runway section. Time to leave the start box and time to reach the goal box were recorded. Upon reaching the goal box the rat was allowed access to the filled food cup for 15 sec. before being removed to a holding cage for a 60 sec. intertrial interval. All rats were given 30 training trials - 20 on day 1 and 10 more trials the following day. Extinction contingencies were initiated 60 sec. after the last training trial. Extinction trials were identical to training trials except that no food was present in the food cup in the goal box. The criterion for extinction was two consecutive trials on which an animal took longer than 60 sec. to reach the goal box.

Three treatment conditions were used in this study. In the CRF condition all animals received all 30 training trials, as well as all extinction trials in Room B. In the variable condition on day 1 of training animals received 16 training trials in Room B and 4 trials in Room A. On day 2 these animals were given 8 trials in Room B and 2 trials in Room A. The particular trials that occurred in Room A were randomly determined with the restriction that no 2 consecutive trials could occur in that room. For these animals all extinction trials occurred in Room B. The final condition (EXPOSURE) was run the same way as the variable condition except that instead of receiving actual training trials in Room A these animals received only exposure to the Room A context.

Table 3 summarizes the results of this experiment. As can be seen, animals in the 3 treatment conditions did not differ significantly in terms of running speed on the last training trial. However, animals in the variable condition required more trials to reach the extinction criterion than animals in the other groups. An analysis of these data confirmed that the variable animals took significantly more trials to extinguish than did either the CRF or exposure group animals. These results suggest that when the memory of training contains varied contextual representations, subsequent extinction is prolonged.

In the second study (Gordon and Wittrup, experiment 2, in preparation) we attempted to duplicate the effects of varied training by cueing the memory of runway training in a second context. This experiment contained one group (CRF) that was treated identically to the corresponding group in experiment 1. Aside from this group three additional treatment groups were

EXPERIMENT 1

GROUP	MEAN TOTAL RUNNING TIME (SECS)		MEAN TRIALS TO EXTINCTION
	X̄ TRIALS 1-5	TRIAL 30	
CRF	6.40	3.53	5.50
VARIABLE	10.23	4.63	8.75
EXPOSURE	13.58	4.04	5.14

Table 3. Acquisition and Extinction Results as a
Function of Treatment Condition.

run. In the cue-A group animals received 16 training trials
in Room B and 4 cueing treatments in Room A on the first training day. Cueing involved confining animals in the runway
startbox for 5 sec. then transferring them to a holding cage
in the same room for an additional 30 sec. On day 2 these
animals received 8 training trials in Room B and 2 cueing
trials in Room A. Extinction was carried out in Room B and the
procedure was the same as that used in the prior study. The
third condition (Cue-B) was run the same way as the Cue-A
condition except that the cueing trials occurred in Room B
rather than in Room A. Finally, in the confine A condition
animals also were treated the same as the Cue-A animals except
that simple exposure to the Room A context was substituted for
the cueing treatments in that room.

The results of this study are represented in Table 4. As in
the first experiment the various treatment groups did not differ significantly in terms of running times on the last training trial. However, the Cue-A animals did take significantly
more trials to reach the extinction criterion than did any
other group. None of the remaining groups differed from each

EXPERIMENT 2

GROUP	MEAN TOTAL RUNNING TIME (SECS)		MEAN TRIALS TO EXTINCTION
	X̄ TRIALS 1-5	TRIAL 30	
CRF	8.28	2.30	3.50
CUE B	8.70	3.07	3.75
CUE A	7.14	2.49	9.86
CONFINE A	7.55	2.24	4.50

Table 4. Acquisition and Extinction Results as a
Function of Treatment Condition.

other in terms of this measure. The failure of the Cue-B animals to exhibit increased resistance to extinction shows that
the greater resistance in the Cue-A animals was not due simply
to the interspersing of cueing trials among training trials.
Apparently cueing must occur in a context other than the training context for prolonged extinction to result. Also, the fact
that the Confine-A animals extinguished rapidly indicates that
this prolonged extinction depended on animals being cued in
context A and not on simply being exposed to this alternate
context during training.

Together these studies demonstrate that training animals in one
context and then cueing in another affects extinction performance in the same way as actually training animals in two contexts. This is, of course, what one would predict if the
malleability hypothesis is correct. Under varied training
conditions we assume that an animal's training memory would

come to contain representations of multiple contexts. We
would predict that the same should be true when animals are
trained in one context and cued in another. During acquisition
trials the training context should become represented as part
of the training memory. However, the alternate context present at the time of cueing or retrieval also should become represented in the memory, so that for these animals the training memory also should contain representations of multiple contexts. For this reason, we would expect the subsequent performance of these animals to be similar if not the same as the
animals trained in multiple contexts.

Cue-induced Interference Effects

Our original hypothesis was that contextual information present
when an animal retrieves a training memory will become represented as part of that memory. The three sets of experiments
described thus far all were designed to test one implication
of this hypothesis. This implication is that the retrieval
context should control subsequent retention performance in the
same way as the training context originally controlled such
performance. In other terms, the retrieval context should begin to function as if it had been the context present at the
time of learning. Clearly this prediction is supported by the
findings we have described above. However, this is not the
only testable implication that can be derived from this hypothesis.

It is well established, for example, that when animals acquire
more than one response in the same context, there is a tendency
for these responses to interact with each other when the animal
subsequently returns to that context. One example of this type
of interaction occurs in what has been termed the "memory interference paradigm." In this paradigm animals are trained
to perform conflicting responses in the same context. When
animals later are tested for retention of the more recently
acquired response they tend to perform more poorly than animals trained on that response alone (c.f. Spear, 1971). In
effect, training on the first of the two responses interferes
with retention of the second response. Furthermore, just as
in proactive interference paradigms involving human verbal retention, the interference effect in animals tends to increase
as the retention interval is increased, (e.g. Gordon, Frankl
and Hamberg, 1979).

The fact that this phenomenon depends on an animal acquiring
both responses in the same context, suggests that the common
context plays a key role in producing the effect. One interpretation of the phenomenon, for example, is that the same

context becomes represented in the memory of both responses at the time these responses are acquired. Thus, at the time of testing when this common context is present, both memories are activated simultaneously and the animal is forced to choose between two incompatible responses on the test. (See Spear, 1971.)

Regardless of the specific explanation applied to the memory interference phenomenon, it is clear that such interference does occur when conflicting responses are learned in the same stimulus context. Thus, one implication of the memory malleability hypothesis is that interference also should occur when one response is acquired in a given context and the memory of a conflicting response is retrieved in that context. In this case, we would expect that the memory for both responses would contain some representation of the common context just as if both responses had been acquired in that context. Recently, we have completed an experiment designed to test this prediction (Gordon, Mowrer, McGinnis and McDermott, in preparation).

In this experiment, the basic interference paradigm consisted of passive avoidance learning followed by active avoidance learning, then by a retention test. Passive avoidance was trained in the same avoidance apparatus we have described previously. During training each animal was placed in the start chamber of the apparatus and, after 3 sec., the door separating the chambers was opened. If an animal crossed into the black chamber within 60 sec. after the door was opened, the chamber door closed and a gridshock of 1 sec. duration was applied to the floor of the black chamber. A successful avoidance response consisted of remaining in the white start chamber for 60 sec. after the chamber door was opened. Following either a gridshock or a successful avoidance each animal was removed to a holding cage for a 30 sec. intertrial interval. Training continued until an animal had achieved 2 consecutive 60 sec. suppressions in the white chamber plus 5 additional trials. The procedures for active avoidance acquisition and for retention testing were the same as those used earlier in our studies involving avoidance learning and retention. In this study an interval of 10 min. separated passive and active avoidance acquisition and an interval of 1 hour elapsed between active avoidance learning and the test.

The experiment itself involved 6 different treatment conditions. Three of these conditions (B-B-B, A-B-A and NC) were used to demonstrate that interference normally occurs in this paradigm only when conflicting responses are acquired in the same context. Thus, in condition B-B-B, the acquisition of both

responses and retention testing all took place in Room B. In
condition A-B-B, passive avoidance learning occurred in Room A,
while both active avoidance training and retention testing occurred in Room B. Animals in condition NC received the same
treatment as the B-B-B animals with one exception. These animals received non-contingent shocks in a large chamber in
Room B rather than experiencing passive avoidance training. The
number, sequence and temporal spacing of these shocks for a
given animal were the same as had been experienced by a randomly chosen animal from the B-B-B condition during passive avoidance training.

The remaining 3 conditions (T-T, T-NT, and NT-T) were run to
determine if interference can be produced by having animals retrieve the passive avoidance memory and acquire the active
avoidance memory in a common context. Thus, animals in all of
these conditions acquired passive avoidance in Room A and received active avoidance training and retention testing in Room
B. In condition T-T all animals experienced a constant tone
during passive avoidance acquisition in Room A. Then, approximately 3.5 min prior to active avoidance acquisition, these
animals were brought to Room B for a cueing treatment. This
treatment involved placing the animals in a holding cage and
initiating the tone that had accompanied passive avoidance
learning. The duration of the tone cue was 15 sec. The remainder of the cue - active avoidance training interval was
spent in the holding cage in Room B. No tone was present in
Room B during active avoidance training or at the time of testing. The remaining conditions were run almost identically to
this T-T condition except for a single change in each group.
In the T-NT group, animals were treated the same as the T-T
animals except that no tone was administered prior to training
in Room B. In the NT-T group, rats received the tone cue prior
to training in Room B, but did not have a tone present during
passive avoidance acquisition.

An analysis of acquisition data revealed that there were no
significant differences among the groups in terms of either
passive or active avoidance acquisition rates. This result
suggests that the various treatments administered prior to
active avoidance learning had little differential effect on
avoidance acquisition, itself. Figure 5 summarizes the response latencies on test trial 1 as a function of treatment
condition. It should be reiterated that short latencies reflect good retention of the active avoidance response, while
longer latencies indicate interference with the retention of
this response. Statistical analyses of these data revealed
several important results. First, it is clear that retention
of active avoidance was excellent in both the NC and A-B-B

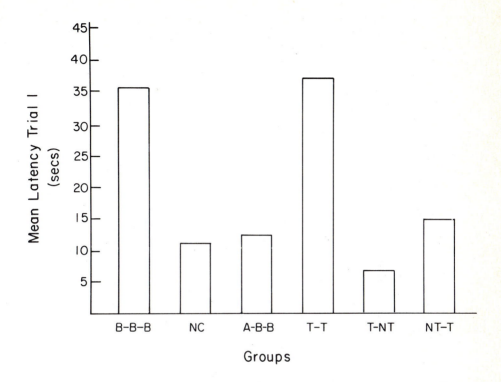

Figure 5.

groups. In other words, if no conflicting response is acquired (NC group) or if a conflicting response is acquired in a different context from that in which active avoidance is learned, retention of the avoidance response is good. No differences between the NC and A-B-B groups were found. On the other hand, the B-B-B animals exhibited significantly longer latencies than either the NC or the A-B-B animals. In the B-B-B condition retention of active avoidance was interfered with by the prior learning of the passive avoidance response. These results confirm that for memory interference to occur it is normally necessary for the conflicting responses to be acquired in the same context.

Of particular interest are the test scores of the T-T animals. These animals showed significantly poorer retention of active

avoidance than either the NC or the A-B-B animals. As a matter
of fact, animals in the T-T and B-B-B conditions did not differ
in terms of retention test latencies. This finding is particularly important in light of the test performance exhibited by
the NT-T and T-NT animals. Animals in these conditions did not
differ from animals in the NC and A-B-B groups. However, both
the NT-T and T-NT animals exhibited significantly better retention of active avoidance than did either the T-T or the B-B-
B rats. These results indicate that memory interference does
occur when animals are cued for one memory and acquire a second
memory in the same context. This effect does not occur without
cueing (c.f. group T-NT) nor does it occur when a cueing treatment is unrelated to the prior conflicting memory (c.f. group
NT-T).

Certainly, these findings are consistent with the prediction
made from the malleability hypothesis. It appears that memory
interference can occur when one memory is retrieved and another
memory is acquired in the same context. This prediction was
based on the idea that a training memory should include representations of a given context whether the memory is acquired or
retrieved in that context. As long as this memory includes such
representations, regardless of how those representations are
established, the memory should produce interference. This, in
fact, does appear to be the case.

The Relationship Between "Adding Contextual Representations" and Higher-order Conditioning

We have shown in a variety of situations that when a cueing
treatment is presented in a novel context, the novel context
subsequently comes to control the learned behavior related to
that cue. This control exists even though the context never
was present when the learned behavior was originally acquired.
Given this description it occurred to us early in this work
that the present phenomenon might be interpretable as a special
case of higher-order conditioning. It is possible, for example,
to view the cueing treatment as a first order CS that has come
to elicit some response during original acquisition. The particular response elicited by the cueing treatment could be a pre-
paratory motor response, a change in some motivational state or,
in the terms we have been using, the activation of the training
memory. In any event the presentation of the cueing treatment
in a previously novel context would mean that the cue would
elicit its response in the presence of these "neutral" contextual stimuli and these previously neutral stimuli would come
to elicit the same response through higher-order conditioning.
In this view the novel context would be acting as a second-

order CS capable of eliciting some response that would affect retention performance.

In general this appears to be a plausible explanation of how control of retention performance transfers to a new context at the time of cueing. We have, however, discovered certain problems with such an interpretation. The first of these problems is that higher-order conditioning is not known to be a particularly robust phenomenon when only one conditioning trial is used. Yet, in the studies we have described a single exposure to cueing in a novel context seems to have strong effects on subsequent retention performance in that context. This does not, of course, preclude an explanation in terms of higher-order conditioning, but it does seem to stretch the plausibility of such an interpretation.

A second problem with the higher-order conditioning explanation arises from some experiments recently completed by Robert Mowrer in our laboratory. Mowrer was interested in the effects of cueing in an irrelevant context. By "irrelevant" it is meant that the cueing context is unrelated to either the training or the test context. The principle question addressed in these experiments was whether cueing in an irrelevant context would affect the ability of the training context to control subsequent retention. The condition of primary interest was one in which animals were trained to avoid in Room A, were given a cueing treatment in Room B and then were tested for retention back in Room A. Mowrer found that animals in this condition performed more poorly on the test than did animals trained and tested in Room A but given no cueing treatment. In other words, cueing in the irrelevant context actually <u>decreased</u> the capacity of the training context to control test performance.

Such a finding is difficult to explain if one assumes that cueing acts simply to elicit a response that can become associated with the irrelevant context. Why would the formation of a new association be expected to detract from the capacity of the original context to control responding? We know of nothing in the conditioning literature that would predict such a finding. Mowrer, however, has suggested at least two possible explanations for these results within a memory retrieval framework (Mowrer and Gordon, in preparation).

One possibility is that the addition of contextual representations to an existing memory actually expands the number of representations in that memory. Thus, at the time of recall it takes longer for an animal to search through the memory to determine if the test stimuli match stimuli represented in the training memory. This increased retrieval time is then reflected in poorer test performance than would occur if the

memory had fewer representations. The second possibility is
simply that a given memory has a set size. Therefore, the
addition of new contextual representations to a memory may lead
to some loss of the original representations from that memory.
Obviously such a loss would hinder the control of the original
training context over retention performance, since that context
would no longer be represented in the training memory.

One final problem with explaining the cueing effect in terms of
higher-order conditioning involves the circumstances under which
the effect occurs most strongly. We know that higher order
conditioning is most effective when the second order CS precedes
the first order CS by a brief time. This would suggest that in
our paradigm exposing the animals to a novel context and then
cueing would be more effective than cueing followed by exposure
to the novel context. Although most of our studies have involved overlapping exposure to these events, we do have some
data which speak to these predictions concerning order of presentation. We have found in certain of our control conditions,
for example, that when animals are exposed to a novel context
and then given a cueing treatment in a different context, the
novel context fails to establish control over subsequent responding (c.f. Gordon, McCracken, Dess-Beech and Mowrer, 1981).
On the other hand, pilot work in our laboratory recently has
looked at the effectiveness of exposing animals to a cueing
treatment in one context and then exposing animals immediately
to a novel context. These preliminary studies suggest that such
a procedure does result in the novel context beginning to
function like the training context.

Obviously, these results only suggest what effects the order of
exposure to cueing treatments and novel contexts might have. To
this point no formal experiments assessing this variable have
been completed. It is notable, however, that this preliminary
evidence indicates that the strongest cueing effect occurs when
cueing precedes exposure to the novel context. According to a
higher-order conditioning explanation, this sequence of events
should result in the weakest cueing effects because this sequence would actually constitute a backward conditioning trial.

Clearly, we need to do additional experiments to assess the
higher order conditioning explanation more directly. It should
be particularly fruitful to explore the effects of cue-context
exposure order in producing the kind of cueing effects we have
reported. However, based on the preliminary data we have and
based on the observations we have made so far, attempts to explain these cueing effects in terms of higher-order conditioning appear unpromising.

Conclusion

We began this series of experiments in an effort to determine if stored memories become malleable when they are cued or retrieved. Specifically, we sought to discover if contextual stimuli present at the time of cueing would become represented in, or incorporated into, the memory being cued. We believe that the data reported in this chapter support an affirmative answer to this question. Clearly, contextual stimuli present when a training memory is cued do begin to function as if they had been present at the time of training. In other words, the cueing context begins to function as if it had been represented in the original memory for training. Importantly, these data do not appear amenable to other interpretations such as an explanation in terms of higher-order conditioning.

Such data suggest to us that an animal's memory of a training experience is not necessarily fixed at the time of learning. They suggest that the content and possibly even the structure of an animal's memory may continue to evolve and change as long as the animal continues to retrieve and use that memory. These data imply that retrieval may be more than the mechanism by which we access stored information. It may be an important time for the reprocessing and updating of previously stored information.

The view that the memories of animals are malleable has a number of implications for the study of memory processing in infra-humans. Most obvious, however, concerns the study of retrieval and our ability to predict retention performance. Traditionally, the study of memory retrieval has centered on the relationship between training and testing experiences. We have been interested in what stimuli are present at the time an animal learns and in how an animal represents those stimuli in a training memory. Our predictions concerning retention performance have been based almost entirely on the degree of similarity between test stimuli and stimuli present during learning.

The present work indicates that an understanding of training and test conditions may be insufficient to allow us to predict retention performance reliably. Prediction of retention performance may, in addition, depend on our ability to determine if an organism has had previous retrieval opportunities, and if so, what information was available to the organism at the time of those opportunities. Furthermore, reliable prediction may depend on our ability to understand the rules by which memories change at the time of retrieval. Without such an understanding our explanations of memory processing in animals remain incomplete.

References

1   Bartlett, S. C. Remembering: a study in experimental and social psychology. Cambridge, England: Cambridge University Press, 1932.

2   Bransford, J. D. & Franks, J. J. Sentence memory: A constructive vs. interpretive approach. Cognitive Psychology, 1972, 3, 193-209.

3   Calhoun, W. H. Central nervous system stimulants. In E. Furchtgott (Ed.), Pharmacological and Biophysical agents and behavior. New York: Academic Press, 1971.

4   Cole, W. G. & Loftus, E. F. Incorporating new information into memory. American Journal of Psychology, 1979, 92, 413-425.

5   DeVietti, T. L., Conger, G. L. & Kirpatrick, B. R. Comparison of the enhancement gradients of retention obtained with stimulation of the mesencephalic reticular formation after training or memory reactivation. Physiology and Behavior, 1977, 19, 549-554.

6   Feldman, D. T. & Gordon, W. C. The alleviation of short-term retention decrements with reactivation. Learning and Motivation, 1979, 10, 198-210.

7   Gordon, W. C. Susceptibility of a reactivated memory to the effects of strychnine: a time-dependent phenomenon. Physiology and Behavior, 1977, 18, 95-99. (a)

8   Gordon, W. C. Similarities between short-term and reactivated memories in producing interference in the rat. American Journal of Psychology, 1977, 90, 231-242. (b)

9   Gordon, W. C. Mechanisms for cue-induced retention enhancement. In N. E. Spear & R. R. Miller (Eds.), Information processing in animals: Memory mechanisms. Hillsdale, N. J.: Lawrence Erlbaum Associates, 1981.

10 Gordon, W. C. & Feldman, D. T. Reactivation induced interference in a short-term retention paradigm. Learning and Motivation, 1978, 9, 164-178.

11 Gordon, W. C., Frankl, S. E. & Hamberg, J. M. Reactivation-induced proactive interference in rats. American Journal of Psychology, 1979, 92, 693-702.

12 Gordon, W. C., McCracken, K. M., Dess-Beech, N., &
    Mowrer, R. R. Mechanisms for the cueing phenomenon: The
    addition of the cueing context to the training memory.
    Learning and Motivation, 1981, 12, 196-211.

13 Gordon, W. C., Mowrer, R. R., McGinnis, C., & McDermott,
    M. J. Cue-induced changes in the contextual control of
    proactive interference. In preparation.

14 Gordon, W. C. & Spear, N. E. The effects of strychnine on
    recently acquired and reactivated passive avoidance memo-
    ries. Physiology and Behavior, 1973, 10, 1071-1075. (a)

15 Gordon, W. C. & Spear, N. E. The effect of reactivation of
    a previously acquired memory on the interaction between
    memories in the rat. Journal of Experimental Psychology,
    1973, 99, 349-355. (b)

16 Gordon, W. C. & Wittrup, S. M. Effects of cueing on re-
    sistance to extinction in the runway. In preparation.

17 Hertel, P. T. & Ellis, H. C. Constructive memory for
    bizarre and sensible sentences. Journal of Experimental
    Psychology: Human Learning and Memory, 1979, 5(4), 386-
    394.

18 Lewis, D. J. Psychobiology of active and inactive memory.
    Psychological Bulletin, 1979, 86, 1054-1083.

19 Loftus, E. F. The malleability of human memory. American
    Scientist, 1979, 67, 312-320.

20 Loftus, E. F. & Zanni, G. Eyewitness testimony: The in-
    fluence of the wording of a question. Bulletin of the
    Psychonomic Society, 1975, 5, 86-88.

21 Misanin, J. R., Miller, R. R. & Lewis D. J. Retrograde
    amnesia produced by ECS after reactivation of a consolidated
    memory trace. Science, 1968, 160, 554-555.

22 Mowrer, R. R. & Gordon, W. C. Effects of cueing in an
    irrelevant context. In preparation.

23 Riccio, D. C. & Ebner, D. L. Postacquisition modification
    of memory. In N. E. Spear & R. R. Miller (Eds.), Infor-
    mation processing in animals: Memory mechanisms.
    Hillsdale, N. J.: Lawrence Erlbaum Associates, 1981.

24 Spear, N. E. Forgetting as retrieval failure. In
   W. K. Honig & P. H. R. James (Eds.), Animal memory.
   New York: Academic Press, 1971.

25 Spear, N. E. The processing of memories: Forgetting and
   retention. Hillsdale, N. J.: Lawrence Erlbaum Associates,
   1978.

26 Sutherland, N. S. & Mackintosh, N. J. Mechanisms of animal
   discrimination learing. New York: Academic Press, 1971.

27 Wittrup, S. M. & Gordon, W. C. The alteration of a training memory through cueing. American Journal of Psychology,
   In press.

ANIMAL COGNITION AND BEHAVIOR
Roger L. Mellgren, editor
© North-Holland Publishing Company, 1983

ON THE THEORY OF GENE-CULTURE CO-EVOLUTION
IN A VARIABLE ENVIRONMENT

H. Ronald Pulliam

State University of New York

Though some social scientists have appreciated the interactions between genetic and cultural evolution (e.g., Campbell, 1965 and Durham, 1976), the genetical evolution of behavior and the cultural evolution of human societies have been, for the most part, treated as distinct fields of inquiry. In their recent book Genes, Mind, and Culture, Charles Lumsden and E. O. Wilson (1981) argue, however, that genetic and cultural evolution are tightly coupled one to another. The basic premise of Lumsden and Wilson's argument is both simple and compelling. An individual's chances of survival and reproduction depend on his or her behavior, but the decision to adopt one behavior or another is influenced by both the individual's genotype and his or her social environment. Thus, the social environment influences the relative Darwinian fitness of individuals and, thereby, influences the course of genetic evolution. Furthermore, the genetic composition of a population influences the social environment and, thereby, influences the course of cultural evolution. Starting from this simple premise, Lumsden and Wilson construct a complicated mathematical model of the process of gene-culture co-evolution and then use their model to deduce hitherto hidden features of the human mind and of human societies.

The purpose of this paper is to extend the analysis of Lumsden and Wilson by addressing what I see as three serious shortcomings of their model of gene-culture co-evolution. First, their model is so complex as, on the one hand, to discourage all but the most mathematically adept readers. In this paper, I attempt to develop a simplified version of the Lumsden and Wilson model which retains the biological and cultural richness of their original model but is nonetheless easier to understand and more amenable to analysis. Secondly, despite the fact that many authors have argued that culture is an adaptation to a rapidly changing environment, Lumsden and Wilson consider gene-cultural co-evolution only in a constant environment where a particular behavior always has the same fitness consequences.

I allow the environment to change between generations and show
that this substantially changes the conclusion that can be
drawn concerning the importance of genetic constraints on behavior. Finally, Lumsden and Wilson consider only the situation where an individual's decision to adopt one behavior or
another is influenced by the relative numbers of behavior users
in the social environment and not at all by the relative success of individuals that adopt different behaviors. I argue
that humans selectively imitate those individuals that they
perceive as successful, and that this greatly alters the course
of gene-culture co-evolution.

The Lumsden-Wilson Model

Lumsden and Wilson envision a co-evolutionary circuit according to which genes and culture interact to influence individual
decisions. In turn, individuals decisions ultimately effect
both cultural and genetic changes. In the words of Lumsden
and Wilson, "The epigenetically guided actions of the individual members create the cultural patterns, but the patterns
influence the actions and, ultimately, the frequencies of the
underlying genes themselves" (p. 265). Lumsden and Wilson
make an important contribution to social theory by specifying
the numerous steps involved in going from genes to culture and
back to genes again.

At the heart of the Lumsden and Wilson model of gene-culture
co-evolution is a submodel of the life cycle (Figure 1). Each
individual begins life as a zygote with a certain genotype
inherited from his or her parents. During early development,
each individual is exposed to alternative modes of behavior,
or culturgens as they are called by Lumsden and Wilson. A representation of each culturgen is stored in each individual's
long-term memory and each individual evaluates the culturgens
according to innate biases specific to his or here genotype
and to his or her perceptions of the usage and evaluations
of the same culturgens by other individuals in his or her own
or parental generations. In the words of Lumsden and Wilson,
the juveniles "record impressions of how many adults use each
of the two culturgens" and during "exploration and play they
access the culturgens still further by means of lessons, games,
practice, and conversations" (p. 267).

After the juvenile socialization period, each individual enters
a prereproductive forage period during which time he or she
has many opportunities to make decisions concerning culturgen
usage and at each decision point, may reevaluate the culturgens
and may shift usage, as illustrated by the arrows between
culturgens in Figure 1. The decisions made during this prereproductive period determine the amount of resources gathered

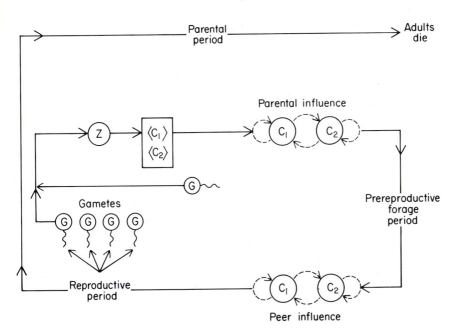

Figure 1. The co-evolutionary circuit (adapted from
Lumsden and Wilson 1981). Gametes (G) carry genetic
information that controls the epigenetic development
of the zygotes (Z). Developing individuals are exposed
early in life to alternative behaviors (C_1 and C_2) which
they subsequently evaluate under parental and peer influences. The behavior finally adopted determines success during the prereproductive "forage" period, which,
in turn, limits individual contributions to the gene
pool of the next generation. Adults live long enough
to enculturate the new generation but die before the
next reproductive phase.

and, thereby, the individual's subsequent Darwinian fitness.
According to Lumsden and Wilson, to "gather a resource means
either literally to accumulate food in a direct manner or else
to secure a richer subsequent harvest through the establishment
of territory, attainment of rank in a hierarchy, formation of
economic and political alliances, or other devices" (p. 270).

The first phase of adult life, is the reproductive period during which time the resources gathered in the prereproductive
forage period are converted into gametes which contribute to
the zygotes of the next generation according to a fixed
fertility rule involving random mating. Again, according
to Lumsden and Wilson, the "variation in fertility due to
alternative culturgen usage is the sole cause of variation in
absolute fitness among the genotypes in the population"
(p. 270). After reproduction, the adults live long enough to
guard their offspring and to participate in their socialization. The adults, however, die before the beginning of the
next reproductive phase, making the generations "quasi-discrete".

A central concept in the co-evolutionary circuit is the
socialization of children by members of both their own and
their parent's generations. According to the Lumsden-Wilson
model, at each decision point, each individual reevalutes its
culturgen usage according to a transition matrix that gives
the probability that an individual adopts culturgen i given
that he or she last preferred culturgen j. The actual values
of the transition probabilities used in making a particular
decision depend on both the individual's genotype and on his
or her social environment. The genotypic influence is via
the specification of innate biases that determine the transition probabilities prior to specific social influence.

The Lumsden and Wilson models are greatly complicated by the
fact that each individual may change his or her mind about
culturgen usage many times. The result is a very complicated
set of equations describing the time between decision points
and the integration of innate biases and social influence
into each decision. In order to simplify their model, Lumsden
and Wilson assume that the prereproductive forage period is
long enough that the distribution of behavioral decisions is in
a steady state for all but a negligible fraction of the prereproductive period. Still the equations are so complex as
to defy simple interpretation and analytic solution.

In the models that follow, I assume that each individual makes
only one decision regarding culturgen usage. That decision
may be influenced by both the behavioral usage patterns of the
adult generation and the peer generation but once it is made
it is fixed for life. The result of this single change is to
simplify greatly the mathematics without greatly altering the
model's biological and cultural assumptions.

A Simple Model of Gene-Culture Co-evolution

Following Lumsden and Wilson, I assume a large, randomly mating population with quasi-discrete generations, meaning "the adults persist long enough to enculturate the juveniles but die before the next breeding phase takes place" (p. 266). Also, I assume that the only genetic influence on variation in individual behavior is due to allelic variation at a single autosomal locus with two alleles, A and B.

As do Lumsden and Wilson, I assume that an individual's reproductive success is determined solely by the individual's behavior, though variation in behavior is influenced by both genetic and cultural factors. In the simplest case (and the only one explicitly considered by Lumsden and Wilson), there is but a single habitat, and an individual who uses behavior 1 has fitness W_1 as compared to an individual who chooses behavior 2 who has fitness W_2. Thus, the mean fitness of individuals of genotype ij depends on the fraction (U_{ij}) of such individuals using behavior 1 and the fraction ($1-U_{ij}$) using behavior 2. In particular, the mean fitnesses of the three genotypes are given by the following equations:

$$W_{AA} = U_{AA}W_1 + (1-U_{AA})\ W_2, \tag{1a}$$

$$W_{AB} = U_{AB}W_1 + (1-U_{AB})\ W_2,\ \text{and} \tag{1b}$$

$$W_{BB} = U_{BB}W_1 + (1-U_{BB})\ W_2. \tag{1c}$$

The total number of individuals using behavior 1 in generation t is given by:

$$N_1{}^t = U_{AA}{}^t N_{AA}{}^t + U_{AB}{}^t N_{AB}{}^t + U_{BB}{}^t N_{BB}{}^t, \tag{2}$$

where N_{ij} gives the number of individuals of genotype ij. The fraction of the population at large using behavior 1 in generation t is given by $U^t = N_1{}^t/N_t$, where N_t is the total population size. The fractions $U_{AA}{}^t$, $U_{AB}{}^t$, and $U_{BB}{}^t$ can be thought of as genotype specific functions of the social environment experienced by individuals in generation t. To simplify the model, I assume that all individuals are socialized early in life and make only one decision concerning the fitness determining behavior. In the version of the model discussed, I assume that the behavioral decisions of the generation being socialized are influenced primarily by the behavior usage pattern of the parental generation. This assumption can be easily modified to allow for both parental and peer influence on individual decisions.

Following Lumsden and Wilson, I use the term assimilation
function to relate individual behavioral decisions to the
social environment experienced by the individual. An assimilation function $U_{ij}{}^t$ specifies the probability that an individual of genotype ij in generation t adopts behavior 1.
In general, this probability is thought to be influenced by
the usage patterns of both the parental and peer generations.
In the very simplest case, however, the probability of adopting behavior 1 is given by a constant value $U_{ij}{}^o$, regardless of
the social environment experienced by the individual. Throughout this paper, $U_{ij}{}^o$ is referred to as the "innate bias" towards adopting behavior 1. In the extreme case where $U_{ij}{}^o$ =
1.0 and there is no social influence, it can be said that behavior 1 is genetically determined for individuals of genotype
ij. Furthermore, in any case where $U_{ij}{}^o > 0.5$, it can be said
that individuals of genotype ij have a genetic predisposition
in favor of adopting behavior 1.

In contrast to an individual with a fixed probability of adopting a particular behavior, a saturable trend watcher (Lumsden
and Wilson, p. 135) is more likely to adopt whichever behavior
is most prevalent in his or her social environment. For the
example discussed here, I assume that individuals in generation
t are influenced only by the usage pattern, U^{t-1}, of their
parental generation. In particular, I use the function

$$U_W{}^t = A_1 - [A_2 \tan^{-1}(1-2\ U^{t-1})] \tag{3}$$

to specify the probability that a saturable trend watcher
adopts behavior 1. For all examples presented, I have set
$A_1 = A_2 = 0.5$. As shown in Figure 2, with these parameter
values, a trend watcher is equally likely to adopt behavior 1
or behavior 2 if the parental generation is split 50:50 in behavior usage; however, if most of the parental generation uses
behavior 1, a trend watcher adopts behavior 1 with a probability of almost 90%.

Innate bias and trend watching can be combined in various
degrees by specifying a parental socialization factor, α_{ij},
which averages the two. In general, the probability that an
individual of genotype ij in generation t adopts behavior 1
is given by the assimilation function

$$U_{ij}{}^t = \alpha_{ij}U_W{}^t + (1-\alpha_{ij})\ U_{ij}{}^o. \tag{4}$$

The parental socialization factor can take on values from 0.0
to 1.0 and specifies the relative sensitivity of the adoption

decision to the usage pattern of the parental generation.
Figure 2 illustrates assimilation functions for various values
of U_{ij}^{o} and αij.

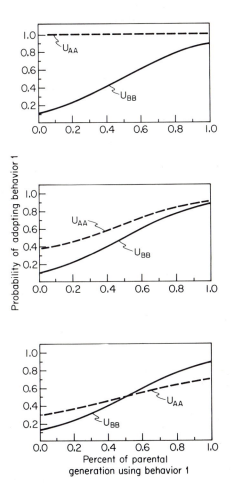

Figure 2. Assimilation functions. The probability
that an individual adopts a particular behavior
is taken to be a function of both the social environment experienced by the individual and the individual's genotype. Each assimilation function
depicted is of the form $U_{ij} = \alpha_{ij} U_w + (1-\alpha_{ij})U_{ij}$
(see text for details). In all three cases $_{AA}$
= 1.0. In Case A, U_{BB}^{o} = 1.0 and α_{BB} = 0.0; in case
B, U_{BB}^{o} = 1.0 and α_{BB} = 0.7; and in case C, U_{BB}^{o} =
0.5, and α_{ij} = 0.5.

Under the assumption of random mating and approximate Hardy-Weinberg equilibrium, the mean fitness of an individual in the population is

$$\overline{W}^t = p_t^2 W_{AA}^t + 2p_t(1-p_t) \, W_{AB}^t + (1-p_t)^2 W_{BB}^t, \tag{5}$$

where p_t is the frequency of gene A in generation t. The gene frequency in the next generation is given by the standard equation

$$p_{t+1} = (p_t^2 W_{AA}^t + p_t (1-p_t) \, W_{AB}^t)/\overline{W}^t. \tag{6}$$

Finally, the usage of behavior 1 in the next generation is given by

$$U_1^{t+1} = p_t^2 U_{AA}^{t+1} + 2p_t (1-p_t) \, U_{AB}^{t+1} + \\ (1-p_t)^2 \, U_{BB}^{t+1}. \tag{7}$$

Since the fitness terms (W_{ij}) depend on usage frequencies and the total usage depends on gene frequencies, equations 6 and 7 are tightly coupled justifying the term gene-culture co-evolution.

Lumsden and Wilson assume a constant foraging environment in which an individual's behavioral decision "determines both the strategy of exploitation and the reproductive success" of foragers (p. 270). I assume that the environment may change and that the behavior that results in greater fitness in one environment results in lower fitness in the second environment. In the examples to follow, the environment changes between generations according to the Markov transition matrix

$$T = \begin{bmatrix} T_{11} & T_{12} \\ T_{21} & T_{22} \end{bmatrix}, \tag{8}$$

where T_{ij} is the probability of a transition from environment j to environment i during the time between two generations, and the probability of no change is $T_{jj} = 1-T_{ij}$.

The Evolution of Assimilation Functions

Using equations 6 and 7, the evolution of assimilation functions can be studied by specifying an assimilation function for each genotype and following changes in gene frequency through time. In the case of a constant environment, equations 6 and 7 can be solved directly, giving the conditions for a rare gene to increase in frequency. In the case of a varying environment, the evolution of assimilation functions can be studied using computer simulation.

Assuming no dominance (i.e., $U_{AB} = (U_{AA} + U_{BB})/2$) and following the convention that in a constant environment individuals using behavior 1 have greater mean fitness, gene A increases in fitness so long as $W_{AA}{}^t > W_{BB}{}^t$, which is the case if $U_{AA}{}^t > U_{BB}{}^t$. Thus, if one genotype has a probability of using behavior 1 which is strictly greater than that of the other genotype (e.g. Figures 2a and 2b), genes resulting in more genetically determined behavioral responses (higher $U_{ij}{}^0$ and lower α_{ij}) are favored by natural selection.

The dynamics of gene-culture co-evolution in a constant environment can be analyzed with a simple graph whose axes are gene frequency in generation t (p_t) and usage frequency in the parental generation (U^{t-1}). Using equations 6 and 7, for any point (p_t, U^{t-1}) on this graph, the changes p and U can be plotted as a single change vector indicating the direction and magnitude of co-evolutionary change. This is illustrated in Figure 3 using the assimilation functions shown in Figure 2a and the fitness values $W_1 = 1.50$ and $W_2 = 0.50$. As for any situation in a constant environment where U_{AA} is strictly greater than U_{BB}, the result is complete fixation of gene A which promotes greater usage of the more fit behavior. This result can be generalized as the statement that gene-culture co-evolution in a constant environment favors lower responsiveness to the socialization and greater genetic control of behavior.

This result is similar to that of Lumsden and Wilson who argue that the "tabula rasa state, in which no innate bias exists in culturgen choice, is shown to be unstable" (p. 304) and in "all cases examined, directed cognition due to genetically biased epigenetic rules replaced undirected cognition". I argue that this generality holds only in a constant environment in which one behavior always has greater mean fitness than the other.

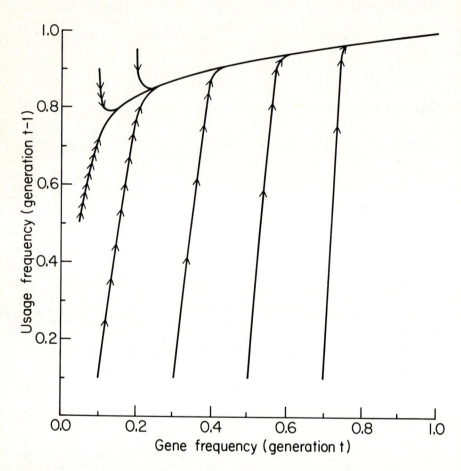

Figure 3. Gene-culture co-evolution in a constant environment. For any gene frequency in generation
t (p_t), and the usage frequency of the appropriate
parental generation (U_{t-1}), the change vector,
(Δp, ΔU) can be calculated. For this example, U_{AA}
and U_{BB} are as given in Case A of Figure 2.

For gene-culture co-evolution in a variable environment, a gene will be selected only if its geometric mean fitness exceeds that of other genes in the population. The geometric mean fitnesses depend on 1) the proportion of the time the environment favors one behavior over the other and 2) the extent to which individuals of each genotype track their changing environments.

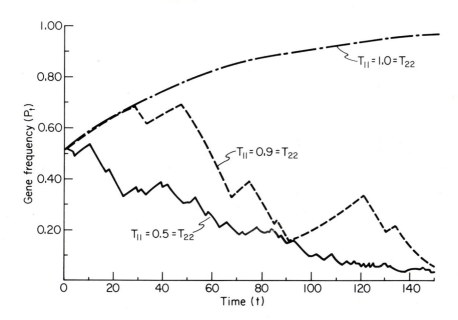

Figure 4. Gene frequency change for gene-culture co-evolution in a variable environment. In each example, the genotypes contrasted are those shown in case A of Figure 2. For $T_{ii} = 1.0$, the environment is constant and genes favoring more fixed behavioral responses are favored. In a variable environment ($T_{ii} = 0.9$ and 0.5), however, genes favoring greater responsiveness to the social environment are favored, on average.

Figure 4 compares gene-culture co-evolution in constant and variable environments. In all three examples, all parameter values except habitat autocorrelation, are the same. As in the last example, in a constant environment, gene A which promotes greater usage of behavior 1 gradually replaces gene B which promotes greater sensitivity to the social environment. Gene-culture co-evolution in a variable environment is shown

in the other two examples, where the habitat autocorrelations
($T_{11} = T_{22}$) are 0.9 and 0.5, respectively. In both cases the
gene promoting more variable behavioral responses gradually
replaces the gene promoting more fixed responses.

For the two examples of gene-culture co-evolution in a variable
environment shown in Figure 4, the usage frequencies did not
track environmental changes but rather gradually declined to
0.5. This suggests that the evolution of more variable behavioral responses was not due to social learning per se.
This can be shown to be the case by comparing assimilation
functions such as those shown in Figure 2c, which differ in
their responsiveness to the social environment but not in
their mean predisposition towards either behavior. Either of
the assimilation functions shown in Figure 4c will replace the
function U_{AA} in Figure 4b. However, when two functions such
as those in Figure 4c are compared directly, the usage frequencies of all three genotypes converge to 0.5. Once this
has happened, individuals of all three genotypes are equally
likely to adopt either behavior and, therefore, are equally
fit. For example, a genotype which uses behavior 1 with
probability 0.5 regardless of the social environment is equally
fit as a trend watcher genotype that uses behavior 1 with probability 0.5 when the parental generation is split 50:50 but
otherwise uses whichever behavior is more frequently used by
the parental generation.

In order for usage frequencies to track changes in the environment in an adaptive manner, there must be social transmission
of some information about which behavior is better adapted
to current conditions. This can be accomplished in one of two
ways. First, social learning can be highly structured with,
for example, offspring only, or at least primarily, learning
from their parents (e.g., see Cavalli-Sforza and Feldman, 1981
and Boyd and Richerson, 1976). In this situation, parents who
use maladaptive behaviors have fewer offspring who learn from
them. This leads to a gradual shift in usage frequency without
any necessity of actual perception of the relative fitness
consequences of the behaviors in question (Pulliam and Dunford
1981). Alternatively, offspring might be more likely to learn
from particularly successful adults of the parental generation,
regardless of their biological relatedness to those adults.
This is most likely to occur if the costs and benefits of the
various behaviors can be perceived and compared.

Selective imitation of successful adults can be modeled in the
following simple manner. Social learning is viewed as a two
step process. First, offspring are exposed to the behaviors
of the parental generation and acquire tentative usage

probabilities $(V_{ij}{}^t)$ according to the assimilation functions previously given in equations 3 and 4. The tentative usage probability is, thus,

$$V_{ij}{}^t = \alpha_{ij}U_W{}^t + (1-\alpha_{ij})\ U_{ij}{}^o,$$

and, if there were no further evaluations of the alternative behaviors, this would be the probability of using behavior 1. However, the new generation does not merely acquire behaviors passively from the parental generation; rather, they partially perceive and compare the relative costs and benefits of the behaviors in question. Perhaps, this is accomplished through peer contact and discussions of the successes and failures of the parental generation.

The magnitude of selective imitation is specified by the geno-type specific constants S_{AA}, S_{AB} and S_{BB}. If in the parental generation, behavior 1 is the more successful behavior, the probability that an individual of genotype ij adopts behavior 1 is given by

$$U_{ij}{}^t = S_{ij} + (1-S_{ij})\ V_{ij}{}^t.$$

If however, behavior 2 is the more successful genotype, the usage probability for behavior 1 is given by

$$U_{ij}{}^t = (1-S_{ij})\ V_{ij}{}^t.$$

The selective imitation constants can range from 0.0, where there is no selective imitation, to 1.0, where each individual adopts with probability 1.0 whichever behavior was more bene-ficial to the parental generation.

Natural selection favors selective imitators. Even with very modest values of selective imitation, in the range of 0.1 to 0.2, selective imitators quickly replace non-selective imitators. Furthermore, selective imitation greatly enhances the value of the social transmission of behavioral preferences from the parental generation. Recall that in the absence of selective imitation, selection is neutral with respect to a genotype that always chooses behavior 1 with probability 0.5 versus a trend watcher that chooses whichever behavior is used most frequently by the parental generation.

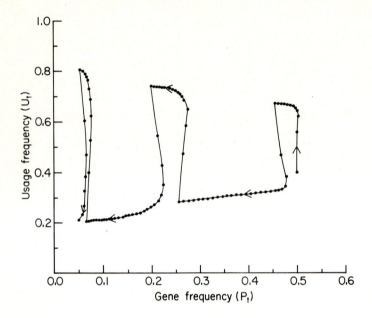

Figure 5. The influence of imitating successful
   individuals on gene-culture co-evolution. The
   parameter values used here and in Figure 6 are
   $U_{AA}^O = 0.5$, $\alpha_{AA} = 0.0$, $S_{AA} = 0.2$, $U_{BB}^O = 0.5$,
   $\alpha_{BB} = 1.0$, and $S_{BB} = 1.0$. Notice that social
   imitation of successful individuals allows the
   usage pattern to track environmental changes in
   an adaptive manner, even as gene frequencies
   change.

Figure 5 shows gene-culture co-evolution in a variable environment for the situation where all three genotypes have the same
parameter values for innate bias ($U_{AA}^O = U_{BB}^O = 0.5$) and
selective imitation ($S_{AA} = S_{BB} = 0.2$) but differ in their
parental socialization factors ($\alpha_{AA} = 0.0$ and $\alpha_{BB} = 1.0$).
Gene B quickly increases in frequcy and eventually completely
replaces gene A. Furthermore, usage frequencies track the
environmental changes in an adaptive manner. Figure 6 indicates that the selective advantage of the trend watching genotypes is due to their greater ability to track environmental
changes. After Gene B is fixed in the population, individuals

have greater than an 80 percent chance of adopting whichever behavior did better in the parental generation.

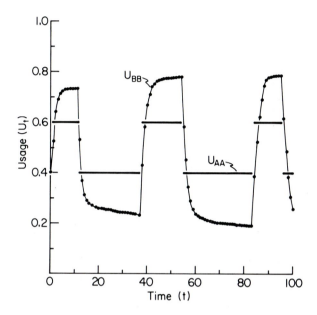

Figure 6. Tracking a variable environment. The two genotypes differ in their responsiveness to the social environment ($\alpha_{AA} = 0.0$ and $\alpha_{BB} = 1.0$) though individuals of both genotypes imitate successful individuals in the parental generation ($S_{AA} = 0.2 = S_{BB}$). As depicted in Figure 5, the greater tracking ability of genotype BB, results in a gradual fixation of gene B in the population.

Discussion and Conclusions

The model presented above suggests that natural selection favors genetically fixed behavioral responses in a constant environment where the same response always confers greater fitness than other possible responses. However, in a variable environment, where one behavior works best in some generations and another works best in other generations, natural selection may favor less fixed behavioral responses. In particular, in a variable environment, natural selection may favor social transmission of behavior between generations. Furthermore, the value of social transmission of behavior is greatly enhanced by the ability to imitate selectively those individuals of the

parental generation that are particularly successful. Even a
moderate ability at selective imitation allows substantial
intergenerational tracking of environmental changes.

Lumsden and Wilson concluded that natural selection normally
favors strong innate biases in behavioral choices. I have
suggested that this conclusion does not generally hold for
evolution in rapidly changing environments. Did the environment faced by early hominids change rapidly enough to favor
social transmission of behavior over innately fixed behavior?
In particular, did the hominid environment change so much more
rapidly than that of other species so as to explain the greater
role of social transmission for humans than for other species?

Rose (1980) suggests that the rapid increase in hominid brain
size over the last few million years is best explained as an
unprecedented "mental arms race". According to this hypothesis,
both the requirements of ecological adaptation and intraspecific
competition favor increased generalized calculating ability,
which, in turn, allows organisms to generate novel behavioral
responses to unanticipated fitness contingencies. Since among
the most complex of problems faced by an intelligent animal are
those generated via intraspecific competition, the greater the
mental abilities of one's competitors, the greater the utility
of increased personal calculating ability. Thus, according to
Rose's hypothesis what has distinguished hominid evolution is
an unprecedented feedback loop whereby the evolution of intelligence created the need for yet greater intelligence.

If Rose's hypothesis is essentially correct, the hominid social
environment may have indeed changed very rapidly. If the
evolution of intelligence effectively accelerated the pace of
environmental change, it would also have increased the utility
of the social transmission of behavior. Moreover, the socially
transmitted information would itself have become yet more
weaponry to be employed in the mental arms race.

The genetic evolution of hominid intelligence may have quickly
passed into a self-accelerating period of gene-culture co-
evolution. According to the model of gene-culture co-evolution
developed here, during this period, hominid mental evolution
was probably characterized by a progressive eroision of genetic
constraints on behavior. However, given a limited source of
genetic variability, this self-acceleration phase may have been
relatively short-lived, quickly passing to a phase of purely
cultural evolution with little or no subsequent genetic change.
Just how far gene-culture co-evolution may have proceeded towards the evolution as a true human tabula rasa will no doubt
be debated for some time to come.

References

1 Boyd, R. & Richerson, P. J. A simple dual inheritance model of the conflict between social and biological evolution. Zygon, 1976, 11, 254-262.

2 Campbell, D. T. Variation and selective retention in socio-cultural evolution. In H. R. Barringer, G. I. Blanksten & R. W. Mack, eds., Social change in developing areas. Cambridge, Mass.: Schenkman Publishing Co., 1965.

3 Cavalli-Sforza, L. L. & Feldman, M. W. Cultural transmission and evolution: a quantitative approach. Monographs in Population Biology (16), Princeton, N.J.: Princeton University Press, 1981.

4 Durham. W. The adaptive significance of cultural behavior. Human Ecology, 1976, 4, 89-121.

5 Lumsden, C. J. & Wilson, E. O. Genes, mind, and culture: the coevolutionary process. Cambridge, Mass., Harvard University Press, 1981.

6 Pulliam, H. R. & Dunford, C. Programmed to learn: an essay on the evolution of culture. New York: Cambridge University Press, 1980.

7 Rose, M. R. The mental arms race amplifier. Human Ecology, 1980, 8, 285-293.

ANIMAL COGNITION AND BEHAVIOR
Roger L. Mellgren, editor
© *North-Holland Publishing Company, 1983*

THE EFFECTS OF THE AVAILABILITY OF LINGUISTIC LABELS ON CONCURRENT DISCRIMINATION LEARNING BY AN ASL RESPONDING CHIMPANZEE

George H. Kimball, N. Jack Kanak
University of Oklahoma

and

Roger S. Fouts
Central Washington University

Since the early 1970's, most of the research reported on non-human primates use of symbolic "language" has concentrated on describing or demonstrating linguistic attributes deemed essential or intrinsic in human language use. As a starting point, this was probably reasonable given the Gardners' (Gardner and Gardner, 1969, 1971) initial reports of the success of Project Washoe. Prior to that time, all previous efforts to teach chimpanzees to "talk" had been failures (e.g., the Kelloggs and Hayes), and the inability of chimps to utilize vocal language had been conveniently ascribed to "insufficient cognitive capacities." The obvious success of Project Washoe, and the subsequent reports of other projects, utilizing both American Sign Language (ASL) (Fouts, 1973, 1974, 1976, 1977; Fouts, Chown, Kimball and Couch 1975; Fouts et al., 1976), and other symbolic systems (Rumbaugh, 1977; Premack, 1972, etc.) triggered attempts to define "language" in such a way as to exclude all other species "utterances," yet not exclude some human "language users" as well. In retrospect, this was probably a very fortuitous event, in that the emerging chimpanzee literature illuminated how ill-defined and myopic our definitions of human language really were. Obviously, we, as a species, were much more sophisicated in our symbolic representation, but our rigorous, "scientific," operational definitions of "language" were primarily a continuum of the ridiculous ("it's something humans do, but no other species does") to the sublime (Chomsky's "ideal" linguist, a category to which <u>no</u> human in the world truly qualifies).

When these studies first appeared, "human uniqueness" supporters analyzed the reported utterances, and categorized lists of elements the chimpanzees "had demonstrated" or "failed to demonstrate." For instance, they could <u>use</u> and <u>comprehend</u> the names of objects, but they could not "<u>reconstitute</u>" their language elements into new or unique "utterances." Therefore, they <u>do not</u> use language like a human (Bronowski and Bellugi, 1976).

After many series of experiments by the various projects attempting to attend to the specific, ever changing definitions of human language by setting-up various clever paradigms to try and demonstrate these capabilities were present in the chimpanzee, it became apparent that some of the scientific community perceived this body of research as attempts to describe a non-human species as "human language users." This was not the intent. Of course they are not human language users. They are not human. Chimpanzees will never utilize their communicating capabilities in a strictly human manner (if such a capacity can ever be defined).

Nevertheless, scientific friend and foe alike must acknowledge the amazing cognitive capacities of these remarkable creatures. The chimpanzee truly "uses" the acquired symbolic linguistic system to communicate with us about concepts, knowledge, and perceptions of the world, in a manner that is sophisticated beyond any previous acknowledgement of a species other than homo sapien.

During the late 1970's and early 1980's, other emphases began to emerge while the linguistic questions continued to be addressed. Along with the obvious interest in "chimp-to-chimp" signed (or symbols) communication (Gorcyca, Garner and Fouts, 1975; Gardner and Gardner, 1978; Savage-Rumbaugh, Rumbaugh and Boysen, 1978), and cultural transmission of the acquired signs from Washoe to her infant (Fouts, Kimball, Davis and Mellgren, 1979), other investigations began to transcend the specific questions of linguistic content. Regardless of whether or not the chimps used perfect syntax, or produced sophisticated and unique sentences, the fact remained that here was an animal who both comprehends and uses the names of objects to describe things and communicate about them. In what way does this benefit them cognitively? Numerous theories of learning, cognition, retention, etc., abound in the human literature, affording a special place in the cognitive hierarchy to humans by the virtue of our ability to cognitively manipulate these abstract symbols.

In one study, (Buchanan, Gill and Braggio, 1981) Lana (a chimpanzee sophisticated in the Yerkish symbols) demonstrated serial position effects (both primacy and recency effects) in a free recall task of lists of symbols which represent words, which is basically analogous to well-documented human performance on similar tasks (Murdock, 1962). She also showed above-chance clustering on recall of categorized lists.

Another study (Savage-Rumbaugh, Rumbaugh, and Boysen, 1978) investigated the relationships between chimpanzees internal representation (symbolization of the objects), tool use, and cooperative interanimal (chimp-to-chimp) communication required to accomplish specific tasks to acquire foods. Basically, one chimpanzee saw which of several sites was "baited." This chimp could not acquire the food without the appropriate tool (i.e., a wrench to unscrew some lug nuts). It was necessary for him to communicate to another chimpanzee what tool (i.e., keys, money, wrench, etc.) would be appropriate to acquire the food. The second chimp would read the utterance, select the tool, and pass it to the "director" chimp, who would then procure the food and share with chimp #2. When their keyboard (Yerkish symbols) was not functioning, they adopted the strategy of either cycling through all the tools, or repeatedly offering the same tool. Clearly, they were "using" their communication system both proficiently and accurately.

In the past, investigations of discrimination learning phenomena have demonstrated many comparable learning capabilities and strategies between man and the higher primates, as well as illuminating some differences. Analogous higher-order capabilities (which at one time, like language, were thought to be species-specific to man) such as the acquisition of a "learning set" or "learning to learn" have been demonstrated in many animal studies, as well as comparable performances across species using complex concurrent discrimination procedures.

In one of the earliest studies, Harlow (1949) showed that once appropriate reaction sets had been formed in monkeys, the sets may be transferred from one pair of discrimination objects to another, making it possible for the subjects to meet a strict criterion for formation of a discrimination with a minimum amount of specific training. This "learning set" phenomenon in primates was subsequently demonstrated in many experiments (Riopelle and Copeland, 1954; Meyer, 1951). Thune (1950) and Meyer and Miles (1953) showed that human subjects also improve their rate of acquisition in the course of learning several lists. This improvement was termed "learning to learn" and was considered analogous to the learning set phenomenon observed in monkeys.

Analysis of discrimination learning capacities has also provided some comparative data. The "serial discrimination" procedures introduced by Riopelle, Harlow, Settlage and Ades (1951) made possible a more direct comparison between human discrimination learning and animal discrimination learning. The serial discrimination procedure corresponds to the procedures employed in human verbal learning experiments (i.e., training on several object pairs in succession). Unlike the

more typical (successive) discrimination learning procedure
utilized in animal experiments, in which a given object pair
is presented on several trials in immediate succession, the
serial discrimination task involves training on several other
object pairs interposed between the successive trials of an
individual problem. One form of this serial procedure is partly analogous to serial verbal learning, i.e., the object pairs
are presented in a constant order on each run through the
training set. Another form of the procedure is more analogous
to paired associate learning and is essentially identical to
verbal discrimination paradigms, i.e., using varied orders of
presentation on successive runs. Since Leary (1957) has shown
that constant and varied orders of presentation result in
essentially equal performance by monkeys, Bessemer (1971) concluded that it is most appropriate to refer to these procedures
collectively as "concurrent discrimination" learning, using the
terminology suggested by Hayes, Thompson and Hayes (1953a).

When comparisons of these acquisition phenomena across
species between humans and animals have been made, the comparison has inevitably utilized the procedures of animal discrimination experiments. This is necessitated by the fact
that theories postulating human verbal discrimination learning
strategies generally include an overt pronunciation response
of the correct word or item. As is obvious, the ability to
use language, or more specifically the ability to name the object, had long been considered to be unique to humans. Therefore, when comparing across species, the learning tasks were
necessarily simplified and designed to minimize the contribution of any "language" component. It appears that under
these conditions where the linguistic competence of the organism plays little role, humans and the higher primates perform
roughly the same in the acquisition of concurrent lists and in
the acquisition of a learning set.

As an example, in a direct comparison across species of
learning set performance, Hayes, Thompson and Hayes (1953b)
compared eight chimpanzees (age 15 months to 26 years, naive
in learning set) to six human children (age 2 to 7 years, also
with no pre-experimental experience). Even though no verbal
response was necessary, it is possible to assume that the
possession of language, or the ability to name the test objects would provide an additional salient cue, and therefore
would enhance acquisition, relative to whatever "central representative process" the animals might use. The human subjects
did name many of the objects, but there was no indication that
this helped them to learn either the individual problems or the
general principles of the experiment. What their data did
show is that when pre-experimental experience was equated,

humans, chimpanzees, and rhesus monkeys differ little in their ability to acquire sets for object-discrimination learning. Thus, under these conditions where verbal reply was by design unnecessary for task acquisition, performance does not appear to be facilitated by the possession of language. However, for more specific human discrimination learning tasks such as learning a list of words, many predictions can be made concerning the discriminative value of the linguistic elements.

In the current study, an experimental design was selected to allow a direct comparison of the results to a verbal discrimination learning theory that emphasizes the importance of linguistic processes during list acquisition. Typically, verbal discrimination learning tasks are very similar to paired-associate tasks, except with PA tasks, the subject is presented pairs of items, and upon presentation of one item, is expected to recall the other item. In verbal discrimination learning, one item of the pair of items is designated as the correct item and the alternative is the incorrect item. When both items are presented, the subject must recognize and select the "correct" item.

The primary interest of this study was to determine if the linguistic attributes are a salient feature of the stimulus to be discriminated for the chimpanzee, as it appears to be in humans, or if species differences determine different learning strategies independent of the ability to attach labels to the stimulus objects.

Experiment 1

Subject. The subject was a young male chimpanzee, Ally. Ally was slightly over five years of age at the beginning of Experiment 1. He was born at the Institute for Primate Studies, and taken from his natural mother at the age of two months. He was raised in a human home until he rejoined the chimpanzee colony at the Institute at four and one-half years of age. At the beginning of Experiment 1, Ally had a sign language vocabulary of 120 reliable signs.

Lists. The lists used in Experiment 1 are shown in Table 1. Both lists contain 18 pairs of objects for which Ally knew the correct name sign. Each list represented three different experimental manipulations of item relationships - same right, same wrong, and control. The Same Right (SR) condition was composed of 6 of the pairs. In the SR condition the 3 correct items of the 3 pairs were re-paired with 3 new incorrect items, thus forming 6 pairs with 3 repeated correct items.

Table 1

Experiment 1
Lists and Errors Per Pair

Condi-tion	List 1	No. of Errors	List 2	No. of Errors	Reversal	No. of Errors
	Baby-Zipper	2	Leaf-Fork	2	Fork-Leaf	6
	Baby-Insect	3	Leaf-Tobacco	4	Tobacco-Leaf	2
	Hat-Cat	7	Turtle-Shocker	0	Shocker-Turtle	6
SR	Hat-Cup	3	Turtle-Bed	8	Bed-Turtle	3
	Watch-Bird	9	Radio-Car	1	Car-Radio	9
	Watch-Pen	4	Radio-Bib	0	Bib-Radio	8
	Shoe-Banana	3	Mirror-Box	0	Box-Mirror	5
	Pig-Banana	0	Glasses-Box	3	Box-Glasses	4
	Rock-Tie	2	Food-Purse	2	Purse-Food	3
SW	Pillow-Tie	0	Monkey-Purse	2	Purse-Monkey	3
	Blanket-Money	0	Bath-Frog	0	Frog-Bath	3
	Pants-Money	0	Key-Frog	4	Frog-Key	5
	Book-Comb	0	Telephone-Hair	3	Hair-Telephone	2
	Flower-Brush	7	Pin-Fruit	0	Fruit-Pin	4
	Airplane-Button	0	Cage-Tree	9	Tree-Cage	5
C	Drink-Berry	2	Toothbrush-Knife	1	Knife-Toothbrush	7
	Dog-Sweet	1	String-Chair	1	Chair-String	5
	Nut-Pipe	2	Spoon-Ball	8	Ball-Spoon	4

	Total	45		Total	48		Total	84
	Mean	2.5		Mean	2.66		Mean	4.66

Errors Per Condition:

List 1	List 2	Reversal
SR-28	SR-15	SR-34
SW- 5	SW-11	SW-23
C-12	C-22	C-27

Six pairs were also included in the Same Wrong (SW) condition, in which the incorrect items of three pairs were re-paired with new correct items. The remaining six pairs composed the Control (C) condition and were made up of mutually exclusive pairs of correct and incorrect items (each item appeared only once). Using a single subject, the more typical transfer paradigms used to manipulate frequency of occurrence of the items (i.e., where groups of subjects learn one list of items, then transfer to another list in order to determine the degree of interference or facilitation the first list provides) are inappropriate. Other experiments (Ekstrand, et al. 1966) in which the frequency of items within a single list were manipulated by varying the number of right and/or wrong item occurrences have generally supported frequency theory predictions.

Three random orders were used for both of the lists, with the restriction that no more than 2 correct items appeared in succession in a given spatial (left-right) position, and that 2 pairs of the same "treatment" (SR, SW, C) condition never appeared in succession.

<u>Apparatus</u>.   The apparatus is diagramed in Figure 1.   The humans were seated on the floor, therefore the partition was only approximately 1.2 meters in height, and was constructed of plywood. The stimulus box was constructed so it could be

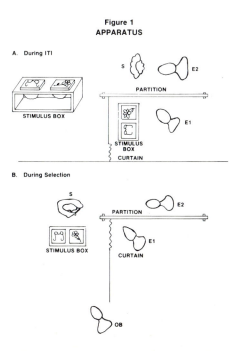

Figure 1
APPARATUS

A. During ITI

B. During Selection

moved behind the partition, and the "exemplars" (objects) were
placed on hinged lids covering the bowls containing the reward.
The stimulus box was also constructed of plywood. In addition
to the apparatus shown in Figure 1, exemplars for each word
on the lists were needed. The exemplar for sweet was a "mint"
(candy) and the exemplar for food was a "monkey biscuit." All
others are self-explanatory.

Procedure. A double-blind procedure was in effect throughout the experiment. As shown in Figure 1, Experimenter 1 (E1)
arranged the objects on the stimulus box, moved the box into
view, and pulled the curtain to prevent observation. Experimenter 2 (E2) then allowed Ally to move around the partition
and view the objects. When Ally made his response, E2 an-
nounced the sign Ally had made, and the Observer confirmed it
if he perceived the same response. In the event of an "inappropriate" response (i.e., naming an object which was something other than the pair of stimulus objects), the Observer
would say "not there" and E2 instructed Ally to choose between
the two stimulus objects. The Observer overruled E2 if he perceived a different choice than E2, or saw a sign that E2 missed.
Neither the Observer nor E2 knew which item was correct, so
when Ally made his choice, E1 would announce "_____ is
correct (or incorrect)" and E2 then raised the lids of both
objects. If Ally's choice was correct, he was allowed to take
the piece of fruit under the correct item. If Ally was incorrect, he was still shown what was under both items but was
not allowed to have the reward under the correct item. Ally
was then taken behind the partition by E2, and E1 arranged the
next pair of stimulus objects. The inter-trial interval (IT1)
was maintained at between 20-30 seconds. The only other timing
restriction imposed was when Ally came around the partition and
viewed the object pair he had only 15 seconds in which to reply.
If Ally failed to respond, or responded inappropriately, after
15 seconds the trial was recorded as a "No Response" (and was
classified as an error). Ally was then shown the correct and
incorrect item, and the Experimenters proceeded to the next
pair on the list. This procedure continued until a criterion
of two consecutive errorless trials of the entire list had
been completed. On an average, three trials (complete runs
through the entire list) per day were run. After a period of
one week following criterion on List 1, Ally began List 2, and
the same procedures were followed.

Immediately following criterion for List 2 (same as List 1
criterion) a further manipulation was performed. This manipulation consisted of a simple reversal of all the pairs of
List 2. All the items that had been previously correct were
now incorrect, thereby effectively switching the SR and SW

treatment conditions (the C condition still contained the same pairs, simply reversed). The first trial of the "Reveral" List commenced the next day with no instruction to Ally. The same procedures used for the acquisition of List 1 and List 2 were maintained throughout the learning of the Reversal List.

Other constraints and controls. The frequency theory's (Ekstrand, Wallace and Underwood, 1966) straightforward pre-mises allow specific predictions of verbal discrimination learn-ing performance to be made under various list manipulations and transfer conditions. Because the single subject design of this study necessitated within-list manipulation of the items rather than transfer between lists, a brief description of frequency theory constructs, and a discussion of particular variables which have been empirically shown to affect acquisi-tion via frequency theory is appropriate.

The frequency theory assumes that the cue for discrimina-tion of the presented item pairs is provided by the subject's perceived difference in the frequency of occurrence between the right and wrong members of a verbal discrimination pair. The theory states that if two members of a verbal discrimina-tion item pair are of equal frequency, then liguistically add-ing a hypothetical "frequency" unit to either member should make the pair more discriminable. Ekstrand et al. assume these units may be added in four ways. First, the representational response (RR) is assumed to add a frequency unit as a function of the subject's perception of the verbal units to be dis-criminated. Thus, in a two-alternative discrimination, the RR would add a unit of frequency to each member of the pair (the ratio of frequency units between Right items and Wrong items would now be R:W, 1:1). Second, it is assumed that the pro-nunciation response (PR) adds another unit of frequency to the item in the pair which is pronounced by the subject (assuming the correct item is picked, the R:W ratio is now 2:1). Third, it is hypothesized that the next response to add a frequency unit is the rehearsal-of-the-correct-response (RCR). The RCR adds a frequency unit via covert or overt pronunciation of the correct alternative during the feedback interval (R:W would now be 3:1). The fourth way frequency is added is through the implicit associative response (IAR). When word associates are present in a list, the perception of one member of the word associate pair is assumed to elicit its associate. Thus, the presentation of the word "king" in a list would presumably add frequency to its primary associate, "queen," via the IAR. Therefore, in this situation, the item pair could have built up a 4:1 discrimination ratio, making the item labelled "4" easier to be picked on subsequent trials.

Ekstrand et al. (1966) also provided for two ways of responding to frequency differences. Rule 1 responding is said to occur when the subject chooses the most frequent alternative, and Rule 2 responding occurs when the subject chooses the least frequent alternative. Usually Rule 1 responding is most appropriate, but it has been shown that in a few "transfer" situations, and in special "mixed-list" manipulations, Rule 2 responding may be more appropriate.

In a review of the literature, Eckert and Kanak (1974) categorized the research generated by frequency theory into three areas: Acquisition, Transfer, and Retention. As is the case with most theories, contradictory results to frequency theory have been demonstrated by manipulating certain aspects of the verbal discrimination task. Due to these contradictory findings, the variables that have been shown to affect acquisition need to be addressed, as well as an explanation of the list manipulations in this study, and our attempts to control for these variables in the experiment utilizing the chimpanzee.

The variables affecting acquisition are numerous. Basic to the frequency theory is the acquisition of frequency units. Subsidiary to the aforementioned responses that add frequency units, is the implication that the subject's discrimination of the frequency unit differential does not depend on the way in which the frequency difference accumulates. Eckert and Kanak (1974) identified four ways frequency has been manipulated to test these assumptions: (1) by permitting the subject to familiarize himself with either the right or wrong item, or both, prior to acquisition of the list, (2) by experimentally manipulating the number of times a particular right or wrong item appears in a list, (3) by selecting words of different normal frequency usage from "normative" sources, and (4) by varying the percentage of occurrence of the response member in the feedback interval. In Experiment 1, all the items on the lists were already familiar to the subject, but to test for equal familiarity with the names of the objects, all the items were presented to the subject, and Ally was asked to name each item by signing. The items were presented in a random order until he had responded to each item ten times. Any item not responded to correctly 100% of the time was labeled a low frequency item, and was re-exposed until 100% accuracy was maintained for the entire list. As mentioned earlier, a "mixed" list was used to vary the number of times the designated correct or incorrect items would appear, to generate predictions based on testing frequency theory with humans. The items originally designated as lower frequency items were also controlled so that they were equally divided as correct and incorrect items. During the feedback interval, both the correct and incorrect item were identified (by a piece of fruit) to the subject.

Another variable affecting acquisition is the "Implicit Associative Response" (IAR). The adding of a frequency unit by association with another word on the list such as "king-queen" is somewhat different when using sign language rather than a verbal response. Studies using deaf subjects (Putnam, Iscoe and Young, 1962; Youniss, Feil and Furth, 1965) manipulated similarity of the words on the basis of the sign commonly used to portray that word. For example, "pretty" and "beautiful" have identical signs and therefore are high similarity words. In Experiment 1, highly similar signs were equally divided between correct and incorrect items and were never paired with each other.

"Imagery" has also been found to be an important variable during acquisition. Several researchers (Paivio and Rowe, 1968; Rowe and Paivio, 1971; and Ulrich and Balogh, 1972) found that list acquisition was facilitated by the use of high image pairs, whether mixed or unmixed lists were used. Since physical exemplars were used on our lists, rather than written words or pictures of the signs, the effect of imagery is distributed across conditions and should be negligible.

"Verbalization of the perceived correct item" is another variable basic to the frequency theory, in that a hypothetical frequency unit is assumed to be added. For the most part, the data regarding the effects of verbalization on the acquisition of a verbal discrimination list are generally consistent with derivations of the frequency theory. However, Wilder and Levin (1973) compared pronouncing vs. pointing as the method of choice to picture pair discriminations in nursery school children, fifth graders, and college students. They found pronunciation facilitated learning for the nursery schoolers, but there was no difference on the fifth graders or college students. In Experiment 1, it was necessary for the chimpanzee to "sign" the name of the objects of his choice, rather than simply pointing at his selection.

"Feedback conditions" have also been a source of theoretical interest. According to the frequency theory predictions, presentation of both correct and incorrect items in feedback would impede learning, as opposed to only presenting the correct item. During the feedback interval in Experiment 1, both the correct and incorrect items were present, and reward of a piece of fruit differentiated the correct from the incorrect item.

Results and Discussion

The results of Experiment 1 (see Figure 2) clearly indicate that Ally did not learn the list as the frequency theory
would predict a human would. An analysis of variance (using
the Shine and Bower, 1971, model for single subject designs)

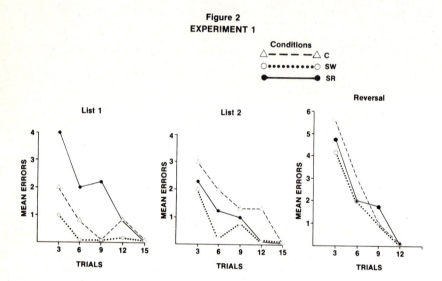

Figure 2
EXPERIMENT 1

Conditions
△— — — —△ C
○•••••••••○ SW
●————————● SR

List 1   List 2   Reversal

MEAN ERRORS / TRIALS

revealed there was a significant ($p < .001$) difference across
treatment conditions with the Same Wrong (SW) condition resulting in less errors on both List 1 and List 2. The Same Right
(SR) and Control conditions flip-flopped across lists implying
that equal information was available in these conditions.

One possible interpretation of the acquisition data for
Lists 1 and 2, suggests that Ally was utilizing what has been
called an "avoidance" type learning strategy (i.e., avoid S-
rather than approach S+). If he was using this type of
strategy, more information would be available to him in the SW
condition, and essentially equal amounts of information would
be available in the SR and C conditions (due to non-repetition
of the S- items). Other research (Harlow, 1959; Cross and
Brown, 1965) has shown that performance of a nonhuman primate
on an object discrimination task is enhanced by prior familiarization of the S- object, as opposed to prior familiarization
of the S+ object or a control condition. Cross and Vaughter
(1966) found the same effect in human children under the age
of 4 1/2 years. D'Amato and Jagoda (1961) have suggested that
mechanisms of inhibition enhance survival in primitive or
developmentally immature organisms, and therefore are adaptive.

The learning of the Reversed List 2 (Reversal Phase) does not appear to follow the same strategy. If Ally persisted in using an avoidance strategy, the old SR condition (which would be the new SW condition) should have been the easiest to learn. However, these reversal phase data show no statistically significant difference across conditions. One possible explanation of this apparent switch of strategy, may be due to the fact that this was not a novel list to be learned (with respect to new items being added). Ally presumably had already learned the "associative" relationships between the pairs of items as an inviolate whole. Simply avoiding the wrong item of each pair, or a "win-stay, lose-shift strategy" (Levine, 1959), could be the most appropriate or parsimonous strategy and would show no difference across conditions.

Another possibility is that the difference noted in List 1 and List 2 acquisition across conditions, could be due to differential saliency of the stimulus objects, or specific object preferences, independent of the potential linguistic information available. Perseverance in responding to preferred objects would obviously influence the effect of the experimentally manipulated conditions. A post hoc analysis of the errors for each pair (see Table 1) indicates at first glance that this might have been initially a rather powerful influence. However, the Reversal Phase again provides interpretative clues. If stimulus saliency (or particular object preference) was the overriding factor in Ally's selections, then those stimuli with the most errors during the learning of the list could be expected to be practically errorless in Reversal after the first or second response. The data for the Reversal Phase do not support this contention, as the three most common errors during List 2 acquisition were not the three least common errors during the Reversal Phase. They were all near the mean of 4.66 errors per pair.

Experiment 2

Although Experiment 1 did demonstrate that a chimpanzee can learn concurrent discrimination lists using ASL as his method of responding, it does not truly prove that his "linguistic" skills facilitated learning the lists. Object preferences paired with a simple avoidance strategy could have been the salient discriminative features, rather than the linguistic attributes in the associative pairings. Experiment 2 attempts to determine the effect knowledge of the name of the objects had upon the rate of list acquisition.

Methods and Procedures

In Experiment 2, Ally learned to discriminate a new list of 18 pairs of objects, all of which he had no sign for in his vocabulary. The treatment conditions remained the same, as well as the procedures, with one exception. For this list, Ally had only to "point" at the object he believed to be correct, rather than attempt to name either item. If his learning strategy was primarily avoidance of the incorrect object, rather than simple item preferences, the basic results by conditions (superiority of the SW condition) should remain approximately the same. Further, analysis of the comparative rate of list acquisition from Experiment 1 to Experiment 2 should give some indication as to the effect the knowledge of the name of the objects has in mediating, or establishing, the associations between the object pairs. Presumably, all other factors being equal, the linguistic tag (available in Experiment 1) would provide an additional discriminable feature to the association, thereby facilitating discrimination and producing more rapid list acquisition.

Results and Discussion

The results of Experiment 2 (see Figures 3 and Table 2) show that this list was obviously more difficult to learn in terms of (1) trials to criterion (Experiment 1-15 trials, Experiment 2-24 trials), (2) total number of errors, (Experiment 1, List 1 and 2 averaged 46.5 errors; Experiment 2, 87 errors), and (3) errors per condition. Again the Shine and Bower (1971) N of 1 ANOVA showed that there was a significant

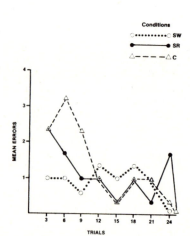

Figure 3
EXPERIMENT 2

Table 2

Lists and Erors Per Pair

Condition

SR	Knob-Ring	3
	Knob-Coil	3
	Handle-Baseplate	8
	Handle-Sil Cock	5
	Bolt-Sheep	4
	Bolt-Pepper Shaker	6
SW	Battery-Toggle Switch	3
	LightBulb-Toggle Switch	5
	Bottle-Clothes Pin	3
	Toy-Clothes Pin	0
	Shell-Shoe Horn	10
	Ink Bottle-Shoe Horn	2
C	Tin Container-Phone Jack	4
	Flower Pot-Film Canister	7
	Eraser-Cork	5
	Compass-Fuse	5
	Toilet Stopper-Glue Tube	6
	Net-Matchbox	8
	Total Errors	87
	Mean	4.8

Errors Per Condition:

SR-29
SW-23
C-35

(p < .05) difference in conditions. Analysis of the first 9 trials showed initial SW superiority (avoidance) again, but in later trials the errors were randomly distributed across conditions. However, concluding that the other lists (Experiment 1) were learned more rapidly due to the linguistic significance of the objects is still somewhat tenuous. It is possible that the difference in trials to criterion could still simply be a function of object familiarity (or meaningfulness) as he obviously had spent more "pre-experimental" time with and around the objects he had name signs for (Experiment 1).

Experiment 3

Experiment 3 was designed to balance more equally the subject's pre-experimental exposure to the stimulus objects, as well as providing more evidence on the effect of linguistic versus non-linguistic mediation of the association between the objects.

Methods and Procedures

In Experiment 3, the 16 pair task (see Table 4) was equally divided between signed (S) and unsigned (US) objects (16 in each condition). All of the objects were unfamiliar to Ally initially. Ally was taught the name sign for 16 of the objects, via the methods described by Fouts (1973). The amount of time and reinforcement necessary before Ally reached reliability on the signs (defined by Gardner and Gardner, 1969, and Fouts 1973, as 15 consecutive errorless responses) was carefully recorded, and yoked to specific objects to which no name signs were taught. The yoked objects (US) were played with, and familiarized, for an identical length of time, and were associated with fruit, tickling, hugging, etc., as their corresponding signed objects (S) were, during acquisition of the name signs.

Table 3

Experiment 3 Design
CORRECT OBJECT

	UNSIGNED	SIGNED
INCORRECT OBJECT — UNSIGNED	Both Objects Unsigned (USB) Unsigned Object--Correct Unsigned Object--Incorrect	Signed Objects Correct (SC) Signed Object--Correct Unsigned Object--Incorrect
INCORRECT OBJECT — SIGNED	Signed Object Incorrect (SI) Signed Object--Incorrect Unsigned Object--Correct	Both Objects Signed (SB) Signed Object--Correct Signed Object--Incorrect

The objects were distributed in a 2 x 2 factorial (see Table 3) with the following limitations and manipulations: (1) S objects slow in reaching the sign-production criterion were paired with US objects that received equivalent extended familiarization time (2) no objects were re-paired within the list, and (3) S and US objects were equally divided as correct (C) and incorrect (IC).

Results and Discussion

The results (see Figure 4 and Table 4) show that this was apparently the most difficult list for Ally to learn. Trials to criterion (30) and total errors (175) were substantially

Table 4

Experiment 3
Lists and Errors Per Pair

Condition			
SI (Signed Object Incorrect)	Feather Duster	Boat	18
	Curler	Horse	14
	Film Reel	Glove	7
	Candle Stick Holder	Paint Brush	6
SC (Signed Object Correct)	Bean	Hoze Nozzle	15
	Wood	Large Piece (Ball)	4
	Cheese	Sifter	16
	Camera	Orange Ring	3
SB (Both Objects Signed)	Letter	Elephant	14
	Bee	Chain	12
	Collar	Bear	9
	Car	Train	7
USB (Both Objects Unsigned)	Moto Piece	Slide Rule	18
	Hose End	Fish Reel	8
	Cork	Pretzel	14
	Screwdriver	Small Piece (Ball)	10
		Total Errors	175
		Mean	10.9

Errors Per Condition:

      SI-45
      SC-38
      SB-42
     USB-50

increased, even though this list contained only 16 pairs of
items as opposed to 18 (in Experiment 1 and 2). (Experiment 1:
Lists 1 and 2, 15 trials to criterion, 46.5 total errors;
Experiment 2: 24 trials to criterion, 97 total errors). Again,
the N of 1 ANOVA showed a significant (p < .01) difference in
errors per condition.

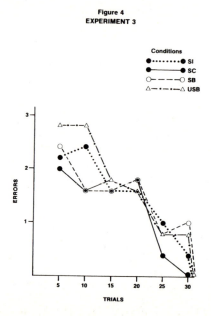

Figure 4
EXPERIMENT 3

Experiment 3 was designed as an attempt to resolve some of the
difficulties involved in the interpretation of the results of
Experiment 1 and 2. These difficulties were a function of the
inability to assess the impact of differences in stimulus
familiarity (and presumably meaningfulness) due to the unequated
pre-experimental exposure of the objects to the subject.

The compound list design in Experiment 3 (utilizing both Signed
and Unsigned objects) does however produce some constraints
that need to be considered in an interpretation of performance.

Because this was, in effect, the fifth list Ally had learned over a period of time, it is possible the increased errors in discrimination could have been due to proactive interference. On the other hand, since this was a novel list, with no items repeated from any previous list, the learning of this list is more analogous to a typical A-B, C-D non-specific transfer paradigm, so the influence of proactive interference should be small.

If, as it seems in Experiment 1 and Experiment 2, Ally's primary strategy for discrimination learning is avoidance, with any available linguistic tags potentially used as mediators in the object associations, the Signed Object Incorrect (SI) and the Signed Object Both (correct and incorrect) (SB) conditions should be initially superior to the Signed Object Correct (SC) and Both Objects Unsigned (USB) conditions. Assuming Ally perceives the linguistic tag as an additional or salient discriminable stimulus for the pair (as indicated in the trials to criterion of Experiment 2 compared to Experiment 1), the SI and SB conditions would initially provide the most information for an avoidance strategy because a signed object occupies the S- position. However, persevering in attending to the linguistic element could produce decremental results across time in the SI condition, due to it being an ineffectual strategy. Accordingly, the absence of discriminable linguistic features should make the USB condition hardest to learn initially. Linguistic mediation in combination with avoidance should favor most effective learning in the SB and SC conditions, where avoiding the incorrect response would lead to selection of the correct object (which would be a signed object).

Had Ally's strategy been strictly avoidance, with no attention to linguistic mediation, there should be no differences in acquisition across conditions because there are no repaired S- items (resulting in essentially equal available information) and all items were equated for familiarity prior to acquisition.

Analysis of the first 15 trials in greater detail (see Figure 5) appears to support the avoidance strategy accompanied with linguistic mediation (after the 15th trial, all conditions approached criterion at approximately the same rate). Initially, the SI condition was equivalent to the SC condition. The predicted decrement in performance (increased errors) caused by attending to the linguistic elements in the SI condition is apparent in trials 6-9, while the conditions containing a "signed" object in the correct position (SC and SB do not show this decrement.

Figure 5
EXPERIMENT 3

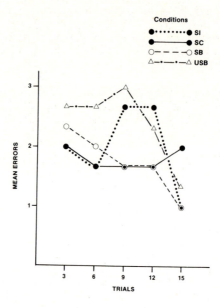

Conditions

●·······● SI
●————● SC
○— — —○ SB
△—·—·—△ USB

MEAN ERRORS

TRIALS

Additional evidence of Ally's attention to the linguistic elements is provided by both analysis of the early errors within the SI condition, and observational data (signing the name of the incorrect object) collected during the experiment. If the increasing number of errors in the SI condition were caused by attention to the linguistic tags of the incorrect objects rather than simple item preferences, the effect should be generalized (randomly distributed) across all pairs within the SI condition. Analysis of the first 10 trials shows that "boat" was missed 7 times, while "horse," "glove," and "paintbrush" were missed 5 times each. In addition, during These 22 errors, Ally <u>named</u> the incorrect object 10 times, even though pointing <u>was all</u> that was necessary. Only <u>once</u> did he name the (S-) object of the pair, and he then chose the correct (S+) (unsigned) object.

The observational data that were recorded by the observer, show that attention to the linguistic element persevered a long time. As late as trials 25, 26, and 29 Ally still signed "boat," "glove," and "elephant" and chose these incorrect items. On another trial, he signed "boat," was shown that the "feather duster" was correct, yet still was observed re-hearsing "boat" while returning around the partition for the ITI.

Conversely, the pair he missed the least was "camera"--"orange ring" (SC condition). On Trial 1, he signed and picked "camera," and thereafter (17 of the first 19 trials) signed "camera" (making only 3 errors, once signing "camera" and choosing the "orange ring").

Ally was also observed rehearsing the names of the objects in the SC and SB conditions many times, occasionally rehearsing the incorrect item as well as the correct item (in the SB condition).

Finally, Ally also occasionally attempted to "name" an unsigned item (for example he called the "feather duster" a "flower").

Obviously, Ally considered the linguistic tag to be an important feature of the stimulus objects to be discriminated. The results of Experiment 3 in conjunction with the superiority of rate of acquisition of the Experiment 1 lists over the Experiment 2 list, indicate that knowing the name of an object can facilitate discrimination learning in a chimpanzee. It is important to note that having acquired a rudimentary linguistic communication system does not seem to change the species tendency toward learning strategies based on an avoidance principle (rather than approach). Rather, linguistic labeling apparently enhances the discriminability of what to avoid. It is also important to point out that related research has shown this sort of phenomenon to be age or maturity dependent in humans.

As mentioned earlier, Harlow (1959) showed that the performance of nonhuman primates given information trials to a nonrewarded object (S-) will, in subsequent two-object discrimination trials, be superior to animals given comparable information trials with a rewarded object (S+). This phenomenon has been called the Moss-Harlow effect (Moss and Harlow, 1947). Cross and Vaughter (1966) subsequently demonstrated that the Moss-Harlow effect was apparently age-dependent for human children. Children under the age of 4 1/2 years showed improved performance in discrimination learning when presented with prior negative information trials (like nonhuman primates), while children older than 4 1/2 showed superior acquisition after positive information trials (replicating Vaughter and Cross, 1965). This led the authors to conclude that the emerging role of language facilitates the transfer of learning strategies from a more primitive or "associative" mode of learning to a more "cognitive" mode (e.g., as described by

Kendler and Kendler, 1962, and White, 1965). It is also hypothesized by these authors that the reason the nonhuman primate
remains in an avoidance strategy of learning rather than a more
cognitive (approach) mode, could be because the symbolic manipulation of language units is not possible.

Kendler and Kendler (1962) characterized this cognitive transition period as that period when the human child begins to
have the capability to be guided by a mediating response to the
presented stimulus, as well as by the stimulus itself. Kendler
and Kendler further feel that the usual, but not only, mediation
is linguistic in nature. Even though children can label cues
prior to the transition point, they apparently do not utilize
these cues when problem solving. Reese (1962) states that
younger children have a "mediational deficiency" because overlearning of a word must take place, before that word can be used
to mediate.

Kendler and Kendler (1962) also state the act of mediating is
part of a "horizontal process" (represented by a series of S-R
behavioral chains). Ontogentic development then establishes
linguistic "vertical" (referring to the fact that the horizontal
chains are occurring simultaneously) connections between horizontal processes. Older children can, given this vertical
connection, use the word they generate to select stimulus and
response components from different behavioral chains. In so
doing, their behavior supposedly becomes "conceptual" (Kendler,
1964).

It remains to be seen whether chimpanzees, as they mature and
their "language" vocabularies mature and expand, demonstrate
these traits in cognition which have been labeled "more cognitive" or "less primitive." As more chimpanzee subjects (who
have comprehended and produced ASL since infancy) become available, and as their vocabularies and environmental interactions
mature, further light may be shed on the cognitive consequences
of language acquisition by a non-human species, as well as providing evidence to the appropriateness or generalizability of
these theories to humans.

Acknowledgements

The authors would like to acknowledge and thank the following individuals for their contribution to this study. Thanks to Dr. Joe B. Couch and Dr. Roger Mellgren for their assistance in the design, implementation, and interpretation of these data. Additional thanks to Dr. W. B. Lemmon and the staff of the Institute for Primate Studies at the University of Oklahoma, and the numerous undergraduate and graduate students who assisted in observing, filming, and data collection. And special thanks to Glenda Smith for the preparation of this manuscript.

References

1 Bessemer, D. W. & Stolinitz, F. Retention of discrimination and an analysis of learning set. In A. Schrier & F. Stollinitz (Eds.), Behavior of Non-Human Primates. New York: Academic Press, 1971, 1-57.

2 Bronowski, J. & Bellugi, U. Language, name and concept. Science, 1970, 168, 669-673.

3 Buchanan, J. P., Gill, T. V., & Braggio, J. T. Serial position and clustering effects in a chimpanzee's "free recall," Memory and Cognition, 1981, 9, 651-660.

4 Cross, H. A. & Brown, L. T. Discrimination reversal learning in squirrel monkeys as a function of number of acquisition trials and prereversal experience. Journal of Comparative and Physiological Psychology, 1965, 59, 429-451.

5 Cross, H. A. & Vaughter, R. M. The Moss-Harlow effect in preschool children as a function of age. Journal of Experimental Child Psychology, 1966, 4, 28-284.

6 D'Amato, M. R. & Jagoda, H. Analysis of the role of overlearning in discrimination reversal. Journal of Experimental Psychology, 1961, 45-50.

7 Eckert, E. & Kanak, N. J. Verbal learning: A review of the acquisition, transfer, and retention literature through 1972. Psychological Bulletin, 1974, 81, 582-607.

8 Ekstrand, B. R., Wallace, W. P., & Underwood, B. J. A frequency theory of verbal discrimination learning. Psychological Review, 1966, 73, 566-578.

9 Fouts, R. S. Acquisition and testing of gestural signs in four young chimpanzees. Science, 1973, 180, 978-980.

10 Fouts, R. S. Language: Origin, definitions and chimpanzees. Journal of Human Evolution, 1974, 3, 475-482.

11 Fouts, R. S. Comparison of sign language projects and implications for language origins. In Harnad, Steklis & Lancaster (Eds.), Origins and Evolution of Language and Speech. New York, New York Academy of Sciences, 1976, 589-591.

12 Fouts, R. S. Artificial and human language acquisition
in the chimpanzee. In D. Hamburg & J. Goodall (Eds.),
Perspective on Human Evolution V. Berkley, California,
1977. (Burg Wartenstein Symposium, No. 62, July, 1974,
Gloggnitz, Austria.)

13 Fouts, R. S., Chown, W., Kimball, G. H., & Couch, J. B.
Grammatical production of novel prepositional phrases
by a chimpanzee using American Sign Language. Paper
presented at the Psychonomic Society Conference in Denver,
Colorado, November, 1975.

14 Fouts, R. S., Chown, W., Kimball, G. H., & Couch, J. B.
Comprehension and production of American Sign Language by
a chimpanzee (Pan). Paper presented at the XXI Inter-
national Congress of Psychology in Paris, France, July,
1976.

15 Fouts, R. S., Kimball, G. H., Davis, D. L., & Mellgren, R.
Washoe and her children. Paper presented at the Psycho-
nomics Society Conference in Phoenix, Arizona, November,
1979.

16 Gardner, R. A. & Gardner, B. T. Teaching sign language
to a chimpanzee. Science, 1969, 165, 664-672.

17 Gardner, B. T. & Gardner, R. A. Two way communication
with an infant chimpanzee. In Schrier and Stollnitz
(Eds.), Behavior of Nonhuman Primates. New York:
Academic Press, 1971, 117-183.

18 Gardner, B. T. & Gardner, R. A. Emergence of language.
Paper presented at the American Association for the Advancement of Science, Washington, D.C., 1978.

19 Gorcyca, D., Garner, P. H., & Fouts, R. S. Deaf children
and chimpanzees: A comparative sociolinguistic investigation. Paper presented at the Speech Communication
Association Meetings in Houston, Texas, December, 1975.

20 Gorcyca, D., Garner, P. H., & Fouts, R. S. Deaf children
and chimpanzees: A comparative sociolinguistic investigation. In N. Ribhie Key (Ed.), Nonverbal Communication
Today: Current Research; In Joshua A. Fishman (Editor-in-
Chief), The Hague: Mouton Publishers, 1977.

21 Harlow, H. F. The formation of learning sets. Psychological Review, 1949, 56, 51-65.

22 Harlow, H. F. Learning set and error factor theory. In
    S. Koch (Ed.), Psychology A Study of a Science, Vol. 2,
    New York: McGraw-Hill, 1959, 492-537.

23 Hayes, K., Thompson, R., & Hayes, C. Discrimination learn-
    ing set in chimpanzees. Journal of Comparative and
    Physiological Psychology, 1953, 46, 105-107 (a).

24 Hayes, K., Thompson, R., & Hayes, C. Discrimination learn-
    ing set in chimpanzees. Journal of Comparative and
    Physiological Psychology, 1953, 46, 99-104 (b).

25 Kendler, H. H. The concept of the concept. In A. W.
    Melton (Ed.), Categories of Human Learning. New York:
    Academic Press, 1964.

26 Kendler, H. H. & Kendler, T. S. Vertical and horizontal
    processes in problem solving. Psychological Review, 1962,
    69, 1-16.

27 Meyer, D. R. Food deprivation and discrimination reversal
    learning by monkeys. Journal of Experimental Psychology,
    1951, 41, 10-16.

28 Meyer, D. R. & Miles, R. C. Intralist-interlist relations
    in verbal learning. Journal of Experimental Psychology,
    1953, 45, 109-115.

29 Moss, E. & Harlow, H. F. The role of reward in discrimi-
    nation learning in monkeys. Journal of Comparative and
    Physiological Psychology, 1947, 40, 333-342.

30 Paivio, A. & Rowe, E. J. Noun imagery frequency and mean-
    ingfulness in verbal discrimination. Journal of Experi-
    mental Psychology, 1968, 76.

31 Premack, D. Teaching language to an ape. Scientific
    American, 1972, 227, 92-99.

32 Putnam, V., Iscoe, I., & Young, R. Verbal learning in the
    deaf. Journal of Comparative and Physiological Psychology,
    1962, 55, 843-846.

33 Reese, H. W. Verbal mediation as a function of age level.
    Psychological Bulletin, 1962, 59, 502-509.

34 Riopelle, A. J., Harlow, H. F., Settlage, P. H., & Ades,
    H. W. Performance of normal and operated monkeys on visual
    learning tests. Journal of Comparative and Physiological
    Psychology, 1951, 44, 283-289.

35 Riopelle, A. J. & Copeland, E. L. Discrimination reversal
    to a sign. Journal of Experimental Psychology, 1954,
    48, 143-145.

36 Rowe, E. J. & Paivio, A. Word frequency and imagery effects
    in verbal discrimination learning. Journal of Experimental
    Psychology, 1971, 88, 319-326.

37 Rumbaugh, D. M. (Ed.), Language learning by a chimpanzee.
    New York: Academic Press, 1977.

38 Savage-Rumbaugh, E. S., Rumbaugh, D. M., & Boysen, S.
    Linguistically mediated tool use and exchange by chimpan-
    zees (pan troglodytes). The Behavioral and Brain Sciences,
    1978, 4, 539-554.

39 Shine, L. C. & Bower, S. M. A one-way analysis of variance
    for single subject designs. Educational and Psychological
    Measurement, 1971, 31, 105-113.

40 Thune, L. E. The effect of different types of preliminary
    activities on subsequent learning of paired-associate
    material. Journal of Experimental Psychology, 1950, 40,
    423-438.

41 Ulrich, J. R. & Balogh, B. A. Imagery and meaningfulness
    of right and wrong items in verbal discrimination learn-
    ing. Psychonomic Science, 1972, 29, 68-70.

42 Vaughter, R. M. & Cross, H. A. Discrimination reversal
    performance in children as a function of prereversal
    experience and overlearning. Psychonomic Science, 1965,
    2, 363-364.

43 White, S. H. Evidence for a hierarchical arrangement
    of learning processes. In L. P. Lipsitt & C. C. Spiker
    (Eds.), Advances in Child Development and Behavior.
    Vol. 2, New York: Academic Press, 1965.

44 Wilder, L. & Levin, J. R. A developmental study of pro-
    nouncing responses in the discrimination learning of words
    and pictures. Journal of Experimental Psychology, 1973,
    15, 278-286.

45 Youniss, J., Feil, R. N., & Furth, H. G. Discrimination of
    verbal material as a function of intrapair similarity
    in deaf and hearing subjects. Journal of Educational
    Psychology, 1965, 56, 184-190.

ANIMAL COGNITION AND BEHAVIOR
Roger L. Mellgren, editor
© North-Holland Publishing Company, 1983

TOWARD A COMPARATIVE PSYCHOLOGY OF MIND

Gordon G. Gallup, Jr.

State University of New York at Albany

Introduction

Mind has been an elusive phenomenon in psychology. Until recently the concept of mind was eschewed by most psychologists because as traditionally conceived it was devoid of any empirical status. Yet in our day to day discourse we use the term freely. Statements such as "What's on your mind?", or "I can't make up my mind" are commonplace, even among psychologists who disavow the existence of mind. Coming into their first course in psychology many students think psychology is the study of mind, but usually within the first week they discover that psychologists feel that mind is somehow outside the realm of science. For instance, I used to tell students in introductory psychology that no one ever heard, saw, tasted, or touched a mind. So while minds may exist, they are inaccessible. Obviously, judging from the title of this chapter, I have "changed my mind." However, there are psychologists who talk as if mental events did not exist, and even contend that external contingencies provide a complete account of behavior. Someone once characterized the history of psychology and the rise of behaviorism by saying that psychology first lost its soul, then it lost its consciousness, and now it seems to be in danger of losing its mind.

In a sense, what I will attempt in this chapter is a resurrection of mind in the context of comparative psychology. In order to salvage the concept of mind, and anticipate reactions from my critics, at least three things have to be accomplished. First, it has to be given some fairly solid operational anchor points. In other words, a set of procedures and empirical markers must be established with some degree of consistency and consensus for purposes of defining and identifying instances of mind. In order to be useful it would also have to be shown that the concept of mind will serve to integrate and organize existing information. Finally, in order to gain scientific credibility one would have to demonstrate that the formulation permitted the derivation of specific, testable predictions about how the presence or absence of mind ought to effect behavior. In the last analysis the value of cognitive interpretations of behavior rests on the extent to which they have heuristic value.

Self-Awareness

What I intend to show is that self-awareness subsumes both consciousness and mind. Indeed, the tendency to treat these as separate entities has probably confused rather than clarified issues about cognition. In my view self-awareness, consciousness, and mind are all a reflection of the same process (Gallup, 1982). But before attempting an integration of these phenomena I will need to start with an extended review of self-awareness.

Whereas self-awareness is pretty much taken for granted in humans, the problem of trying to assess this capacity in animals is formidable. Some psychologists feel that the question of self-awareness is simply not amenable to objective study in other organisms (e.g., Gardiner, 1974; Klüver, 1933), others feel that humans are unique in their capacity for self-conception (e.g., Buss, 1973; Kinget, 1975), and there are even those who argue that apparent instances of self-aware behavior in both humans and nonhumans can be subsumed by a simple reinforcement analysis (e.g., Epstein, Lanza and Skinner, 1981).

However, the fact that most of us do take self-awareness in humans for granted has made us rather careless and sloppy in terms of our thinking about this issue (Gallup, 1975). In order to get a solid conceptual as well as an empirical handle on self-awareness, we need to define the concept in such a way as to permit an objective analysis. To that end, I would propose the following definition. An organism is self-aware to the extent that it can be shown capable of becoming the object of its own attention. Now I will show how this can be translated into the world of tangibles.

One way to assess an organism's capacity to become the object of its own attention is to confront it with a mirror. Ironically, my interest in mirrors stems from a course requirement. As a graduate student I took a course in experimental psychology which required that each student conduct an independent research project. Quite by accident, one morning I found myself contemplating what kind of research problem to tackle while shaving in front of a mirror. It occurred to me that the question of whether animals might be capable of recognizing

themselves in mirrors was an intriguing and potentially re-
searchable issue. A review of the literature (see Gallup,
1968) showed that people had confronted animals with mirrors
from time to time and recorded their responses, but systematic work was lacking and no one had developed a clear idea
of how to objectively determine whether an animal had succeeded in correctly deciphering mirrored information about itself.

Mirrors have unique psychological properties. In front of a
mirror any visually capable organism is ostensibly an audience
to its own behavior. In principle, mirrors enable organisms
to see themselves as they are seen by others. However, the
literature clearly shows that most animals react to themselves
in mirrors as if they were seeing other animals. A variety
of species respond to their own images by attempting to attack
the reflection, court the reflection, and in general treat the
reflection much as they would a conspecific. Some species of
primates even attempt to reach behind the mirror in an apparent
attempt to contact the "other" animal. But unlike humans, most
animals seem incapable of recognizing the dualism implicit in
mirror surfaces, and even after prolonged exposure fail to
discover the relationship between their behavior and the reflection of that behavior in a mirror.

Self-Recognition in Humans

In humans self-recognition of one's own reflection is common
place, but not universal. The capacity to recognize one's
self in a mirror appears to be subject to maturational and
experiential constraints. In the first place, the ability to
correctly interpret mirrored information about the self presupposes prior experience with mirrors. For instance, people
born with congenital visual defects who undergo operations in
later life which provide for normal sight, respond just like
nonhumans and initially react to themselves in mirrors as
though they were seeing other people (von Senden, 1960).
Small children also show other-directed behavior to mirrors
and are notorious for reacting to their reflection as if confronted by a playmate. Social behavior in response to mirrors
begins at about six months of age, but the average child does
not start to show reliable signs of self-recognition until 18
to 24 months (for a critical review of this literature see
Gallup, 1979b).

Some people never learn to recognize themselves. There are
instances of profoundly retarded children, adolescents, and
even comparably handicapped adults who have had a life-time
of exposure to mirrors, but persist in treating the reflection

in much the same way that you or I would treat another person
(e.g., Harris, 1977). Thus, self-recognition would appear to
require a level of intellectual functioning which exceeds that
of even some humans. It is also interesting to note that prolonged mirror gazing has been associated with the onset of
schizophrenia in man (e.g., Ostancow, 1934), and some schizo-
phrenics appear to have lost their capacity for self-recognition as evidence by talking to the reflection and laughing at
their images in mirrors (e.g., Wittreich, 1959).

Self-Recognition in Chimpanzees

After pondering the possibility of an evolutionary discontinuity in the ability to decipher mirrored information about
the self, I decided that a fair assessment of this capacity
would require that animals be given extended exposure to
mirrors. In the natural habitat of most species, mirror surfaces are rarely encountered and when such encounters occur,
exposure is typically very brief. Therefore, in the initial
study I gave a number of group-reared, preadolescent chimpanzees individual exposure to themselves in mirrors for eight
hours a day over a period of ten consecutive days (Gallup,
1970). Invariably, their first reaction to the mirror was to
respond as if they were seeing another chimpanzee, and they
all engaged in a variety of social gestures while watching the
reflection. After about two days, however, the tendency to
treat the image as a companion began to disappear. This was
replaced by the emergence of what seemed to be a self-directed
orientation. That is, rather than show species typical social
behavior and respond to the mirror as such, they began to use
the mirror to respond to themselves. They started to use the
mirror to groom and inspect parts of their bodies they had
not seen before, and progressively began to experiment with
the reflection by way of making faces, looking at themselves
upside down, and assuming unusual postures while monitoring
the results in the mirror.

In addition to time sampling the behavior of chimpanzees confronted with mirrors, we also kept careful records of the
amount of time they spent watching the reflection. Prior to
the development of a self-directed orientation, all the chimpanzees showed an avid interest in the mirror, but with the
appearance of mirror mediated, self-referenced activities this
interest diminished.

While I was convinced that the chimpanzees had discovered the
dualism implicit in mirrored surfaces and had succeeded in
correctly deciphering mirrored information about themselves,
I was concerned that my colleagues might not find my subjective impressions very compelling. So in an attempt to provide

a more convincing demonstration of self-recognition, I devised an unobtrusive and more rigorous test. After the last day of mirror exposure, each chimpanzee was anesthetized, removed from its cage, and taken to another room. While completely unconscious, the chimpanzee was placed on a table, and I carefully painted the uppermost portion of an eyebrow ridge and the top half of the opposite ear with a bright red, odorless, non-irritating, alcohol soluble dye (rhodamine B-base). After the marking procedure was complete, the animal was returned to its cage and allowed to recover in the absence of the mirror.

The significance of this technique is three-fold, and these details have unfortunately been overlooked by investigators attempting to adapt the approach to assessing self-recognition in children (see Gallup, 1979b). First, the chimpanzees had no way of knowing about the application of red marks since the procedure was accomplished under deep anesthesia. Second, the dye was carefully selected so as to be free from any tactile or olfactory cues. We even wore the dye on our own skin for several days just to be sure. Finally, the marks were strategically placed at predetermined points on the chimpanzee's face so that upon recovery it would be impossible for the dye to be seen without a mirror.

Following recovery from anesthesia, all subjects were observed to determine the number of times any marked portion of the face was touched in the absence of the mirror. The mirror was then reintroduced as an explicit test of self-recognition. Upon seeing their painted faces in the mirror, all the chimpanzees showed repeated mark-directed responses, consisting of attempts to touch and inspect marked areas on their eyebrow and ear while watching the image. In addition, there was over a three-fold increase in viewing time, suggesting that there was something about seeing themselves with red marks on their faces which greatly enhanced their visual attention to the mirror. Several chimpanzees also showed noteworthy attempts to visually examine and smell the fingers which had been used to touch these facial marks. I suspect that you would respond pretty much the same way, if upon awakening one morning you saw yourself in the mirror with red spots on your face.

In an attempt to eliminate any doubt about the source of these reactions, several comparable chimpanzees which had never seen themselves in mirrors were also anesthetized and marked. However, when presented with the mirror they showed no mark-directed responses, patterns of self-directed behavior were completely absent, and the dye was ignored. Marked chimpanzees without prior mirror experience responded as if they had

been suddenly confronted with another chimpanzee and proceeded
to gesture, vocalize, and threaten their reflection. These
data strongly suggest that self-recognition had been learned
by the former animals sometime during the previous ten days
of mirror exposure. We now know that chimpanzees are capable
of evidencing self-recognition on a mark test after as few as
only four days of mirror exposure (see Suarez and Gallup,
1981).

Self-Recognition in Other Primates

These findings have been replicated with chimpanzees a number
of times in different laboratories by different investigators
(e.g., Gallup, McClure, Hill and Bundy, 1971; Hill, Bundy,
Gallup and McClure, 1970; Lethmate and Dücker, 1973; R. L.
Thompson, personal communication; S. P. Spitzer, personal
communication; Suarez and Gallup, 1981). Orangutans have also
been tested for self-recognition and show results which are
largely indistinguishable from those obtained from chimpanzees.
Initially, patterns of other-directed behavior diminish over
the first few days of mirror exposure. Self-directed responses
begin to appear on about the third day. When tested for self-
recognition, orangutans respond just like chimpanzees and
attempt to use the mirror to locate and investigate the marks
on their faces (Lethmate and Dücker, 1973; Suarez and Gallup,
1981).

In general, primates, of which there are over 190 species, can
be divided into about five fairly distinct taxonomic categories.
The smallest and most "primitive", such as tree shrews, are
classified as prosimians. Next are New World monkeys, consisting of such familiar South American species as squirrel
monkeys and capuchins, followed by Old World monkeys, such as
baboons and rhesus monkeys, then apes (gibbons, siamangs), and
great apes (chimpanzees, orangutans, gorillas). Whether humans can continue to justify occupying a separate category
is a matter which could be subject to some debate in light of
the striking behavioral (e.g., Gallup, Boren, Gagliardi and
Wallnau, 1977) and biological similarities between humans and
great apes (e.g., Yunis and Prakash, 1982).

With the exception of humans, chimpanzees, and orangutans, all
other primates tested thus far have failed to recognize themselves, even after extended exposure to mirrors. Spider
monkeys, capuchins, squirrel monkeys, stumptailed macaques,
rhesus monkeys, crabeating macaques, pigtailed macaques,
liontailed macaques, mandril, olive, and hamadryas baboons,
two species of gibbons, and even gorillas have been systematically tested using these procedures but none have shown

any indication that they were capable of realizing that their behavior constituted the source of the behavior depicted in the mirror (see Gallup, Wallnau and Suarez, 1980; Gallup, 1982). Thus, in the process of trying to resolve one apparent evolutionary discontinuity, I appear to have run up against another.

An obvious counter charge in the face of these differences might be that monkeys and lesser apes simply have not been given an ample opportunity. If as Epstein et al. (1981) claim, the development of self-recognition is simply a matter of discovering the appropriate mirror contingencies, then extended exposure to mirrors should eventually provide for self-referenced behaviors in the presence of the reflection. But how much exposure is enough? Would a month suffice, two months, a year? Eventually you reach the point where continued exposure to a mirror ceases to be a viable excuse.

In one instance, I gave a crabeating macaque five months, or over 2,400 hours of mirror exposure (Gallup, 1977a). But to no avail. Throughout the period she failed to show any signs of self-recognition, and continued to respond to her image as if confronted by another monkey. On the test of self-recognition she appeared oblivious to the marks on her face, and merely threatened the reflection. Benhar, Carlton and Samuel (1975) ran a series of five experiments with olive baboons in an attempt to find self-recognition, and even went so far as to try to teach one baboon to recognize itself with raisin rewards. However, despite their continued efforts they found no evidence that baboons could learn to correctly interpret mirrored information about themselves, and conclude that the "ability to distinguish between 'self' and 'others' seems to be a quality specific to man and apes"(p. 207). Robert L. Thompson (personal communication) has conducted a number of studies in an attempt to find self-recognition in pigtailed macaques, but, like the rest of us, has also had no success. In one instance, he kept a pigtailed macaque in a cage containing a polished metal mirror for an entire year, but the monkey still gave no evidence of self-recognition and social responses directed toward the reflection continued at a diminished rate for the duration of the period. The most extended exposure to date has been documented by M. Bertrand (personal communication) who made systematic observations of a home-reared pigtailed macaque which has had daily access to mirrors for seven years, but still seems incapable of recognizing its own reflection, and even now occasionally makes social responses to the image. In stark contrast with other

primates, chimpanzees and orangutans often begin to show
evidence of self-directed behavior in as few as 2 or 3 days of
mirror exposure (Gallup, 1970; Suarez and Gallup, 1981).

Absence of Self-Recognition in Gorillas

By far the most surprising failure to find self-recognition
involves gorillas, which in addition to chimpanzees and
orangutans are the only other primates classified as great
apes. This may seem particularly puzzling in the face of the
anecdotal reports by Patterson (e.g., 1978) who claims that
the gorilla named "Koko" can recognize herself in a mirror.
Koko has ostensibly been taught sign language, and Patterson
contends that since Koko can label her reflection in a gestural
mode she recognizes her own image. But this can hardly be con-
strued as evidence of self-recognition. A child held in front
of a mirror and told its name over and over might eventually
learn to associate the name with the reflection, but this
would not mean it had learned to recognize itself (see Gallup,
1979). Labeling the reflection qualifies as nothing more than
a demonstration of associative learning.

In spite of Patterson's impressions, and the impressions she
may have created, there have been at least three subsequent
attempts to demonstrate self-recognition in gorillas under
experimental conditions, with as yet no success. In a comparative study of self-recognition in all three species of great
apes (Suarez and Gallup, 1981), we attempted a more rigorous
assessment than had been used previously. One possible account
of the gorilla's failure to show self-recognition might be that
they are simply not interested in themselves and lack the in-
clination or motivation to touch and inspect superimposed body
marks. If this were true, the absence of mark-directed behavior on a test of self-recognition would be inconclusive.
To assess this possibility, in addition to applying pink dye
to facial features, we also marked each of four gorillas on
the wrist to see if they would respond to marks that could be
seen directly. After recovery from anesthesia all the gorillas
showed an avid interest in their marked wrists, which they
repeatedly touched and inspected, but not one of them used
the mirror to respond to comparable marks on the face. As
further evidence that gorillas are oblivious to the source
and significance of what they see in a mirror, unlike chimpanzees who show an abrupt increase in visual attention to the
mirror when allowed to confront themselves with marks on their
faces, the gorillas showed no change in viewing time.

Our overwhelming impression of gorillas is that their behavior
in response to mirrors most closely resembles that of preadolescent rhesus monkeys (see Gallup et al., 1980). Throughout the sixteen days of mirror exposure we provided, the gorillas never showed self-directed behavior at any time. Although
the sceptic might again charge that 16 days of exposure was
insufficient, Ledbetter and Basen (1982) recently completed a
study in which a pair of captive gorillas were given almost
400 hours of exposure to themselves in mirrors extending over
a period of several months, yet they report no evidence of self-
recognition. J. Lethmate (personal communication) has now
tested as many as a dozen gorillas for self-recognition, and
consistent with the rest of us finds little or no evidence
that they are capable of realizing that their behavior is the
source of the behavior depicted in the mirror.

It is especially important to note that the performance of
gorillas on a variety of more traditional psychological tasks
(e.g., learning set formation, discrimination, discrimination
reversal, oddity, delayed response) is largely indistinguishable from chimpanzees and orangutans (see Suarez and Gallup,
1981 for a review of this literature). Thus, one would be
hard pressed to argue that the gorilla's failure to recognize
itself in a mirror is a simple consequence of mere species differences in learning ability or concept formation.

Comparative studies of high-resolution chromosomes, protein
structure, and even DNA itself (e.g., Yunis and Prakash, 1982)
show that all three of the great apes and humans are closely
related, with humans and chimpanzees being the most similar,
closely followed by gorillas. Orangutans were apparently the
first to diverge from the protohominoid line, and are the
least similar biochemically to the rest of us. Therefore, with
gorillas being intermediate between orangutans and chimpanzees,
but more like chimpanzees, it is curious that among great apes
they constitute the cognitive exception to self-awareness.

In one sense, however, the gorilla's inability to recognize
itself in a mirror only causes conceptual problems to the
extent that you suffer from the misconception of evolution
as a progressive process. Evolution does not occur by design,
it is a byproduct of selection, and the raw material for such
selection are genetic accidents. Differential or nonrandom
reproduction is what drives all evolutionary change. Whereas
we are all closely related, humans did not evolve from chimpanzees, nor did gorillas evolve from orangutans. We are
related by virtue of sharing a common ancestor. It is worth
reiterating in this context that there are even some humans
who are incapable of recognizing themselves in mirrors.

When it comes to the study of psychological phenomena there is
a very real danger in letting ourselves become intellectually
shackled by taxonomic categories based largely on gross mor-
phology (Gallup, 1982). As detailed in a subsequent section,
I strongly suspect that the key to the conceptual difference
between gorillas and other great apes lies in their social
structure and lack of interspecific competition. It is also
interesting to note that the gorilla brain is structurally
the least lateralized of all great ape brains (LeMay and
Geschwind, 1975).

The Paired Exposure Technique

However, since it is admittedly difficult to deal with negative
instances, we have felt compelled to try to devise alternative
ways of making the identity of the reflection more explicit in
a continuing effort to explore the possibility of self-recognition in other primates. In a recent study (Gallup et al.,
1980) we attempted to capitalize on the rhesus monkey's well
developed capacity for individual recognition. Individual
recognition, or the capacity to recognize and respond differentially to conspecifics, is what underwrites practically all
stable social relations. For example, the formation and maintenance of a dominance hierarchy would be impossible without
the capacity to identify and differentiate among members of
the same species. It occurred to us that we might be able to
contrive a situation in which individual recognition could be
used to facilitate self-recognition. By giving familiar cagemates paired access to a common mirror, each member of the
pair should be able to correctly identify and recognize the
reflection of its companion, thereby posing the question in a
more direct way about the source of the remaining unfamiliar
animal they see in the mirror.

First, we were able to independently establish that indeed
rhesus monkeys could recognize the reflections of cagemates,
and we have tried this technique now with several pairs of
monkeys. As yet, however, we have failed to find even the
slightest indication of self-recognition. In fact, we now have
a pair of rhesus monkeys, housed together since infancy, which
have had continuous access to themselves in the same mirror
since 1978, but as time goes on they tend to show more rather
than less social behavior to the reflection. Our intention
is to continue to maintain these animals under such conditions
to see what effect, if any, a life time of exposure to mirrors
may have. That should satisfy even the most caustic critic.
Whereas it is true that the absence of evidence is not evidence
of absence, this view needs to be tempered by the fact that the
same holds true for ghosts, ESP, Santa Claus, and the tooth
fairy.

I might also point out that failing their initial attempt, Ledbetter and Basen (1982) also tried our paired exposure technique with gorillas, but the results were disappointing. While both gorillas were able to recognize their partner's reflection in a common mirror, and used the mirror to watch one another, they never used the mirror to respond to themselves.

Self-Recognition as an Index of Self-Awareness

While it is well established that monkeys can learn to use mirrored cues to both contact and manipulate objects they cannot otherwise see (e.g., Brown, McDowell and Robinson, 1965), they appear incapable of learning to sufficiently integrate features of their own reflection to use mirrors to respond to themselves. It is important to emphasize that it is not because they are incapable of learning to respond to mirrored cues. When looking at the reflection of a human or a piece of food they can detect both the 180° reversal and the dualism as it pertains to objects other than themselves, and do learn to respond appropriately by turning away from the mirror to gain more direct access to the object of the reflection (Tinklepaugh, 1928). Yet they completely fail to interpret correctly mirrored information about themselves.

Taken collectively, my interpretation of the inability of most species to recognize themselves in mirrors is that they lack an essential cognitive category for processing mirrored information about themselves (Gallup, 1977b). It is not simply a consequence of perceptual-motor deficits. Rhesus monkeys have little or no difficulty learning to respond to mirrored cues. Moreover, all of the perceptual-motor skills needed to recognize and respond differentially to conspecifics are part and parcel of those required for self-recognition. On most tests of psychological functioning in primates, such as those that assess learning and concept formation, one finds that the differences among most primates (including humans) are matters of degree, not kind. For instance, chimpanzees form learning sets more rapidly than rhesus monkeys, but with extended training rhesus monkeys can achieve the same level of performance (Rumbaugh and McCormack, 1967). Yet when it comes to self-recognition extended experience does not make up the difference. If primates are ordered in terms of their ability to perform traditional tasks, one finds a continuum or gradation among species. But the capacity to recognize oneself in a mirror represents a dramatic departure from the pattern of data derived from other performance measures. Somehow there appears to be a step-wise change as you move from monkeys and lesser apes to two great apes and man, akin to a psychological void of sorts.

What is missing, I think, is a sense of self. One of the
unique features of mirrors is that the identity of the observer and his reflection in a mirror are isomorphic with one
another. Some philosophers contend that self-awareness is a
logical impossibility. How can the self be aware of itself
without raising the specter of an infinite regress of per-
ceptive selves? If consciousness is equivalent to pointing
at something, then by analogy the index finger cannot point
at itself. Yet mirrors circumvent the philosopher's dilemma
(see Gallup, 1979a). In front of a mirror the index finger can
point at itself. In front of a mirror both the subject and
object of perception are indeed one and the same.

Since the identity of the observer and his reflection in a
mirror are the same, if you do not know who you are, how can
you possibly know who it was you were seeing when confronted
with yourself in a mirror? The capacity to correctly infer
the identity of the reflection is predicated on an identity
on the part of the organism making that inference. Thus, the
monkey's inability to recognize his own image and his tendency
to continue to react as if he were seeing another monkey may be
due to the absence of a sufficiently well integrated self-
concept. The other evidence in support of the chimpanzee's
capacity for self-awareness is their ability to impute mental
states to others (e.g., Premack and Woodruff, 1978). I will
elaborate on this point later in the chapter.

Is it possible to be self-aware in other modalities? The
answer, of course, is yes. People can learn to recognize their
own voices, and perhaps even their own odors. But that is not
the context in which this question is typically posed. The
argument goes something like this. While some species may be
incapable of recognizing their visual images in mirrors, maybe
they could evidence self-recognition in some other modality.
In my opinion, this is a mistaken conception of self-awareness.
This would imply the existence of separate selves (e.g., a
visual self, olfactory self, etc.). One's sense of self
transcends modality specific information, and as such requires
(or presupposes) the capacity for intermodal equivalence.
Your sense of self does not fade, become fundamentally altered,
or disappear when you close your eyes, hold your nose, or
cover your ears.

It is true that some animals respond differentially to their
own cues as compared to those of conspecifics. Consider bats.
Most bats are primarily reliant on echolocation for purposes
of navigation, prey identification, and prey capture. Some
species of bats live in large underground caves, with a single
cave often containing hundreds or even thousands of bats. For

purposes of flying about within the cave an individual bat must
be capable of differentiating his own echoes from those of
other bats, otherwise his sophisticated biological sonar system
would be hopelessly jammed. But differential responses to an
animal's own sound or odor hardly constitute hard evidence of
self-awareness in light of differential familiarity and
habituation effects. As long as the bat can reliably differentiate his echo from the echoes of other bats it is not necessary that he realize that it is in fact his own.

Knowledge of Self Presupposes Knowledge of Others

A prevailing view of self-concept formation in humans is that
the sense of self emerges out of a social milieu. Self-awareness, according to this view, is a byproduct of social interaction with others. For instance, according to G. H. Mead
(1934) in order for the self to emerge as an object of conscious inspection it requires the opportunity to see yourself
as you are seen by others. This approach to self-concept
formation was first introduced by Cooley (1912) who is credited
with what is known as the "looking glass theory" of self.
Cooley felt that the primary information we have about our
selves is represented in terms of reflected appraisals that we
get from other people. For example, if others treat you as
good boy, or as a bad boy, you will come to see yourself as
such.

If knowledge of self requires knowledge of others as the social
interactionists claim, then it occurred to me that chimpanzees
reared in social isolation and deprived of the opportunity to
interact with others might be deficient in their capacity for
self-recognition. In order to test this hypothesis we compared
the performance of group-reared chimpanzees with those that had
been separated from their mothers at birth and reared in cages
which prevented them from seeing other chimpanzees (Gallup,
McClure, Hill and Bundy, 1971). Unlike their socially maintained counterparts, the isolates all showed continued and
unabated interest in the mirror over successive days of exposure. On the test of self-recognition, following the application of red dye and recovery from anesthesia, all of the
group housed chimpanzees showed an abrupt increase in viewing
time and repeatedly touched marked areas on themselves while
watching the reflection. The isolates, on the other hand,
showed no change in viewing time and the number of mark-directed responses on the test was equivalent to the number of
times normal animals touched such marks before even having
seen the mirror.

Consistent with the interactionist position, therefore, it
would appear that self-recognition requires prior social

experience. Perhaps the most striking thing about social interaction as a precondition for self-recognition is that the nature of such experience need not be restricted to members of the same species. Chimpanzees reared in human homes who have never had the opportunity to interact with other chimpanzees, still show obvious signs of self-recognition in response to mirrors (e.g., Hayes, 1951). However, human reared chimpanzees do appear to differ from their more normal counterparts. They act as if they were human. For instance, it is reported that when Washoe, who was reared with humans, was first exposed to other chimpanzees she referred to them in "sign language" as black bugs (Linden, 1974). It is well established that being reared in social isolation can have devastating effects on primate sexual behavior. Human reared chimpanzees, however, are actually more impaired than those raised in abject social isolation (Rogers and Davenport, 1969). One possible account of this anomaly might be that chimpanzees reared by humans think they are human.

There is also an informal but widely publicized experiment on another home reared chimpanzee named "Vicki" (Hayes and Nissen, 1971). Among other things Vicki was taught to sort stacks of photographs into a human and an animal pile. One day unbeknown to Vicki her tutors placed her own picture in the stack, and it is reported that when she came to her photograph she picked it up and without hesitation placed it on the human pile. As I have pointed out elsewhere (Gallup, 1977b), there are at least two ways to interpret such behavior. First, maybe humans are not the only ones to appreciate the similarities between chimpanzees and humans. Alternatively, maybe Vicki thought she was human.

Even humans may not be immune from such effects. Temerlin (1975) reports that as a result of rearing a chimpanzee in his home, his son, who related to the chimpanzee much as he would a playmate, began to develop doubts about his own species identity. One also gets the impression that the same effect led to the eventual demise of the first systematic attempt to study a home reared chimpanzee. The Kelloggs (1933) acquired an infant chimpanzee named "Gua" and reared it with their son "Donald". Just as chimpanzees may be humanized by people, Donald began to show deliterious effects of interacting with Gua. When most children start to learn language, their first words are often "daddy" or "mommy". Not Donald. His first word was "Gua". Shortly after that Donald's language development began to deteriorate, and eventually he stopped talking altogether. Much to his mother's distress and probable chagrin, he began to gnaw on trees and make food barks at meals.

One gets the impression that finally Mrs. Kellogg declared the experiment a hazard to her son's well being, and Gua was sent to a zoo (see Desmond, 1979 for more detail).

In the face of such data which show not only that social interaction is necessary for self-conception, but that the nature of such interaction can have a profound impact on how one conceives of oneself, the orangutan might appear to be an exception. Orangutans can recognize themselves in mirrors. Yet, under natural conditions, except when it comes to sex, adult orangutans are by and large solitary creatures (Maple, 1980). Although this might appear to be at odds with the foregoing, it is not. The solitary behavior of orangutans is maturationally bound. Infant orangutans would perish without the almost continuous presence of a protective, lactating adult female. I suspect that the early mother-infant relationship in orangutans is what may underwrite subsequent self-awareness. We will come back to a possible account of why orangutans are so solitary in a latter context.

Will Reinforcement Suffice?

Although I have taken an opportunity to comment on several previous occasions (Gallup, 1981, 1982) about the operant challenge to my work on self-recognition, since the issue continues to be seen by some as substantive I will offer a few additional observations. Through a series of shaping procedures Epstein et al. (1981) were able to teach pigeons to peck at a blue dot on the breast which could only be seen in a mirror. On the basis of this demonstration they argue that the concept of self-awareness is unnecessary, and by analogy that a reinforcement analysis will subsume the behavior of chimpanzees and humans in response to mirrors.

The issue is not one of self-recognition per se, but rather whether cognitive phenomena can be subsumed by an analysis of environmental contingencies. However, to demonstrate that you can model the behavior of one species by reinforcing a series of successive approximations to what looks like the same routine in another species, does not provide, in a logically compelling way, a demonstration that the behavior of the former species necessarily arose in the same way.

If self-referenced behavior in the presence of a mirror is merely a matter of learning contingencies appropriate to mirrored space, why is it that monkeys and gorillas do not show self-recognition? Conversely, why do pigeons have to be taught, through an intricate shaping regime, to show mark-directed responses, while chimpanzees show such behavior in the

absence of any tutoring? After all, if it is merely a matter
of reinforcement, rhesus monkeys should do much better than
pigeons. Indeed, if, as Skinner would have us believe, even
language acquisition is simply a matter of appropriate reinforcing contingencies, why is it that a group of operant
psychologists (Terrace, Petitto, Sanders and Bever, 1979) have
concluded that in spite of years of careful training chimpanzees cannot master the fundamentals of symbolic communication?
After all, if it is merely a matter of reinforcement, even a
rat should be capable of mastering the subtleties of syntax
and sentence construction in a gestural mode.

Rather than being an issue of substance, it seems to me that
what we have here is a paradigm clash. The controversy will
not be resolved by who is right or wrong. It is a question
of which conceptual framework will prevail. As I indicated
earlier, the utility of any cognitive approach to behavior has
to be measured in part by the extent to which it has heuristic
value. What operant psychologist in his "right mind" would try
to teach a pigeon to peck at a blue dot on itself in a mirror
without a cognitive psychologist to guide him?

E. W. Menzel, Jr. (personal communication) poses the question
this way. If we were to accept the logic of Epstein et al.
then it would follow that "Inasmuch as my teachers somehow or
other got me to perform a few problems in the calculus, I am
the intellectual peer of Newton and Leibnitz, and the latter
individuals not only did not invent the calculus but also
acquired it in the same fashion as I did." Similarly, and I
am paraphrasing Menzel, if the readers of this chapter sub-
scribe to the idea that what I am writing can best be described
in terms of environmental contingencies, then they should confine their evaluation of this material to such matters, and refrain from the tendency to involve themselves in mentalistic
issues such as whether what I write is important, original,
intelligent, or what have you.

Let's return to the question of why pigeons have to be taught
to peck at blue marks on themselves in front of a mirror. In
defense of their use of conditioning procedures, Epstein et al.
argue that pigeons do not show mark-directed behavior spontaneously because they lack the necessary response repertoire.
The logic of this excuse is flawed on several accounts. First,
just as chimpanzees will inspect and remove foreign matter
from themselves in the absence of mirrors, pigeons also show
frequent self-referenced responses in the form of elaborate
preening behavior. So just what is it about the pigeon response repertoire that they lack? Is it really a question
of response repertoires? Monkeys and chimpanzees have

overlapping response repertoires, yet monkeys seem completely
incapable of recognizing themselves in mirrors. Is it that
pigeons need to be taught to respond to mirror cues? Monkeys
can learn to respond to mirror cues, but they do not show self-
recognition.

Not only is the use of conditioning procedures faulty in light
of the logic of Epstein et al., but it seriously contaminates
their data and subverts the rationale for requiring mark-
directed behavior as a test of self-recognition. By teaching
an organism the criterion response, it ceases to be a criterion
of anything other than the effectiveness of the contingencies
that gave rise to its existence. By analogy, if I were to
teach a pigeon to peck the correct answers on an answer sheet
to the Graduate Record Examination, and it achieved a combined
verbal and quantitative score of 1500, would it qualify for
admission to the graduate program at Harvard? Or, would this
demonstrate that performance on the GRE is simply a byproduct
of environmental contingencies? The answer, to both questions
is of course "no". I have no doubt that pigeons could be
taught to do this. But, having done so, all it would illustrate is that reinforcement works. That is not too profound.
We have known that reinforcement works ever since Thorndike
formulated the law of effect.

The attempt to model mark-directed behavior is conceptually
flawed in a number of other ways, and the study by Epstein
et al. contains some serious methodological problems. But
these have been detailed elsewhere. I would invite the
interested reader to consult these critiques (Gallup, 1981,
1982).

Mind as a Byproduct of Self-Awareness

Now I will attempt to deal with the question of mind, and
eventually whether minds might exist in species other than our
own. The significance of self-awareness goes well beyond the
mere capacity to recognize one's own image in a mirror. An
organism which is aware of itself is an organism that is
aware of its own existence. In principle at least, once you
can become the object of your own attention you can begin to
think about yourself. From this point, as I see it, there
are two logical next steps. On the one hand, to be aware of
your existence raises the specter of contemplating your own
eventual nonexistence. The unique price that we pay for self-
awareness is the realization of the inevitability of our own
individual demise. The other implication of being aware of
yourself, is that you can begin to make inferences about
mental states in other individuals. Knowledge of self provides

an intuitive basis for achieving knowledge of others. It is
this latter implication that I will focus on. But before pro-
ceeding we need to establish some definitions.

Mind, like most concepts in psychology, is a matter of definition. Definitions are neither true nor false. Definitions
only differ in terms of their utility. Learning, frustration,
hunger, fear, concept formation, and memory are not directly
observable, tangible entities, but they have proven useful
when subject to rigorous empirical definition. Mind has been
sidestepped by psychologists for want of such definition. I
intend to argue that mind is the capacity to use experience
to model the experience of others. But it is important to
acknowledge that experience is a private event. There is no
way for me to experience your experience, let alone the experi-
ence of another species. The only experience I can ever ex-
perience is my own experience. But just because experience
is private does not mean that we have to abandon the concept.
Atoms and molecules are useful (indeed, indispensible) concepts in chemistry and physics, but they have never been
experienced directly. In principle, we ought to be able to
devise means of inferring and mapping experience, and use it
just like the physicist uses the concept of an atom to organize
data and generate predictions.

I have suggested that mind is represented by the ability to
monitor one's own mental states (Gallup, 1982). In terms of
a computer analogy, a mind can be thought of as a subroutine
which monitors and modulates particular features of the system
at large, and on that basis makes inferences about the operation and disposition of similar systems. Thus if I define
self-awareness as the ability to become the object of your own
attention, consciousness as being aware of your own existence,
and mind as the ability to monitor your own mental states,
then it is obvious that these are not mutually exclusive
cognitive categories. Indeed, I think they are all part and
parcel of the same process.

First, I am assuming that an organism which can become the
object of its own attention, and is therefore aware of itself,
is also aware of its own mental states. Organisms with the
capacity to monitor their own mental states ought to also be
able to reflect on such states, and even simulate such states.
All of which is to say that in the sense in which I am using
the term, evidence of mind must be evidence of introspection.
Introspection presupposes self-awareness. Being unable to
become the object of your own attention, you should be rendered
incapable of introspection for the simple reason that there
would be nothing to introspect about. This is not to deny the

existence of feelings which serve to energize behavior in a large number of species, but lacking the ability to monitor such states I would argue that they are mindless.

Since I will attempt to apply this concept to nonhumans, some people may object to equating mind with introspection on the grounds that introspection requires language. However, I am not restricting the term to verbal reports. You do not have to be able to name something in order to be aware of it. The capacity to differentiate between feelings of hunger, anger, and fear does not require language. One of Premack's (1972) principle conclusions as a result of trying to teach the chimpanzee named "Sarah" language using geometric forms, was that language did not provide Sarah with any new concepts. Rather, language merely allowed her to express what she already knew. The existence of such things as self-awareness, cross-modal perception, and tool fabrication in linguistically naive chimpanzees strongly suggests that many of the traits heretofore thought to be a byproduct of language are in fact preverbal (see also Desmond, 1979). Indeed, the so-called linguistic relativity hypothesis may, at least in some instances, be 180° out of phase with reality. Rather than language imposing limits on thought, thought may impose limits on language.

Now, in order to make a case for the resurrection of mind I need to show how the capacity to monitor one's own mental states can be operationalized. My second assumption is that an individual with a mind should be tempted to suppose that mental states also exist in other individuals. For instance, when we see people in situations that are similar to those we have encountered, we tend to assume that their experience is similar to our own. While it is probably true that no two people experience the same event in exactly the same way, there is pretty good consensus among people as to such things as color differentiation, temperature, distance, and loudness. This also applies, but perhaps to a lesser extent to such states as fear, jealousy, rage, intentionality, happiness, remorse, sympathy and the like. Therefore, in principle, I can use my experience to map your experience with some degree of accuracy.

Using your own experience to model a conspecific's experience opens up some extraordinary possibilities. Based on such an intuitive knowledge of others, you could begin to anticipate and even influence what other people might do. By withholding information, providing misinformation, or distorting information, deception surfaces as a major factor to be contended

with in social interaction and produces concomitant pressure
for the development of detection, grudging, counter-deception
and the like.

My ability to monitor my own mental states gives me a means of
achieving an intuitive knowledge of your mental states. Knowledge of self, in other words, provides for an inferential
knowledge of others. Thus, humans not only attribute purpose,
intent, and various other mental states to one another, but
there is a primitive and almost irresistable tendency to
generalize such attributions to species other than our own
(e.g., "the dog is angry", "the baby monkey wants his mother",
"the rat is trying to escape"). Clearly, the tendency to
impute mental states to others presupposes the capacity to
monitor such states on the part of the individual making such
inferences. Therefore, evidence of anthropomorphism becomes
prima facie evidence of mind. It is ironic, as Premack and
Woodruff (1978) point out, that the process of becoming a behaviorist in effect requires that one learn to suppress this
tendency. Anthropomorphism is a pervasive characteristic of
human existence, which requires explanation in its own right.
In fact, some people even anthropomorphize with machines (e.g.,
computers) and other inanimate objects.

Based on this formulation, not only should mindless species
be incapable of introspection, but because of their resulting
inability to attribute mental states to others, they should
fail to exhibit evidence of such things as gratitude, grudging,
sympathy, empathy, attribution, intentional deception, and
sorrow. As a means of operationalizing the concept, Table 1
summarizes some of these empirical markers of mind. Note,
however, that I have listed two instances of each trait:
hard-wired analogs and examples which entail self-awareness.
Only the latter qualify as evidence of mind because they require making inferences, attributions, and/or assumptions
about states of mind in other individuals.

Mind in Other Primates

Are there species other than humans who have minds? According
to the model I have outlined, the emergence of self-awareness
is equivalent to the emergence of mind. Notice that all
instances of mind listed in the right-hand column of Table 1
have been drawn from research on chimpanzees. Not only is this
consistent with the theory, but the theory is formulated in
such a way as to be testable. Organisms failing to show
evidence of being able to become the object of their own
attention should likewise fail to evidence anything other than
hardwired analogs to most of the traits listed in Table 1.

Table 1

Evidence of Mind
(adapted from Gallup, 1982)

Traits	Hard-Wired Analogs	Instances of Mind
Attribution	Unlearned reactions to conspecific threat postures and predators	Attribution of intent[1] and responsibility[2]; anthropomorphism[3]
Deception	Mimicry	Intentional distortion and/or withholding of information[4]
Reciprocal Altruism	Alarm calls	Reciprocal aid giving[5]; selectively withholding aid from cheaters and stealers[4]
Empathy	Responses to appeasement gestures reactions to infant distress calls	Providing solace to injured conspecifics[2,6]
Reconciliation		Preferential contact between opponents following an aggressive encounter[6]
Pretending	Injury feigning; death feigning	Certain forms of deception[7]

[1]Hebb, 1979
[2]Goodall, 1968
[3]Premack and Woodruff, 1978
[4]Woodruff and Premack, 1979
[5]Savage-Rumbaugh, Rumbaugh and Boysen, 1978
[6]de Waal and van Roosmalen, 1979
[7]Menzel, 1975

As would follow from their capacity to recognize themselves in mirrors, chimpanzees show clear indications of mind. The evidence includes anecdotes, field observations, and laboratory research. Starting with some examples of the former, Hebb (1979) described an exchange between two adult females housed in adjacent cages, where one took a mouthful of water and spit

it at the other only to miss. In spite of the fact that the
chimpanzee who was spit at suffered no untoward consequences,
she reacted by threatening and screaming at her neighbor.
This could be interpreted to mean that she was reacting to the
perceived intentions of the other animal. Clearly, she was
not reacting to the mere stimulus consequences of the act.
An associationist might argue, however, that this was simply
an instance of generalization from prior occasions in which
spitting had been more accurate. However, if I am right, then
a chimpanzee who has had prior spitting experience but has
never been spit at, should react in an equivalent manner on
the very first occasion. Think about it.

Köhler (1927) described how he routinely punished his chimpanzees for eating fecal material, which is technically referred
to as coprophagia and is common among captive chimpanzees. He
reported that as a consequence of this, not only did chimpanzees show signs of agitation and appear to act guilty when
discovered with feces on their faces or in their mouths, but
reacted similarly when confronted by Köhler even if a cagemate
had recently indulged in this forbidden act. While the reaction to Köhler by guilty chimpanzees could easily be explained in associationistic terms, reactions based on the
misbehavior of cagemates seems more compatible with an introspective account, and is clearly consistent with how humans
react to similar situations.

As a result of extensive field observations, Goodall (1968,
1971) has described numerous behavioral episodes among chimpanzees which are compatible with this account. For example,
she has reported instances of what, in the present context,
would appear to be attribution of responsibility. She found
that when two infants were playing together, if one was hurt
by the other, rather than reprimanding the aggressor, the
mother of the victim would often attack the offender's mother.
Blame for the incident, therefore, seems to have been placed
on the perpetrator's mother and not the playmate. Comparable
attributions are similarly differentiated on the basis of
perceived social maturity among humans.

Goodall also reports instances in which subordinate chimpanzees
appear to painstakingly avoid looking at or attempting to take
obscure or otherwise unnoticed fruit in the presence of more
dominant animals, only to indulge themsvles once the others
leave. Although an associationist might argue that while
chimpanzees are probably punished for trying to eat before
more dominant animals have had their fill, the studied attempts
to avoid looking at the food clearly fits an introspective

strategy derived from inferences about how others might react to different kinds of information.

Finally, Goodall describes the development of an intriguing tactic used by a preadolescent chimpanzee to pry his mother away from hours of ceaseless termite fishing. Failing repeated attempts to distract the old female and get her to follow him away from the termite mound, on one occasion the young male resorted to kidnapping his infant brother, a strategy which proved highly effective and was repeated by the youngster from one season to the next as a means of breaking the female's preoccupation with termiting.

Not unrelated to this observation in terms of the capacity it implicates, is Goodall's widely publicized description of a subordinate male who one day picked up a pair of empty kerosene cans from around her camp, and climbed a hill overlooking an area where more dominant males had assembled to eat bananas which had been set out as bait. Having reached the top of the hill, the subordinate chimpanzee began to charge down the hill running directly at the other males while banging the two kerosene cans together. The other animals, startled by the clamor, dropped their bananas and ran for cover. Not only did this prove to be a highly effective technique for intimidating physically superior animals and gaining access to food, but as a consequence the innovator rapidly rose to a position of dominance within the group. While an associationist would probably feel comfortable giving an account of how such behavior might be maintained, it would prove extraordinarily difficult to explain the initial appearance of such a tactic based exclusively on operant principles. It would follow from the present analysis, however, that the chimpanzee struck upon the use of kerosene cans to augment his intimidation display in the first place by introspectively anticipating the probable reaction of other animals.

Premack and Woodruff (1978) have done some ingenious research which bears directly on the question of mind. They pose the issue by asking if the chimpanzee has a "theory of mind", in the sense of imputing mental states to others. Their approach differs from the model I have outlined in that they do not concern themselves with questions of mindlessness, self-awareness, hard-wired analogs, introspection, precursors to mind, or the issue of species limits. But they have shown the chimpanzees do appear to make attributions and inferences about mental states in humans. For instance, if shown the videotape of a person shivering, the chimpanzee will choose among an array of different photographs the one depicting the person activating a familiar heater. Similarly, if shown the videotape of an actor trying to get out of a locked cage, play a

familiar phonograph that is not plugged in, or attempting to
wash down a floor with a hose not attached to the faucet, the
chimpanzee will choose photographs which provide correct
solutions to each of these problems.

In a more recent series of studies, Woodruff and Premack (1979)
have provided compelling evidence on behalf of the chimpanzee's
capacity to engage in deliberate deception. Using a paradigm
which involved the exchange of information in human-chimpanzee
dyads about the location of hidden incentives, they found that
when the human and chimpanzee were required to compete rather
than cooperate for the goal, chimpanzees would selectively
withhold information or even provide misinformation about the
location of the incentive as a means of misleading their
opponent. Woodruff and Premack conclude that the ability to
provide either accurate or inaccurate information, depending
upon the context and characteristics of the other member of the
pair, constitutes evidence of intentionality.

Under less contrived conditions, Menzel (1973, 1975) found
that chimpanzees could convey complex information to one another about the direction, probable location, and relative
desirability or undesirability of different hidden objects
that only one chimpanzee had been allowed to see. As an
elegant illustration of the dynamics of mind operating under
natural conditions, Menzel found that after brief exposure to
this kind of problem in a large outdoor enclosure, the unin-
formed chimpanzees began to extrapolate the location of hidden
food based on the leader's initial movements and would often
run ahead of him in an attempt to abscound with the incentive.
In instances of highly prized items, informed chimpanzees were
then observed attempting to ostensibly mislead others by moving initially in directions unrelated to the location of the
incentive. In time, the followers reacted by abandoning an
extrapolation strategy and reverted to keeping the leader under
continuous scrutiny. It would appear, therefore, that chimpanzees make use of intricate introspective strategies based
on inferences about mental states among one another using
subtle gestures, postures, and eye movements.

The model outlined here explains why, almost a decade ago,
Menzel (1973) concluded that while mentalistic terms do not
necessarily explain what chimpanzees do, they often result in
accurate predictions and succinct descriptions of such behavior. If the present analysis is correct, based on their
ability to recognize themselves in mirrors, orangutans should
also exhibit behaviors consistent with a capacity to attribute
mental states to others based on introspection. And that is
a testable proposition. Indeed, given the orangutan's ability

to correctly decipher mirrored information about itself, it is tempting to speculate that the common ancestor to great apes and man had a mind. If orangutans have minds, it is also tempting to suppose that one reason why they are so solitary under natural conditions is because they have discovered that they cannot trust one another. Rape is a common occurrence among orangutans (Maple, 1980).

Are There Any Other Minds?

Connor and Norris (1982) recently published a paper which was provocatively entitled "Are Dolphins Reciprocal Altruists?". In it they systematically review a large amount of observational as well as anecdotal evidence involving instances of mutual assistance and resource sharing among dolphins. All of the evidence converges in a compelling way to suggest that a variety of cetaceans may engage in patterns of reciprocal altruism which are uncontaminated by kin selection and other forms of hard-wired assistance. They even cite instances of helping behavior between members of different genera (e.g., pilot whales and dolphins). Connor and Norris suggest that such remarkable convergence between dolphins and humans (although we are both mammalian forms, we have been diverging ever since the Paleocene) may have been a consequence of strong predatory pressures associated with the exploitation of an aquatic mode for dolphins and the occupation of savannahs by protohumans, which in both cases prompted sophisticated psychological strategies based on mutual interdependence. They conclude by stating that "If, indeed, the unsubstantiated stories of dolphins pushing humans ashore are true, they must be viewed in the same context as humans pushing stranded dolphins back to sea."

Aid giving and mutual assistance which is independent of the degree of genetic relatedness among the participants is evidence of mind (see Table 1). If the present formulation is correct, then in light of these findings with dolphins, they should show evidence of self-awareness. In fact, I have already outlined a procedure which could be used to assess self-recognition in dolphins (Gallup, 1979a), the results of which would constitute an intriguing test of my model.

As evidenced by their peculiar and elaborate grieving behaviors in response to the death of conspecifics (Douglas-Hamilton and Douglas-Hamilton, 1975), elephants are another class of big brained, potentially self-aware mammals which may employ introspectively based social strategies.

Why has a Theory of Mind Been so Long in Coming?

As illustrated by the fact that most of the traits listed in
Table 1 have hard-wired parallels, mind represents a subtle
advantage. The presence of mind is not obvious for at least
three reasons. In the first place, any distinction which
might exist between species in terms of the presence or absence
of mind is masked by the fact that, because of our uncanny
predisposition to anthropomorphize, we treat many species as
if they had minds. The absence of mind is also not obvious
because we take mind for granted. Because we have minds it is
difficult to imagine what it would be like without a mind.
Someone once said that the fish will be the last one to discover water. By the same token we are so immersed in mind
that we have been oblivious to it.

The best illustration of what it would be like to be without
a mind is a phenomenon called "blindsight" (see Humphrey,
1980; Weiskrantz, Warrington, Sanders and Marshall, 1974).
Blindsight is a condition in which destruction of major areas
of the visual cortex produces patients who think they are
blind. It was commonly assumed that indeed they were blind
until Weiskrantz et al. decided to probe the apparent inability to see in several of their patients who had sustained
injury to the visual cortex. To their amazement, they discovered that if such patients could be persuaded to guess
about the location of objects in visual space they showed a
high degree of accuracy. In fact, such people can not only
correctly identify the position of objects in space but they
can even "guess" their form. These patients, therefore, are
still capable of processing certain kinds of visual information, but they are not aware of it. In effect, they have
been rendered mindless in the visual modality. My guess is
that if they had not had a prior history of normal vision,
they would probably not be as encumbered by their present
handicap (see section on "Consciousness").

Finally, the presence or absence of mind is also obscured by
the fact that it has been to the distinct advantage of
organisms during the course of evolution to act as if they had
minds, and as a consequence the behavior of many species contains a variety of hard-wired analogs (Gallup, 1982). For
instance, take a baby duckling's initial reaction to the
silouette of a hawk. It is not necessary that the duck have
any insight into the significant danger posed by hawks or
their ostensible intentions. The only thing that counts is
that they respond appropriately during their first encounter
with a hawk. Indeed, a first encounter would preclude the
effective use of attributions. The same is true of injury

feigning in some species of ground nesting birds. The female
incubating a clutch of eggs, will respond to an approaching
predator with a distraction display which consists of leaving
the nest and acting as if she were injured by dragging a wing
while moving away from the nest. This frequently has the
effect of distracting the predator away from the nest, and precludes egg and/or chick predation. But there is no reason to
assume that the act of injury feigning is predicted on an
introspective capacity to model a predator's likely reaction
to sighting an injured bird, or that injury feigning occurs in
an intentional, deliberate way. Rather, this is simply a hard-
wired instance of deception which capitalizes on the fact that
some predators are highly attracted to injured prey. The
literature on the behavior of animals under natural conditions
is literally filled with a variety of hard-wired analogs (see
Gallup, 1982 for more examples).

Evidence of Mind

Premack and Woodruff (1978) have outlined some ingenious
experimental procedures for assessing an organism's ability
to make attributions about mental states in others. But,
if, as I have argued, organisms have evolved in many instances
to act as if they had minds, how does one distinguish between
hard-wired analogs and legitimate states of mind?

I have previously outlined several ways to disentangle these
two alternatives (Gallup, 1982). In general, the least suspect
instances of mind are those that are the most biological arbitrary. Because organisms have been selected over the course of
evolution to act in ways that maximize their net genetic representation in subsequent generations, to qualify as compelling
evidence of mind, apparent instances of attribution or imputa-
tion should involve individuals other than kin. For example,
members of the same species may be relatively indifferent to
pain and distress among one another (it is not uncommon for a
wounded chicken to be pecked to death by its cagemates), but
because of kin selection this might not be the case among close
relatives. With a few exceptions involving ecological inter-
dependencies (e.g., predator-prey relations), the ideal case
would be one involving members of different species. Interspecific instances of apparent mental inference are less likely to be genetically contaminated than those involving conspecifics. Because of possible problems engendered by stimulus
generalization, it would also be helpful if the individuals
involved were morphologically dissimilar, as in the case of
porpoises and humans.

For similar reasons, the more biologically irrelevant the imputed mental state the better (e.g., the desire to watch television, as opposed to an apparent "desire" to eat or evade predation). Moreover, hard-wired instance of mind should be relatively independent of experience (as in the case of a baby duckling's reaction to the sillouette of a hawk), whereas legitimate instances of mind should take into account the unique characteristics of the individual participants, the particular situation at hand, and the relationship of both of these elements to prior experience. Hard-wired instances of mind are more likely to be species-specific and stereotyped, while genuine states of mind will involve improvisation and relatively unique and even opportunistic attributions which can be varied to meet the demand of different contexts.

Premack and Woodruff's (1978) account of a chimpanzee's reaction to a human attempting to play a disconnected phonograph qualifies as a relatively pure instance of mind. It involves (1) imputing a mental state to another species, it is (2) clearly neutral as far as biological predispositions are concerned, and it is (3) predicated on prior experience with phonographs. However, in the last analysis a single instance of mind will not suffice. To make a strong case for mind requires convergent validation involving multiple instances which cut across the categories listed in Table 1.

Consciousness

I have suggested previously (Gallup, 1977) that there may be two dimensions to conscious experience. Consciousness, I argued, has bidirectional properties. I can be aware of objects and events in the world around me, and I can become the object of my own attention. I can think about myself, contemplate my own existence, and speculate about the mechanisms which may underlie my own behavior. Based on this view, the bidirectional properties of consciousness would translate into awareness and self-awareness. But, once again I am beginning to "change my mind". What I am about to suggest in this section will be perceived by many as radical and even counterintuitive, but it is consistent with the data. Awareness without self-awareness may be a logical impossibility. Unless you are aware of being aware, how could you possibly be aware? Without self-awareness there would be nothing to be aware of.

The ability to monitor the environment and respond to stimuli does not presuppose conscious awareness. A single pithed frog (one whose brain has been scrambled with a steel probe) can still respond appropriately by withdrawing his leg when it is dipped into a vial of acid. People who sleepwalk rarely

bump into things or fall down stairs. Even much of what people
do on a day to day basis does not require a conscious component.
As a relatively experienced typist (of necessity), I do not
have to think about and carefully organize each movement of my
fingers as I type this manuscript. One might argue that these
movements are still being modulated by tactile and propriocep-
tive cues, and I would agree. But it does not require much in
the way of conscious processing on my part.

We become progressively unconscious with respect to many well
learned activities. Responses which have become habitual are
those which are so well established that they no longer require monitoring. Many people have had the experience of
driving on a restricted access road at night and glancing into
their rear view mirror to discover a pair of tail lights moving in the opposite direction, only to suddenly realize that
they cannot remember having seen the car coming toward them in
the first place. Similarly, it is not uncommon for a veteran
smoker to light a cigarette and then discover that there is one
already burning in the ashtray in front of him. To describe
these as unconscious is simply another way of saying that they
are mindless activities. Much of what we do may be mindless.
Indeed, most people are mindless about mind!

This last point is instructive. In a very real sense science
can be thought of as a means of expanding human awareness.
Not only would microscopes and telescopes be obvious instances
of this sort, but the discovery of interrelationships among
events (e.g., smoking and cancer) also qualify. Having a mind
is what provides the potential for such awareness. If major
scientific breakthroughs are represented by new forms of awareness, then it follows that the most significant psychological
issues in the future are those that we are probably the least
aware of. There is a growing consensus among sociobiologists
(see Leak and Christopher, 1982) that we may have been selected
during the course of evolution to be unconscious of some of our
motives. Consider deception. Providing conspecifics with misinformation or withholding information could have adaptive
significance under certain conditions. But deception at someone else's expense is a source of pressure for the development
of sophisticated detection techniques and retaliatory measures.
Therefore, the argument goes, there ought to be concomitant
pressure for self-deception since this would permit a much
more convincing strategy based on misrepresentation. Indeed,
most people are mindless about sociobiology, but behave in ways
that make it look as if they had known about the principles
of sociobiology all along and were consciously acting according
to optimal reproductive strategies based on carefully calculated cost/benefit ratios (see Barash, 1979; Gallup and
Suarez, in press).

Now for the radical implications of this approach to consciousness. If patients who have lost their occipital cortex can
still guess the correct form and location of objects in visual
space, why do we need the concept of consciousness to describe
the behavior of most animals? Unless an animal behaves in ways
that suggest it is aware of its own existence, there is no need
to assume it is aware of what it is doing. Yet this does not
preclude a variety of what might appear to be purposive, deliberate, planned activities. The impression of purpose and
planning, however, may be in most instances a byproduct of the
reproductive contingencies which gave rise to and serve to
maintain the species.

Take the digger wasp as a case in point. The gravid female
digger wasp digs a borrow and lays her eggs on the inside. She
then goes out in search of a particular species of caterpillar,
which she captures and stings. The effect of the venom is not
to kill the caterpillar, but to paralyze it. She then flies
back to the burrow with the caterpillar, and places it next to
the opening. First, she goes inside to check on her eggs, and
then comes back out to drag the caterpillar into the nest as an
ostensible source of food for the larva when they first hatch.
This entire sequence appears very deliberate and rational. Yet
when the sequence is interrupted the illusion of purpose stands
out in relief. If, when the wasp returns with the caterpillar
and goes inside to check on the eggs, you displace it from the
threshold of the burrow by a few centimeters, when she comes
back out of the nest she will retrieve the caterpillar, as if
caught in a closed loop, place it back on the threshold, and go
back in once again to check on the eggs. This subroutine will
continue to recycle as long as you continue to displace the
caterpillar (see Tinbergen, 1958). If the wasp was actually
aware of what it was doing there would be no reason to recycle.
Rather, she should simply drag the displaced caterpillar directly into the nest and complete the episode.

This formulation is not likely to be greeted with much enthusiasm by animal rights activists. The entire animal rights
movement is riddled with inconsistencies, groundless precon-
ceptions, and logical flaws (see Gallup and Suarez, 1980).
Is it, for example, somehow alright to attempt by whatever
means to exterminate rats in the inner cities, but inhumane
and diabolical to administer a brief electric shock to a rat
in the laboratory? The issue of trying to minimize pain and
suffering in animals poses a major paradox. If we admit that
pain is undersirable because pain hurts, then to be logically
consistent we would have to concede that it is alright to kill
animals (as long as the technique is painless), since most
species are oblivious to their own existence and therefore

cannot concern themselves with their eventual demise. Yet, the question of whether an animal lives or dies seems to me to have consequences which are far more profound than whether it experiences an occasional aversive stimulus.

Indeed, in mindless species the experience of pain may be different than it is for you and I. Because of our ability to monitor our own mental states we can be aware of being in pain. For a dog owner to have to extract porcupine quills with a pair of pliers from the nose of an unfortunate pet is an excruiating experience, but another (unrelated) dog witnessing this transaction would not be similarly affected. Pain, hunger, fear, frustration, and thirst serve to activate behavioral sequences which have hard-wired adaptive significance, but lacking the ability to monitor such states they may be unconscious.

Chimpanzees, orangutans, and maybe cetaceans and elephants pose a fundamentally different ethical dilemma. The ability to monitor one's own mental states puts you in a unique cognitive domain. Can we continue to justify incarcerating, killing, and abusing such life forms? When a mind dies a universe blinks out of existence.

Concluding Comments

It is ironic, in the context of the present analysis, that for the better part of this century most psychologists have been unwittingly suppressing the primary evidence of mind by actively condemning introspection and teaching students to refrain from engaging in anthropomorphism because it is unscientific. Anthropomorphism is a natural and probably inevitable byproduct of the ability to be aware of one's own mental states.

However, quite independent of its relationship to mind, it seems to me that anthropomorphism still has a legitimate place in psychology. Imputing mental states to species other than our own is a potential source of testable hypotheses. Morgan's canon is a rule of thumb, but not something which should be used to strip the investigator of all his imagination and creativity. To ask whether animals might experience and/or interpret things in ways which are similar to us has heuristic value. As an explanation it is intellectually bankrupt, but as a means of stimulating research it may be a mistake to dismiss anthropomorphism altogether.

However, having argued that introspection is the basis for mind (anthropomorphism is a consequence of mind), it is encumbent upon me to explain why introspection is so shallow it was legitimately abandoned by psychologists long ago as a

means of generating a meaningful data base. Whereas now most
people accept the fact that the brain is the basis for mental
activity, this was not always the case. One of the most
peculiar features of the relationship between brain and behavior is that the central nervous system operates in such a
way as to be virtually oblivious to its own existence. As a
consequence people used to attribute their own mental states,
and occasionally still do, to different parts of the viscera
(e.g., the heart), probably because of the more explicit
feedback we get from changes in visceral activity and variations in emotional states. Another reason why the discovery
of the relationship between brain and behavior was so long in
coming is that major portions of the central nervous system
and corresponding psychological capacities evolved primarily
to do business with the complexities of the external world. I
suspect that this is also why many psychologists tend to
externalize psychological issues, eschew cognitive states, and
insist on environmental determinants as being the important, if
not the only antecedents of behavior.

Self-awareness does not provide automatic access to all of the
reasons why people do what they do. Self-awareness evolved in
response to unique social pressures. The presence of mind is
a byproduct of strange bedfellows. Human evolution was predicated on a high degree of mutual interdependency and need for
cooperation, but this was coupled with intense interspecific
competition. The advantage of mind is one that provides for
the development of introspective social strategies, such as
reciprocal altruism, cheating, deception, and grudging, which
in turn constitute pressure for further refinement and elabora-
tion of even more sophisticated mental machinery. The human
species is intelligent in large part because of the continuous
subtle and intricate competition which we face among one
another (see Humphrey, 1976). From the standpoint of evolution
we are the source of our own intelligence. If my formulation
is correct, the absence of mind in man would not make much difference in terms of day to day behavior, but should put us at
a distinct disadvantage when it comes to cooperating and competing with others.

As illustrated by the chimpanzee who struck upon banging empty
kerosene cans together to augment his intimidation display,
among species with minds intellectual prowess can supplant
physical prowess as the ultimate criteria for access to limited
resources. But whereas a mind may be a reproductive advantage
in a particular social psychological context, it does not follow that in and of itself a mind will guarantee good psychology. We continue to be mindless about many dimensions of our
existence.

Finally, it is important to comment on the relationship between mind and language. From this perspective language can be thought of as a vehicle for sharing and influencing states of mind. Although through inferences chimpanzees make about intentions and other mental states they can convey information among one another, the significance of language is in terms of achieving much more detailed informational states of mind based on spatial, temporal, and numerical relations. I would argue that the primary difference between human and other minds is in terms of their informational content. On the other hand, it is clear that language is not the precursor to mind.

In commenting upon the probable relationship between the complexities of social living and evolution of intelligence, Humphrey (1976) derives the following conclusion. "What, I think is urgently needed is a laboratory test of 'social skill' - a test which ought, if I am right, to double as a test of 'high level intelligence'. The essential feature of such a test would be that it places the subject in a transactional situation where he can achieve a desired goal only by adapting his strategy to conditions which are continually changing as a consequence partly but not wholly of his own behaviour" (p. 316). All of the empirical markers of mind listed in Table 1, along with the procedures outlined in this chapter, would appear to satisfy the essential features of just such a test.

References

1 Barash, D. The whisperings within. New York: Penguin, 1979.

2 Benhar, E. E., Carlton, P. L. & Samuel, D. A. A search for mirror-image reinforcement and self-recognition in the baboon. In S. Kondo, M. Kawai & S. Ehara (Eds.), Contemporary primatology. New York: Karger, Basel, 1975.

3 Brown, W. L., McDowell, A. A. & Robinson, E. M. Discrimination learning of mirrored cues by rhesus monkeys. Journal of Genetic Psychology, 1965, 106, 123-218.

4 Buss, A. Psychology - Man in perspective. New York: Wiley, 1973.

5 Cooley, C. H. Human nature and the social order. New York: Charles Scribner's & Sons, 1912.

6 Connor, R. C. & Norris, K. S. Are dolphins reciprocal altruists? The American Naturalist, 1982, 119, 358-374.

7 Desmond, A. J. The ape's reflexion. New York: Dial Press, 1979.

8 Douglas-Hamilton, I. & Douglas-Hamilton, O. Among the elephants. New York: Viking Press, 1975.

9 Epstein, R., Lanza, R. P. & Skinner, B. F. "Self-awareness" in the pigeon. Science, 1981, 212, 695-696.

10 Gallup, G. G., Jr. Mirror-image stimulation. Psychological Bulletin, 1969, 70, 782-793.

11 Gallup, G. G., Jr. Chimpanzees: Self-recognition. Science, 1970, 167, 86-87.

12 Gallup, G. G., Jr. Towards an operational definition of self-awareness. In R. H. Tuttle (Ed.), socio-ecology and psychology of primates. The Hague, Netherlands: Mouton, 1975.

13 Gallup, G. G., Jr. Absence of self recognition in a monkey (Macaca inscicularis) following prolonged exposure to a mirror. Developmental Psychobiology, 1977, 10, 281-284. (a)

14 Gallup, G. G., Jr. Self-recognition in primates: A comparative approach to the bidirectional properties of consciousness. American Psychologist, 1977, 32, 329-338. (b)

15 Gallup, G. G., Jr. Self-awareness in primates. American Scientist, 1979, 67, 417-421. (a)

16 Gallup, G. G., Jr. Self-recognition in chimpanzees and man: A developmental and comparative perspective. In M. Lewis & L. Rosenblum (Eds.), Genesis of behavior, Vol. 2: The child and its family. New York: Plenum, 1979. (b)

17 Gallup, G. G., Jr. On pigeons, mirrors, and self-awareness. Paper presented at the meeting of the Psychonomic Society, Philadelphia, November 1981.

18 Gallup, G. G., Jr. Self-awareness and the emergence of mind in primates. American Journal of Primatology, 1982, 2, 237-248.

19 Gallup, G. G., Jr., Boren, J. L., Gagliardi, G. J. & Wallnau, L. B. A mirror for the find of man, or will the chimpanzee create an identity crisis for Homo sapiens? Journal of Human Evolution, 1977, 6, 303-313.

20 Gallup, G. G., Jr., McClure, M. K., Hill, S. D. & Bundy, R. A. Capacity for self-recognition in differentially reared chimpanzees. The Psychological Record, 1971, 21, 69-74.

21 Gallup, G. G., Jr. & Suarez, S. D. On the use of animals in psychological research. The Psychological Record, 1980, 30, 211-218.

22 Gallup, G. G., Jr. & Suarez, S. D. Homosexuality as a by-product of selection for optimal heterosexual strategies. Perspectives in Biology and Medicine, in press.

23 Gallup, G. G., Jr., Wallnau, L. B. & Suarez, S. D. Failure to find self-recognition in mother-infant and infant-infant rhesus monkey pairs. Folia Primatologica, 1980, 33, 210-219.

24 Gardiner, W. L. Psychology: A story of a search. Belmont, California: Wadsworth, 1974.

25 Goodall, J. The behaviour of free-living chimpanzees in
    the Gombe Stream Reserve. Animal Behaviour Monographs,
    1968, 1, 161-311.

26 Goodall, J. In the shadow of man. New York: Dell, 1971.

27 Harris, L. P. Self-recognition among institutionalized
    profoundly retarded males: A replication. Bulletin of
    the Psychonomic Society, 1977, 9, 43-44.

28 Hayes, C. The ape in our house. New York: Harper & Row,
    1951.

29 Hayes, K. J. & Nissen, C. H. Higher mental functions in a
    home-reared chimpanzee. In A. M. Schrier & F. Stollnitz
    (Eds.), Behavior of non-human primates. Vol. 3. New York:
    Academic Press, 1971.

30 Hebb, D. O. Consciousness: Problems of method. Paper
    presented at a symposium on consciousness. Dalhousie
    University, August 1979.

31 Hill, S. D., Bundy, R. A., Gallup, G. G., Jr. & McClure,
    M. K. Responsiveness of young nursery reared chimpanzees
    to mirrors. Proceedings of the Louisiana Academy of
    Science, 1970, 33, 77-82.

32 Humphrey, N. K. The social function of intellect. In
    P. P. G. Bateson & R. A. Hinde (Eds.), Growing points
    in ethology. London: Cambridge University Press, 1976.

33 Humphrey, N. K. Nature's psychologists. In B. D. Josephson
    and V. S. Ramachandran (Eds.), Consciousness and the physi-
    cal world. New York: Pergamon Press, 1980.

34 Kellogg, W. N. & Kellogg, L. A. The ape and the child.
    New York: McGraw-Hill, 1933.

35 Kinget, G. M. On being human. New York: Harcourt Brace
    Jovanovich, 1975.

36 Klüver, H. Behavior mechanisms in monkeys. Chicago:
    University of Chicago Press, 1933.

37 Köhler, W. The mentality of apes. London: Kegan Paul,
    1927.

38 Leak, G. K. & Christopher, S. B. Freudian psychoanalysis
    and sociobiology: A synthesis. American Psychologists,
    1982, 37, 313-322.

39   Ledbetter, D. H. & Basen, J. A.   Failure to demonstrate self-recognition in gorillas.   American Journal of Primatology, 1982, 2, 307-310.

40   LeMay, M. & Geschwind, N.   Hemispheric differences in the brains of great apes.   Brain Behavior and Evolution, 1975, 11, 48-52.

41   Lethmate, J. & Dücker, G.   Untersuchungen zum selbsterkennen im spiegel bei orang-utans und einigen anderen affenarten.   Zietschrift fur Tierpsychologie, 1977, 33, 248-269.

42   Linden, E.   Apes, men, and language.   New York: Penguin, 1974.

43   Maple, T. L.   Orang-utan behavior.   New York: Van Nostrant Reinhold, 1980.

44   Mead, G. H.   Mind, self and society from the standpoint of a social behaviorist.   Chicago: University of Chicago Press, 1934.

45   Menzel, E. W., Jr.   Chimpanzee spatial memory organization.   Science, 1973, 182, 943-945.

46   Menzel, E. W., Jr.   Natural language of young chimpanzees.   New Scientist, 1975 (Jan. 16), 127-130.

47   Ostancow, P.   Le signe du miroir dans la demence precoce.   Annales Medico-Psychologiques, 1934, 92, 787-790.

48   Patterson, F.   Conversations with a gorilla.   National Geographic Magazine, 1978, 154, 438-465.

49   Premack, A. J. & Premack, D.   Teaching language to an ape.   Scientific American, 1972, 227, 92-99.

50   Premack, D. & Woodruff, G.   Does the chimpanzee have a theory of mind?   The Behavioral and Brain Sciences, 1978, 4, 515-526.

51   Temerlin, M. K.   Lucy: Growing up human.   New York: Science & Behavior Books, 1975.

52   Terrace, H. S., Petitto, L. A., Sanders, R. J. & Bever, T. G.   Can an ape create a sentence?   Science, 1979, 206, 891-902.

53   Tinbergen, N.   Curious naturalists.   New York: Doubleday, 1958.

54   Tinklepaugh, O. L.   An experimental study of representative factors in monkeys.   Journal of Comparative Psychology, 1928, 8, 197-235.

55   Rogers, C. M. & Davenport, R. K.   Effects of restricted rearing on sexual behavior of chimpanzees.   Developmental Psychology, 1969, 1, 200-204.

56   Rumbaugh, D. M. & McCormack. C.   The learing skills of primates: A comparative study of apes and monkeys. In D. Stark, R. Schneider & H. J. Kuhn (Eds.), Progress in primatology.   Stuttgart: Fischer, 1967.

57   Savage-Rumbaugh, E. S., Rumbaugh, D. M. & Boysen, S. Symbolic communication between two chimpanzees (Pan troglodytes).   Science, 1978, 201, 641-644.

58   Suarez, S. D. & Gallup, G. G., Jr.   Self-recognition in chimpanzees and orangutans, but not gorillas.   Journal of Human Evolution, 1981, 10, 87-99.

59   Yunis, J. J. & Prakash, O.   The origin of man: A chromosomal pictorial legacy.   Science, 1982, 215, 1525-1530.

60   von Senden, M.   Space and sight: The perception of space and shape in the congenitally blind before and after operation.   Glencoe, Illinois: Free Press, 1960.

61   de Waal, F. B. M., and van Roosmalen, A.   Reconciliation and consolation among chimpanzees.   Behavioral Ecology and Sociobiology, 1979, 5, 55-66.

62   Weiskrantz, L., Warrington, E. K., Sanders, M. K. & Marshall, J.   Visual capacity in the hemianopic field following a restricted occipital ablation.   Brain, 1974, 97, 709-728.

63   Wittreich, W.   Visual perception and personality. Scientific American, 1959, 200, 56-60.

64   Woodruff, G. & Premack, D.   Intentional communication in the chimpanzee: The development of deception.   Cognition, 1979, 7, 333-362.

INDEX